Applied Probability and Statistics

BAILEY · The Elements of Stochastic Processes with Applications to the Natural Sciences

BARTHOLOMEW · Stochastic Models for Social Processes, *Second Edition*

BENNETT and FRANKLIN · Statistical Analysis in Chemistry and the Chemical Industry

BHAT · Elements of Applied Stochastic Processes

BLOOMFIELD · Fourier Analysis of Time Series: An Introduction

BOX and DRAPER · Evolutionary Operation: A Statistical Method for Process Improvement

BROWNLEE · Statistical Theory and Methodology in Science and Engineering, *Second Edition*

BURY · Statistical Models in Applied Science

CHERNOFF and MOSES · Elementary Decision Theory

CHOW · Analysis and Control of Dynamic Economic Systems

CLELLAND, deCANI, BROWN, BURSK, and MURRAY · Basic Statistics with Business Applications, *Second Edition*

COCHRAN · Sampling Techniques, *Second Edition*

COCHRAN and COX · Experimental Designs, *Second Edition*

COX · Planning of Experiments

COX and MILLER · The Theory of Stochastic Processes, *Second Edition*

DANIEL · Application of Statistics to Industrial Experimentation

DANIEL and WOOD · Fitting Equations to Data

DAVID · Order Statistics

DEMING · Sample Design in Business Research

DODGE and ROMIG · Sampling Inspection Tables. *Second Edition*

DRAPER and SMITH · Applied Regression Analysis

DUNN and CLARK · Applied Statistics: Analysis of Variance and Regression

ELANDT-JOHNSON · Probability Models and Statistical Methods in Genetics

FLEISS · Statistical Methods for Rates and Proportions

GNANADESIKAN · Methods for Statistical Data Analysis of Multivariate Observations

GOLDBERGER · Econometric Theory

GROSS and CLARK · Survival Distributions

GROSS and HARRIS · Fundamentals of Queueing Theory

GUTTMAN, WILKS and HUNTER · Introductory Engineering Statistics, *Second Edition*

HAHN and SHAPIRO · Statistical Models in Engineering

HALD · Statistical Tables and Formulas

HALD · Statistical Theory with Engineering Applications

HARTIGAN · Clustering Algorithms

HILDEBRAND, LAING and ROSENTHAL · Prediction Analysis of Cross Classifications

HOEL · Elementary Statistics, *Fourth Edition*

HOLLANDER and WOLFE · Nonparametric Statistical Methods

HUANG · Regression and Econometric Methods

JAGERS · Branching Processes with Biological Applications

Methods for Statistical Data Analysis of Multivariate Observations

R. GNANADESIKAN
Bell Telephone Laboratories
Murray Hill, New Jersey

John Wiley & Sons
New York · London · Sydney · Toronto

Published by John Wiley & Sons, Inc.

Library of Congress Cataloging in Publication Data:

Gnanadesikan, R
Methods for statistical data analysis of multivariate observations.

(Wiley series in probability and mathematical statistics)
Bibliography: p.
Includes index.
1. Multivariate analysis. I. Title.
QA278.G6 519.5′35 76-14994
ISBN 0-471-30845-5

Printed in the United States of America

10 9 8 7 6 5 4 3 2

To the family of my childhood
and
the family of my parenthood

स्वस्ति नो बृहस्पतिर्दधातु। सहायकानां च स्वस्ति।

नमः पूर्वपुरुषेभ्यः। ॐ शान्तिः शान्तिः शान्तिः।

In the manner of the Upanishads, I invoke the blessings of the Vedic Lord of Prayers upon you, my good reader, upon myself and upon my collaborators. I also take this opportunity to pay my obeisance to my ancestors and my predecessors in my field.

Preface

This book had its origins in a General Methodology Lecture presented at the annual meetings of the American Statistical Association at Los Angeles in 1966. A more concrete format for the book emerged from a paper (see Gnanadesikan & Wilk, 1969) presented at the Second International Symposium on Multivariate Analysis held at Dayton in June, 1968. That paper provided an outline of objectives for organizing the material in the present book, although the coverage here is more up to date, extensive, and detailed than the one in the paper. Specifically, the book is concerned with the description and discussion of multivariate statistical techniques and concepts, structured according to five general objectives in analyzing multiresponse data. The methods and underlying concepts are grouped according to these five objectives, and a chapter of the book is devoted to each objective.

The book is intended to emphasize methodology and data-based interpretations relevant to the needs of data analysis. As such, it is directed primarily toward applied statisticians and users of statistical ideas and procedures in various scientific and technological disciplines. However, some issues, arising especially out of the newer techniques described in the book, may be of interest to theoretical statisticians. Also, there are algorithmic aspects of the procedures which numerical analysts may find interesting.

Portions of the material in this book have been used by the author as the basis for a graduate-level series of lectures presented at Imperial College of Science & Technology of the University of London in 1969 and at Princeton University in 1971. Although the book can thus serve as a text, it differs from standard textbooks in not containing exercises. In view of the orientation of the book, the natural exercises would be to analyze specific sets of data by using the methods described in the text. However, rather than setting such exercises, which often tend to be artificial, it would seem to be far more useful to expect the students to use the relevant techniques on any real problems which they encounter either in their own work or in the course of their being consulted for statistical advice on the problems of others. Also, for making the purpose and usefulness of a technique more apparent, illustrative examples are used.

Such examples appear throughout the book and constitute an important facet of the presentation.

The coverage in this book is mainly of relatively recent (i.e., within the last decade) developments in multivariate methodology. When more classical techniques are described, the intention is either to provide a more natural motivation for a recent concept or method or to attempt a more complete discussion. A thorough review of all multivariate techniques is not a goal of the book. Specifically, for instance, no attention is given here to the analysis of multiple time series.

Despite the intention to emphasize relatively recent developments, the book inevitably reflects the fact that it was written over a period of six or seven years that have seen a spate of publications on multivariate topics. For instance, whereas material on cluster analysis written from a statistical viewpoint was relatively sparse when Chapter 4 of this book was conceived, there have been several recent articles and even whole books (e.g., Everitt, 1974; Hartigan, 1975) on this topic.

I am grateful to Bell Telephone Laboratories for its support of my efforts in writing this book and for providing so many important facilities without which the task could not have been undertaken. I also thank Imperial College and Princeton University for providing me with the stimulus and opportunity to organize the material for this book. It is a particular pleasure to acknowledge the many valuable comments of Professor D. R. Cox at the time of these lectures at Imperial College. Thanks are due also to my colleague Dr. J. R. Kettenring for his willingness to use parts of this material in a course that he taught and for his several helpful comments, based partly on the experience. I am deeply indebted to many past and present colleagues for their collaborative research efforts with me, which are reflected in various parts of this book. I wish also to acknowledge the kind permissions of other authors, several journals and publishers (including Academic Press, the American Statistical Association, Biometrics, the Biometrika Trustees, the Institute of Mathematical Statistics, Methuen & Co., Pergamon Press, Psychometrika, Statistica Neerlandica, and Technometrics) to incorporate material published elsewhere.

I am grateful to Mrs. M. L. Culp, Miss D. A. Williams, and Mrs. J. Charles for their careful typing of different parts of the manuscript, and the assistance of Messrs. I. L. Patterson and J. L. Warner in collating the exhibits is gratefully acknowledged.

Finally, I express my deepest sense of gratitude to my wife not only for her valuable comments but also for her constant encouragement during the writing of the book.

R. GNANADESIKAN

Murray Hill, New Jersey
January 1977

Contents

Methods for Statistical

Data Analysis of

Multivariate Observations

CHAPTER 1

Introduction

Most bodies of data involve observations associated with various facets of a particular background, environment, or experiment. Therefore, in a general sense, data are always multivariate in character. Even in a narrow sense, when observations on only a single response variable are to be analyzed, the analysis often leads to a multivariate situation. For example, in multiple linear regression, or in fitting nonlinear models, even with a single dependent variable, one often is faced with correlations among the estimated coefficients, and analyzing the correlation structure for possible reparametrizations of the problem is not an uncommon venture.

For the purposes of the present book, a more limited definition of a multivariate situation is used: multiresponse (or multivariate) problems are those that are concerned with the analysis of n points in p-space, i.e., when each of n persons, objects, or experimental units has associated with it a p-dimensional vector of responses. The experimental units need not necessarily constitute an unstructured sample but, in fact, may have a superimposed design structure, i.e., they may be classified or identified by various extraneous variables. One essential aspect of a multivariate approach to the analysis of such multiresponse problems is that, although one may choose to consider the p-dimensional observations from object to object as being statistically independent, the observed components within each vector will usually be statistically related. Exploitation of the latter feature to advantage in developing more sensitive statistical analyses of the observations is the pragmatic concern and value of a multivariate approach.

Most experimenters probably realize the importance of a multivariate approach, and most applied statisticians are equally well aware that multivariate analysis of data can be a difficult and frustrating problem. Some users of multivariate statistical techniques have, with some justification, even asserted that the methods may be unnecessary, unproductive, or misguided. Reasons for the frustrations and difficulties characteristic of multivariate data analysis, which often far exceed those encountered in univariate circumstances, appear to include the following:

1. It seems very difficult to know or to develop an understanding of what one really wants to do. Much iteration and interaction is required. This is also true in the uniresponse case in real problems. Perhaps in the multiresponse case one is simply raising this difficulty to the pth power!
2. Once a multiresponse view is adopted, there is no obvious "natural" value of p, the dimensionality of response. For any experimental unit it is always possible to record an almost indefinitely large list of attributes. Any selection of responses for actual observation and analysis is usually accomplished by using background information, preliminary analysis, informal criteria, and experimental insight. On the other hand, the number of objects or replications, n, will always have some upper bound. Hence n may at times be less than p, and quite often it may not be much greater. These dimensionality considerations can become crucial in determining what analyses or insights can be attained.
3. Multivariate data analysis involves prodigious arithmetic and considerable data manipulation. Even with modern high-speed computing, many multivariate techniques are severely limited in practice as to number of dimensions, p, number of observations, n, or both.
4. Pictures and graphs play a key role in data analysis, but with multiresponse data elementary plots of the raw data cannot easily be made. This limitation keeps one from obtaining the realistic primitive stimuli, which often motivate uniresponse analyses as to what to do or what models to try.
5. Last, but of great importance and consequence, points in p-space, unlike those on a line, do not have a unique linear ordering, which sometimes seems to be almost a basic human requirement. Most formal models and their motivations seem to grasp at optimization or things to order. There is no great harm in this unless, in desperation to achieve the comfort of linear ordering, one closes one's mind to the nature of the problem and the guidance which the data may contain.

Much of the theoretical work in multivariate analysis has dealt with formal inferential procedures, and with the associated statistical distribution theory, developed as extensions of and by analogy with quite specific univariate methods, such as tests of hypotheses concerning location and/or dispersion parameters. The resulting methods have often turned out to be of very limited value for multivariate data analysis.

The general orientation of the present book is that of statistical data analysis, concerned mainly with providing descriptions of the informational content of the data. The emphasis is on *methodology*—on underlying or motivating concepts and on data-based interpretations of the methods. Little or no coverage is given to distribution theory results, optimality properties, or formal or detailed mathematical proofs, or, in fact, to fitting the methods discussed into the framework of any currently known formal theory of statistical inference, such as decision theory or Bayesian analysis.

The framework for the discussion of multivariate methods in this book is provided by the following five general objectives of analyzing multiresponse data:

1. Reduction of dimensionality (Chapter 2).
2. Development and study of multivariate dependencies (Chapter 3).
3. Multidimensional classification (Chapter 4).
4. Assessment of statistical models (Chapter 5).
5. Summarization and exposure (Chapter 6).

The classification of multivariate methods provided by these five objectives is not intended to be in terms of mutually exclusive categories, and some techniques described in this book may be used for achieving more than one of the objectives. Thus, for example, a technique for reducing dimensionality may also prove to be useful for studying the possible internal relationships among a group of response variables.

With regard to the technology of data analysis, although it is perhaps true that this is still in a very primitive state, some important aids either are available or are under development. Raw computing power has grown astronomically in recent years, and graphical display devices are now relatively cheap and widely available. Much more data-analytic software is to be expected in the near future. Hardware-software configurations are being designed and developed, for both passive and interactive graphics, as related to the needs of statistical data analysis. Graphical presentation and pictorialization are important and integral tools of data analysis. (See Gnanadesikan, 1973, for a discussion of graphical aids for multiresponse data analysis.) A feature common to most of the methods discussed in the subsequent chapters of this book is their graphical nature, either implicit in their motivating ideas or explicit in their actual output and use.

In general, the mathematical notation used conforms to familiar conventions. Thus, for instance, $\mathbf{a}$, $\mathbf{x}$, . . . denote column vectors; $\mathbf{a}'$, $\mathbf{x}'$, . . . , row vectors; and $\mathbf{A}$, $\mathbf{Y}$, . . . , matrices. Whenever it is feasible and not unnatural, a distinction is made between parameters and random variables by using the familiar convention that the former are denoted by Greek letters and the latter by letters of the English alphabet. Most of the concepts and methods discussed are, however, introduced in terms of observed or sample statistics, i.e., quantities calculated from a body of data. Statistics that are estimates of parameters are often denoted by the usual convention of placing a hat (ˆ) over the parameter symbol.

Equations, figures, and tables that occur as part of the main text are numbered sequentially throughout the book. However, no distinction is made between figures and tables when they occur in the context of an example, and both are referred to as "exhibits." Thus Exhibit 5*a* is a table of numbers that appears in Example 5, whereas Exhibits 5*b* and *c* both are figures that are part of the same example.

A bibliography is included at the end of the book, and specific items of it that are directly relevant to a particular chapter are listed at the end of the chapter. An item in the bibliography is always cited by the name(s) of the author(s) and the year of publication. Thus Gnanadesikan (1973), Gnanadesikan & Wilk (1969), Kempthorne (1966), Tukey (1962), and Tukey & Wilk (1966) are specifically relevant references for the present chapter.

CHAPTER 2

Reduction of Dimensionality

2.1. GENERAL

The issue in reduction of dimensionality in analyzing multiresponse data is between attainment of simplicity for understanding, visualization, and interpretation, on the one hand, and retention of sufficient detail for adequate representation on the other hand.

Reduction of dimensionality can lead to parsimony of description, of measurement, or of both. It may also encourage consideration of meaningful physical relationships between the variables, for example, summarizing bivariate mass-volume data in terms of the ratio density = mass/volume.

As mentioned in Chapter 1, in many problems the dimensionality of response, p, is conceptually unlimited, whereas the number, n, of experimental units available is generally limited in practice. By some criteria of relevance, the experimenter always drastically reduces the dimensionality of the observations to be made. Such reduction may be based on (i) exclusion before the experiment; (ii) exclusion of features by specific experimental judgment; (iii) general statistical techniques, such as principal components analysis (see Section 2.2), use of distance functions of general utility, and methods for recognizing and handling nonlinear singularities (see Section 2.3); and/or (iv) specific properties of the problem which indicate the choice of a particular (unidimensional) real-valued function for analysis, e.g., relative weights for assigning an overall grade in matriculation examinations.

The first two of these approaches lead to a reduction of measurement in that the number of variables to be observed is diminished. The last two will not, in general, result in reducing current measurements but may reduce future measurements by showing that a subset of the variables is "adequate" for certain specifiable purposes of analysis. The major concern of the present chapter is the discussion of some specific examples of the third approach in the list above.

From the point of view of description, too severe a reduction may be undesirable. Meaningful statistical analysis is possible only when there has not been excessive elimination. Clearly a dominant consideration in the use of statistical procedures for the reduction of dimensionality is the interpretability of the lower dimensional representations. For instance, the use of principal components per se does not necessarily yield directly interpretable measures, whereas a reasonable choice of a distance function will sometimes permit interpretation.

Circumstances under which one may be interested in reducing the dimensionality of multiple response data include the following:

1. Exploratory situations in data analysis, for example, in psychological testing results or survey questionnaire data, especially when there is ignorance of what is important in the measurement planning. Here one may want to screen out redundant coordinates or to find more insightful ones as a preliminary step to further analysis or data collection.
2. Cases in which one hopes to stabilize "scales" of measurement when a similar property is described by each of several coordinates, for example, several measures of size of a biological organism. Here the aim is to compound the various measurements into a fewer number which may exhibit more stable statistical properties.
3. The compounding of multiple information as an aid in significance assessment. Specifically, one may hope that small departures from null conditions may be evidenced on each of several jointly observed responses. Then one might try to integrate these noncentralities into a smaller-dimensional space wherein their existence might be more sensitively indicated. One particular technique that has received some usage is the univariate analysis of variance applied to principal components.
4. The preliminary specification of a space that is to be used as a basis for eventual discrimination or classification procedures. For example, the raw information per object available as a basis for identifying people from their speech consists, in one version of the problem, of a 15,000-dimensional vector which characterizes each utterance! This array must be condensed as a preliminary to further classification analysis.
5. Situations in which one is interested in the detection of possible functional dependencies among observations in high-dimensional space. This purpose is perhaps the least well defined but nevertheless is prevalent, interesting, and important.

Many problems and issues exist in this general area of transformation of coordinates and reduction of dimensionality. These are problems of concept as to what one hopes to achieve, of techniques or methods to exhibit information that may be in the data, of interpretations of the results of applying available techniques, and of mathematical or algorithmic questions related to implementation. Specifically, if one develops a transformed or derived set of (reduced) coordinates, there is the question of whether these can be given some meaning or interpretation that will facilitate understanding of the actual problem. Similarly, it may or may not be true that derived coordinates, or approximations to these, will be directly observable. Sometimes such observability may occur with gains in efficiency and simplicity of both experiment and analysis.

Another problem in this area is that of the commensurability of the original coordinates and of the effect of this issue on a derived set of coordinates. This is not, apparently, a problem in principle, since there is no difficulty in dealing with functions of variables having different units. However, if the functions are themselves to be determined or influenced by the data, as in principal components analysis, some confusion may exist. An example of the issue involved here is presented in Section 2.2.1.

In looking for a reduced set of coordinates, classical statistical methodology has been largely concerned with derived coordinates that are just linear transforms of the original coordinates. This limitation of concern to linearity is perhaps due at least in part to the orientation of many of the techniques toward multivariate normal distribution theory. More recently, however, techniques have been suggested (Shepard, 1962a, b; Shepard & Carroll, 1966; Gnanadesikan & Wilk, 1966, 1969) for nonlinear reduction of dimensionality.

2.2. LINEAR REDUCTION TECHNIQUES

This section reviews briefly the classical linear reduction methods. First, discussion is provided of principal components analysis, a technique initially described by Karl Pearson (1901) and further developed by Hotelling (1933), which is perhaps the most widely used multivariate method. Second, concepts and techniques associated with linear factor analysis are outlined. Both the principal factor method due to Thurstone (1931) and the maximum likelihood approach due to Lawley (1940) are considered.

2.2.1. Principal Components Analysis

The basic idea of principal components analysis is to describe the dispersion of an array of n points in p-dimensional space by introducing a new set of orthogonal linear coordinates so that the sample variances of the given points with respect to these derived coordinates are in decreasing order of magnitude. Thus the first principal component is such that the projections of the given points onto it have maximum variance among all possible linear coordinates; the second principal component has maximum variance subject to being orthogonal to the first; and so on.

If the elements of $\mathbf{y}' = (y_1, y_2, \ldots, y_p)$ denote the p coordinates of observation, and the rows of the $n \times p$ matrix, $\mathbf{Y}'$, constitute the n p-dimensional observations, the sample mean vector and covariance matrix may be obtained, respectively, from the definitions

$$\bar{\mathbf{y}}' = (\bar{y}_1, \bar{y}_2, \ldots, \bar{y}_p) = \frac{1}{n}\mathbf{1}'\mathbf{Y}', \tag{1}$$

$$\mathbf{S} = ((s_{ij})) = \frac{1}{n-1}(\mathbf{Y} - \bar{\mathbf{Y}})(\mathbf{Y} - \bar{\mathbf{Y}})', \tag{2}$$

where $\mathbf{1}'$ is a row vector all of whose elements are equal to 1, and $\bar{\mathbf{Y}}'$ is an $n \times p$ matrix each of whose rows is equal to $\bar{\mathbf{y}}'$. The $p \times p$ sample correlation matrix, $\mathbf{R}$, is related to $\mathbf{S}$ by

$$\mathbf{R} = \mathbf{D}_{1/\sqrt{s_{ii}}} \cdot \mathbf{S} \cdot \mathbf{D}_{1/\sqrt{s_{ii}}}, \tag{3}$$

where $\mathbf{D}_{1/\sqrt{s_{ii}}}$ is a $p \times p$ diagonal matrix whose ith diagonal element is $1/\sqrt{s_{ii}}$ for $i = 1, 2, \ldots, p$.

A geometric interpretation of principal components analysis is as follows: The inverse of the sample covariance matrix may be employed as the matrix of a quadratic form which defines a family of concentric ellipsoids centered on the sample center of gravity; i.e., the equations

$$(\mathbf{y} - \bar{\mathbf{y}})'\mathbf{S}^{-1}(\mathbf{y} - \bar{\mathbf{y}}) = c, \tag{4}$$

for a range of nonnegative values of c, define a family of concentric ellipsoids in the p-dimensional space of $\mathbf{y}$. The principal components transformation of the data is just the projections of the observations onto the principal axes of this family. The basic idea is illustrated, for the two-dimensional case, in Figure 1. The original coordinates, (y_1, y_2), are transformed by a shift of origin to the sample mean, $(\bar{y}_1, \bar{y}_2)$, followed by a rigid rotation about this origin that yields the principal component coordinates, z_1 and z_2.

Algebraically, the principal components analysis involves finding the eigenvalues and eigenvectors of the sample covariance matrix. Specifically, for obtaining the first principal component, z_1, what is sought is the vector of coefficients, $\mathbf{a}' = (a_1, a_2, \ldots, a_p)$, such that the linear combination, $\mathbf{a}'\mathbf{y}$, has maximum sample variance in the class of all linear combinations, subject to the normalizing constraint, $\mathbf{a}'\mathbf{a} = 1$. For a given $\mathbf{a}$, since the sample variance of $\mathbf{a}'\mathbf{y}$ is $\mathbf{a}'\mathbf{S}\mathbf{a}$, the problem of finding $\mathbf{a}$ turns out to be equivalent to determining a nonnull $\mathbf{a}$ such that the ratio $\mathbf{a}'\mathbf{S}\mathbf{a}/\mathbf{a}'\mathbf{a}$ is

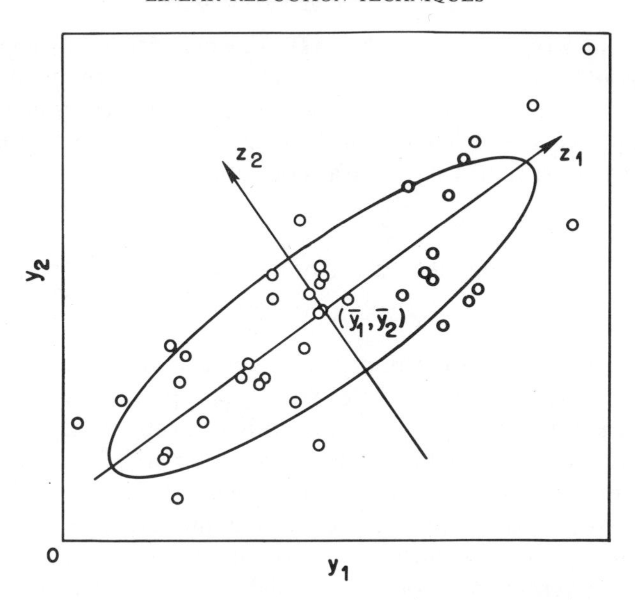

Fig. 1. Illustration of principal components with bivariate data.

maximized. It is well known that the maximum value of this ratio is the largest eigenvalue, c_1, of **S**, and the required solution for **a** is the eigenvector, $\mathbf{a}_1$, of **S** corresponding to c_1.

After the first principal component has been determined, the next problem is to determine a second normalized linear combination orthogonal to the first and such that, in the class of all normalized linear functions of **y** orthogonal to $\mathbf{a}_1'\mathbf{y}$, the second principal component has largest variance. At the next stage, one would determine a third normalized linear combination with maximum variance in the class of all normalized linear combinations orthogonal to the first two principal components. The process may be repeated until p principal components have been determined. The problem of determining the p principal components is equivalent to determining the stationary values of the ratio $\mathbf{a}'\mathbf{S}\mathbf{a}/\mathbf{a}'\mathbf{a}$ for variation over all nonnull vectors, **a**. These stationary values are known to be the eigenvalues, $c_1 \geq c_2 \geq \cdots \geq c_p \geq 0$, of **S**, and the required principal components are provided by $\mathbf{a}_1'\mathbf{y}$, $\mathbf{a}_2'\mathbf{y}$, . . . , and $\mathbf{a}_p'\mathbf{y}$, where $\mathbf{a}_i'$ is the normalized eigenvector of **S** corresponding to the eigenvalue, c_i, for $i = 1, 2, \ldots, p$. The ranked eigenvalues are in fact just the sample variances of the linear combinations of the original variables specified by the eigenvectors.

The above results can also be related to the so-called spectral decomposition (see, e.g., Rao, 1965, p. 36) of the matrix **S**: there exists an orthogonal matrix, $\mathbf{A}$, such that $\mathbf{S}=\mathbf{A}\mathbf{D}_c\mathbf{A}'$, where $\mathbf{D}_c$ is a diagonal matrix with diagonal elements $c_1, c_2, \ldots, c_p$. The columns of $\mathbf{A}$ are the eigenvectors $\mathbf{a}_1, \mathbf{a}_2, \ldots, \mathbf{a}_p$. The principal component coordinates, which for convenience are defined to include a shift of origin to the sample mean, are then specified by the transformation

$$\mathbf{z}=\mathbf{A}'(\mathbf{y}-\bar{\mathbf{y}}), \tag{5}$$

and the principal components transformation of the data is

$$\mathbf{Z}=\mathbf{A}'(\mathbf{Y}-\bar{\mathbf{Y}}). \tag{6}$$

When transformed to the principal component coordinate system, the observations have certain desirable statistical properties. For instance, the sample variance of the observations with respect to the ith principal component is $\mathbf{a}_i'\mathbf{S}\mathbf{a}_i = c_i$, the ith largest eigenvalue of $\mathbf{S}$, for $i=1, 2, \ldots, p$, and the sum of the sample variances with respect to the derived coordinates $=\sum_{i=1}^{p} c_i = \mathrm{tr}(\mathbf{S}) = \sum_{i=1}^{p} s_{ii}$ = sum of the variances with respect to the original coordinates. Furthermore, because of the mutual orthogonality of the representations of the original observations in terms of the principal component coordinates, the sample covariances (and hence the sample correlations) between pairs of the derived variables are all 0. This follows geometrically from the "orthogonal" nature of the two-dimensional configuration of the projections of the observations onto each member of every pair of principal component coordinates. Equivalently, it follows algebraically from the relationship that the sample covariance between the ith and jth principal components coordinates $=$ $\mathbf{a}_i'\mathbf{S}\mathbf{a}_j = c_j\mathbf{a}_i'\mathbf{a}_j = 0$ since $\mathbf{a}_i$ and $\mathbf{a}_j$ (for $i \neq j$) are orthogonal.

The above geometrical, algebraic, and algorithmic descriptions have been presented in terms of the covariance matrix. Clearly, if one standardizes each coordinate by dividing by its sample standard deviation, then the covariance matrix of the standardized variables is just the correlation matrix of the original variables. Thus the above discussion applies to principal components analysis of the correlation matrix.

In light of the current state of the knowledge on numerically stable computational methods, the recommended algorithm for performing the eigenanalysis involved in obtaining the principal components is either the so-called QR method applied to $\mathbf{S}$ or $\mathbf{R}$ (Businger, 1965), or the so-called singular value decomposition technique performed on $(\mathbf{Y}-\bar{\mathbf{Y}})$ or on the standardized form, $\mathbf{D}_{1/\sqrt{s_{ii}}}(\mathbf{Y}-\bar{\mathbf{Y}})$ (Businger & Golub, 1969; Golub, 1968).

If the sample size, n, is not greater than the dimensionality, p, the sample covariance matrix will be singular, corresponding to the fact that all n points will lie on a hyperplane of dimension less than p. Within that linear subspace one can define a dispersion matrix and find its principal components. This will be reflected in the eigenvalue analysis of the singular covariance matrix, in that some of the eigenvalues will be 0. The eigenvectors corresponding to the nonzero eigenvalues will give the projections of the observations onto orthogonal coordinates within the linear subspace containing the observations.

One hope in the case of principal components analysis is that the bulk of the observations will be near a linear subspace and hence that one can employ a new coordinate system of reduced dimension. Generally, interest will lie in the coordinates along which the data show their greatest variability. However, although the eigenvector corresponding to the largest eigenvalue, for example, provides the projection of each point onto the first principal component, the equation of the first principal component coordinate is given by the conjunction of the equations of planes defined by the remaining eigenvectors. More generally, if most of the variability of a p-dimensional sample is confined to a q-dimensional linear subspace, that subspace is described by the $(p-q)$ eigenvectors which correspond to the $(p-q)$ "small" eigenvalues. For purposes of interpretation—detection or specification of constraints on, or redundancy of, the observed variables—it may often be the relations which define near constancy (i.e., those specified by the smallest eigenvalues) that are of greatest interest.

An important practical issue in eigenanalyses is that of judging the relative magnitudes of the eigenvalues, both for isolating "negligibly small" ones and for inferring groupings, if any, among the others. The issue involves not only computational questions, such as the specification of what constitutes a zero eigenvalue, but also questions of statistical inference and useful insight. The interpretation of magnitude and separation of eigenvalues from a sample covariance matrix is considerably complicated by the sampling variation and statistical interdependence, as exhibited even by the eigenvalues of a covariance matrix calculated from observations from a spherical normal distribution. Although there are some tests of significance, which have been proposed as formal inferential aids, a real need exists for data-analytic procedures for studying the configuration of a collection of sample eigenvalues as a whole (see Section 6.2 for further discussion).

Clearly, principal components are *not* invariant under linear transformation, including separate scaling, of the original coordinates. Thus the principal components of the covariance matrix are not the same as those of the correlation matrix or of some other scaling according to measures of "importance." Note, however, that the principal components of the correlation matrix are invariant under separate scaling of the original variables. For this reason, as well as for numerical computational ones, some have urged that principal components analysis always be performed on the correlation matrix. However, for other reasons of a statistical nature, such as interpretation, formal statistical inference, and distribution theory, it often is preferable to work with the covariance matrix. There does not seem to be any *general* elementary rationale to motivate the choice of scaling of the variables as a preliminary to principal components analysis on the resulting covariance matrix.

An important exception regarding invariance occurs when the observations are confined to a linear subspace. In this case, the specification of the singularities is unique under nonsingular linear transformation of the variables. One might expect that, loosely speaking, similar near uniqueness would hold when the data have a "nearly singular" structure.

To conclude the present discussion of linear principal components analysis, an example of application is considered next. The use of the technique will be discussed further in Section 6.4.

Example 1. This example is taken from Blackith & Roberts (1958) and has also been discussed as an application of linear principal components analysis by Blackith (1960) and by Seal (1964). It deals with measurements on 375 $(=n)$ grasshoppers of 10 $(=p)$ characters that were chosen to cover the major areas of the body. The 375 grasshoppers were, in fact, cross-classifiable into eight groups—two species, two sexes, and two color forms. One interest in the data was to study and to characterize basic patterns of growth of the grasshoppers. Suppose that, for $g=1, 2, \ldots, 8$, n_g denotes the number of grasshoppers measured in the gth group, and $\mathbf{X}_g$, $\bar{\mathbf{x}}_g$, and $\mathbf{S}_g$ are, respectively, the matrix of observations, the mean vector, and the covariance matrix for the gth group. The mean vector and the covariance matrix for each group are obtained by using Eqs. 1 and 2 of this chapter. Blackith (1960) reports on a principal components analysis of the pooled 10×10 covariance matrix,

$$\mathbf{S}=\frac{1}{n-8}\sum_{g=1}^{8}(n_g-1)\mathbf{S}_g,$$

where $n=\sum_{g=1}^{8} n_g=375$. The pooled covariance matrix is based on 367 degrees of freedom. Exhibit 1*a*, taken from Blackith (1960), shows the first three eigenvalues and the corresponding three eigenvectors, which, therefore, define the first three principal components. The sum of the three eigenvalues is 16.924, and the first three principal components account for about 99% of the observed variation in 10-dimensional space.

Exhibit 1a. First three principal components for grasshoppers data (Blackith, 1960).

	Variate	Variance		Eigenvalues	
			16.087	0.516	0.321
				Eigenvectors	
1.	Reduced wt. (mg.)	15.7725	1.0000	-0.0678	-0.1056
2.	# Antennal segments	0.5531	0.0523	1.000	-0.1027
3.	Elytron length (mm.)	0.4155	0.0847	0.0694	1.000
4.	Head width (mm.)	0.0138	0.0215	0.0141	0.0155
5.	Pronotal width (mm.)	0.0150	0.0197	0.0146	0.0098
6.	Hind femoral length (mm.)	0.2545	0.0929	0.0928	0.2688
7.	Hind femoral width (mm.)	0.0198	0.0233	0.0024	0.0008
8.	Prozonal length (mm.)	0.0097	0.0110	0.0055	-0.0095
9.	Metazonal length (mm.)	0.0197	0.0150	0.0160	0.0555
10.	Front femoral width (mm.)	0.0015	0.0046	-0.0025	0.0068

Note that the normalization of these eigenvectors has been accomplished by making the largest element 1, instead of the more usual unit Euclidean norm scheme of making the squares of the elements add to 1. This, however, does not interfere with the interpretation of the results. Thus each of the three eigenvectors in Exhibit 1*a* is "close to" a corresponding unit eigenvector with a single nonzero element which is unity: the first to (1, 0, 0, . . . , 0), the second to (0, 1, 0,. . . , 0), and the third to (0, 0, 1, 0, . . . , 0). In other words, the first principal component weights, almost exclusively, the original characteristic measured that has the largest variance, and, similarly, the second and third principal components in turn weight the original characteristics with the second and third largest variances, respectively. The sensitivity of the principal component coordinates to the variances of the original variables implies a critical dependence of the derived coordinates on the choice of scales for observing the original variables. Moreover, in the present example, the two characteristics with largest variances—namely, reduced weight and number of antennal segments—also happen to be measured on different scales from the one (millimeters) used for the remaining eight characteristics. Thus an additional issue here is the effect of the commensurability of the observed responses on the derived principal component coordinates.

One way of handling this difficulty in this example is to omit the two responses measured on very different scales and then perform a principal components analysis on the remaining eight responses. Another approach, which was mentioned earlier, would be to perform the analysis on the correlation matrix instead of the covariance matrix. Exhibit 1*b* shows the eigenvalues (see Seal, 1964) obtained in principal components analyses performed on both covariance and correlation matrices for the full set of 10 responses as well as for the reduced set of 8 responses.

Lines indicating intuitively reasonable separations among the eigenvalues are also shown in Exhibit 1*b*, a dashed line denoting a weak separation and a solid line suggesting stronger separation. The indicated number and location of the separations are seen to be different between analyses performed on covariance matrices and those done on correlation matrices. Thus, both with all 10 responses and with the subset of 8, the principal components analyses of correlation matrices suggest that only the largest eigenvalue is clearly separated from the remaining ones. The analyses based on the corresponding covariance matrices, however, seem to suggest two separations among the eigenvalues.

Seal (1964) provides a reasonable argument for the indicated single separation among the eigenvalues of the 8×8 correlation matrix. Many of the off-diagonal elements of the matrix appear to be essentially the same, thereby indicating a nearly equicorrelational structure among the variables. In the case of the eight responses, all of the correlation coefficients

Exhibit 1b. Eigenvalues for four principal components analyses (Seal, 1964).

	$\mathbf{S}$(10×10)	$\mathbf{R}$(10×10)	$\mathbf{S}$(8×8)	$\mathbf{R}$(8×8)
	16.087	4.802	0.549	3.959
	0.516	0.970	0.145	0.923
	0.321	0.898	0.021	0.867
	0.103	0.852	0.015	0.634
	0.017	0.637	0.009	0.588
	0.012	0.587	0.006	0.501
	0.009	0.499	0.003	0.339
	0.006	0.351	0.001	0.189
	0.003	0.218		
	0.001	0.186		
Total	17.075	10.0	0.749	8.0

appear to be about 0.4. With exact equicorrelational structure, a $p \times p$ correlation matrix will have only two distinct eigenvalues, one being equal to $1+(p-1)r$ and the remaining $(p-1)$ being equal to $(1-r)$, where r is the common value of all the correlation coefficients. An interesting question in the present example is whether the equicorrelation is inherent and experimentally sensible or is induced by the pooling of the covariance matrices from the eight groups of grasshoppers. Pooling several widely different covariance structures may lead to an "average" equicorrelational structure, and the relatively low value of the "common" correlation coefficient (0.4) in the example raises the question of a possible artifactual nature of the observed equicorrelation. Had the covariance matrix within each group been available, a technique described in Section 6.3.2 could have been used to study the appropriateness of the preliminary pooling of the eight covariance matrices in this example.

A somewhat different issue here is the relevance of analyzing the data on the observed scales of measurement rather than transforming the observations before the analysis. An interesting and seemingly appropriate transformation in this case would be to use logarithms of the original observations as the starting point of the principal components analysis. Unfortunately, the raw observations in the example are unavailable, and such an analysis is therefore not possible.

2.2.2. Factor Analysis

The so-called model in factor analysis is

$$\mathbf{y} = \mathbf{\Lambda f} + \mathbf{z}, \tag{7}$$

where $\mathbf{y}$ is a p-dimensional vector of observable responses, $\mathbf{\Lambda}$ is a $p \times q$ matrix of unknown parameters called *factor loadings*, $\mathbf{f}$ is a q-dimensional vector of hypothetical (unobserved) variables called *common factors*, and $\mathbf{z}$ is a p-dimensional vector of hypothetical (unobserved) variables called *unique factors*. [*Note*: To distinguish between $\mathbf{f}$ and $\mathbf{z}$ one needs to impose the condition that each column of $\mathbf{\Lambda}$ has at least two nonzero elements.] Generally, it is further assumed that the components of $\mathbf{z}$ are mutually uncorrelated as well as being uncorrelated with the elements of $\mathbf{f}$. In other words, the covariance matrix of $\mathbf{z}$ is a $p \times p$ diagonal matrix, $\mathbf{\Delta}$, with diagonal elements $\delta_1^2, \ldots, \delta_p^2$, and the cross-covariance matrix between $\mathbf{f}$ and $\mathbf{z}$ is null.

With n observations available on p responses which are being studied simultaneously, the above model may be written as

$$\mathbf{Y} = \mathbf{\Lambda F} + \mathbf{Z}, \tag{8}$$

where $\mathbf{Y}$ is $p \times n$, $\mathbf{F}$ is $q \times n$, and $\mathbf{Z}$ is $p \times n$. The factor-analytic model in

Eq. 7 (or 8), taken together with the above assumptions, specifies the following relationship among the covariance matrices of the different sets of variables involved:

$$\boldsymbol{\Sigma}_{yy} = \boldsymbol{\Lambda}\boldsymbol{\Sigma}_{ff}\boldsymbol{\Lambda}' + \boldsymbol{\Delta}, \tag{9}$$

where $\boldsymbol{\Sigma}_{yy}$ denotes the $p \times p$ covariance matrix of $\mathbf{y}$, and $\boldsymbol{\Sigma}_{ff}$ denotes the $q \times q$ covariance matrix of $\mathbf{f}$. If the q common factors are assumed to be standardized and mutually uncorrelated, then $\boldsymbol{\Sigma}_{ff} = \boldsymbol{I}$, and

$$\boldsymbol{\Sigma}_{yy} = \boldsymbol{\Lambda}\boldsymbol{\Lambda}' + \boldsymbol{\Delta}. \tag{10}$$

Formally, the two cases represented by Eqs. 9 and 10 are indistinguishable. This is due to the fact that one can write $\boldsymbol{\Sigma}_{ff} = \mathbf{T}\mathbf{T}'$, where $\mathbf{T}$ is a lower triangular matrix, and rewrite Eq. 9 as $\boldsymbol{\Sigma}_{yy} = \boldsymbol{\Lambda}^*\boldsymbol{\Lambda}^{*\prime} + \boldsymbol{\Delta}$, where $\boldsymbol{\Lambda}^* = \boldsymbol{\Lambda}\mathbf{T}$. Despite this formal indistinguishability, however, the representations of the data in terms of correlated and of uncorrelated factors would be different. Thus, for purposes of interpretation, it may be important to distinguish between the two cases.

An alternative way of motivating the factor-analytic model, which may be more appealing statistically, is as follows: given p observable variables, $\mathbf{y}$, do there exist $q(<p)$ variables, $\mathbf{f}$, such that the partial correlations between every pair of the original variables upon elimination of the q f-variables are all zero? An affirmative answer to this question may be shown to be equivalent to the factor-analytic model as specified by Eq. 7 (or Eq. 9), for from Eq. 7 it follows that the conditional covariance matrix of $\mathbf{y}$ given $\mathbf{f}$ = covariance matrix of $\mathbf{y} - \boldsymbol{\Lambda}\mathbf{f}$ = covariance matrix of $\mathbf{z} = \boldsymbol{\Delta}$. Hence, from the assumptions concerning $\boldsymbol{\Delta}$, the off-diagonal elements of the conditional covariance matrix, which are the partial covariances between pairs of the y-variables, given $\mathbf{f}$, are all 0, so that the partial correlations between pairs of the elements of $\mathbf{y}$, given $\mathbf{f}$, are also all 0. Conversely, suppose there exists $\mathbf{f}$ such that the partial correlation between every pair of y-variables, given $\mathbf{f}$, is 0. Then, from the definition of partial correlation, it follows that the covariance matrix of the "residuals" from the linear regression of $\mathbf{y}$ on $\mathbf{f}$ is diagonal. The linear regression of $\mathbf{y}$ on $\mathbf{f}$ is $\mathscr{E}(\mathbf{y} \mid \mathbf{f}) = \boldsymbol{\Sigma}_{yf}\boldsymbol{\Sigma}_{ff}^{-1}\mathbf{f}$, where $\mathscr{E}$ stands for expectation, and $\boldsymbol{\Sigma}_{yf}$ denotes the $p \times q$ cross-covariance matrix between $\mathbf{y}$ and $\mathbf{f}$. The conditional covariance matrix of $\mathbf{y}$ given $\mathbf{f}$ = covariance matrix of the "residuals," $\mathbf{y} - \boldsymbol{\Sigma}_{yf}\boldsymbol{\Sigma}_{ff}^{-1}\mathbf{f} = \boldsymbol{\Sigma}_{yy} - \boldsymbol{\Sigma}_{yf}\boldsymbol{\Sigma}_{ff}^{-1}\boldsymbol{\Sigma}_{yf}'$. If this is a diagonal matrix, $\boldsymbol{\Delta}$, then it follows that $\boldsymbol{\Sigma}_{yy} = \boldsymbol{\Lambda}\boldsymbol{\Sigma}_{ff}\boldsymbol{\Lambda}' + \boldsymbol{\Delta}$, where

$$\boldsymbol{\Lambda} = \boldsymbol{\Sigma}_{yf}\boldsymbol{\Sigma}_{ff}^{-1}. \tag{11}$$

Thus the equivalence claimed at the beginning of this paragraph follows.

One hope in using factor analysis is that the number of common factors, q, will be much smaller than the number of original variables, p, thus leading to a parsimony of description which may aid in interpretation and understanding. Formally, the problems to be solved in a factor-analytic approach include (*a*) finding $\mathbf{\Lambda}$ of minimal rank to satisfy the model as summarized by Eq. 9 (or by Eq. 10); (*b*) estimating $\mathbf{\Delta}$; and (*c*) making inferences about $\mathbf{F}$. For present purposes, most of the discussion in the rest of this section is devoted to (*a*). A more extensive and thorough discussion of factor analysis will be found in Harman (1967) and Lawley & Maxwell (1963).

Equation 11 suggests that one way of obtaining $\mathbf{\Lambda}$ is to regress $\mathbf{y}$ on $\mathbf{f}$. However, this is not a feasible direct approach since $\mathbf{f}$ is not observable. Two other methods for estimating the factor loadings—the principal factor and the maximum likelihood methods—are outlined below.

Before discussing these methods, a few additional concepts and terms need to be introduced. First, as a consequence of Eq. 10, one has

$$\text{variance}\,(y_i) = \sigma_{ii} = \sum_{j=1}^{q} \lambda_{ij}^2 + \delta_i^2, \qquad i = 1, \ldots, p,$$

where $\mathbf{\Sigma}_{yy} = ((\sigma_{il}))$ and $\mathbf{\Lambda} = ((\lambda_{ij}))$. The quantity

$$h_i^2 = \sum_{j=1}^{q} \lambda_{ij}^2 \tag{12}$$

is called the *communality* of the ith variable, while δ_i^2 is termed the *uniqueness* of the ith variable ($i = 1, 2, \ldots, p$). It follows that

$$\begin{aligned}\text{total variance} = \text{tr}(\mathbf{\Sigma}_{yy}) &= \sum_{i=1}^{p} \sigma_{ii} \\ &= \sum_{i=1}^{p} (h_i^2 + \delta_i^2) \\ &= \sum_{i=1}^{p} \sum_{j=1}^{q} \lambda_{ij}^2 + \sum_{i=1}^{p} \delta_i^2 \\ &= V + \delta^2, \end{aligned} \tag{13}$$

where $V = \sum_{i=1}^{p} \sum_{j=1}^{q} \lambda_{ij}^2$ is the *total communality*, and $\delta^2 = \sum_{i=1}^{p} \delta_i^2$.

Second, from Eq. 10, it is clear that any orthogonal transformation (rotation) of the factors $\mathbf{f}$ will still satisfy the constraint on the covariance structure as specified by Eq. 10. In fact, any transformation from $\mathbf{f}$ and $\mathbf{\Lambda}$ to $\mathbf{f}_1 = \mathbf{Af}$ and $\mathbf{\Lambda}_1 = \mathbf{\Lambda B}$, where $\mathbf{A}$ and $\mathbf{B}$ are $q \times q$ matrices such that their product $\mathbf{AB}$ is orthogonal, will satisfy the same constraint. In this general transformation, however, although $\mathbf{f}$ may be uncorrelated and standardized, the derived set $\mathbf{f}_1$ need not have either property. If one wishes to remain with standardized uncorrelated factors, the choice of $\mathbf{A}$ to be orthogonal and $\mathbf{B}=\mathbf{A}'$ will suffice. The indeterminacy implied by such transformations in any "solution" obtained for the factor loadings is used to advantage in the so-called practice of *rotating* a preliminary solution to obtain a more interpretable final solution. Further discussion of the issues and procedures involved in rotation is available in Harman (1967).

Without any loss of generality, the original variables may be assumed to be standardized ($\sigma_{ii}=1$ for $i=1, 2, \ldots, p$) so that, with standardized uncorrelated factors, Eq. 10 specifies the following structural representation of the $p \times p$ correlation matrix, $\mathbf{\Gamma}=((\rho_{il}))$:

$$\left.\begin{aligned} &\mathbf{\Gamma}=\mathbf{\Lambda}\mathbf{\Lambda}'+\mathbf{\Delta}, \\ \text{or} \qquad & \\ &\rho_{ii}=1=h_i^2+\delta_i^2 \qquad \text{for } i=1, \ldots, p, \\ \text{and} \qquad & \\ &\rho_{il}=\sum_{j=1}^{q}\lambda_{ij}\lambda_{lj} \qquad \text{for } i\neq l. \end{aligned}\right\} \tag{14}$$

The $p \times p$ matrix $\mathbf{\Gamma}^*$, whose diagonal elements are the communalities h_i^2 and off-diagonal elements are the correlation coefficients ρ_{il} between pairs of the observed variables, is called the *reduced correlation matrix.* This matrix plays an important role in the principal factor method of determining $\mathbf{\Lambda}$ and has the following properties: (i) as a consequence of Eq. 14, if every $\delta_i^2>0$, the diagonal elements of $\mathbf{\Gamma}^*$ are all less than 1; (ii) the rank of $\mathbf{\Gamma}^*=$ the minimum number of linearly independent factors required for reproducing the correlations among the observed variables $=$ the dimensionality of the factor space; (iii) if $h_i^2=1$ for all i, the rank of $\mathbf{\Gamma}^*=p$, and no reduction of dimensionality is accomplished by factor analysis.

The previously stated problem (a) of factor analysis may now be restated as follows: given the intercorrelations among a set of p observed responses, choose the set $\{h_i^2\}$ so as to minimize the rank of $\mathbf{\Gamma}^*$.

The two methods of determining $\mathbf{\Lambda}$ under the model of Eq. 10 (or, equivalently, Eq. 14) are described next.

The Principal Factor Method. This method was proposed by Thurstone (1931)† and more fully described by Thomson (1934). It should not be confused with the principal components method described in Section

† Thurstone credits Walter Bartky with the mathematical solution associated with the technique. It may be of interest that this is the same Bartky, an astronomer, who is also said to have been the originator of the ideas of sequential sampling.

2.2.1, and the similarities and differences of the two techniques are discussed later.

From Eq. 12 it follows that for any factor, f_j, its contribution to the communality of the variable y_i is λ_{ij}^2. Hence the contribution of f_j to the communalities of all p observed responses is

$$V_j = \sum_{i=1}^{p} \lambda_{ij}^2 = \boldsymbol{\lambda}_j'\boldsymbol{\lambda}_j, \tag{15}$$

where $\boldsymbol{\lambda}_j$ denotes the jth column of $\boldsymbol{\Lambda}$, and $j = 1, 2, \ldots, q$. The total communality defined by Eq. 13 is, of course, $V = \sum_{j=1}^{q} V_j$.

The principal factor method involves, as the first stage, choosing the coefficients, $\lambda_{11}, \ldots, \lambda_{p1}$, of the first factor f_1 so as to maximize the contribution of f_1 to the total communality subject to the constraints on the correlation structure as summarized by Eq. 14. In other words, we wish to choose $\boldsymbol{\lambda}_1$ so as to maximize $V_1 = \boldsymbol{\lambda}_1'\boldsymbol{\lambda}_1$ subject to the $p(p+1)/2$ constraints implied by $\boldsymbol{\Gamma}^* = \boldsymbol{\Lambda}\boldsymbol{\Lambda}'$.

The constrained maximization turns out to be equivalent to finding the eigenvalues and eigenvectors of $\boldsymbol{\Gamma}^*$ (see Harman 1967, for details); in fact, the maximum value of V_1 is the largest eigenvalue of $\boldsymbol{\Gamma}^*$, and the required maximizing value of $\boldsymbol{\lambda}_1$ is just proportional to the corresponding eigenvector. Thus, if γ_1 is the largest eigenvalue of $\boldsymbol{\Gamma}^*$ and $\boldsymbol{\alpha}_1$ is the corresponding eigenvector, which is normalized so that $\boldsymbol{\alpha}_1'\boldsymbol{\alpha}_1 = 1$, then

$$\boldsymbol{\lambda}_1 = \sqrt{\gamma_1} \cdot \boldsymbol{\alpha}_1, \tag{16}$$

and the maximum value of $V_1 = \boldsymbol{\lambda}_1'\boldsymbol{\lambda}_1 = \gamma_1$. This "solution" to the problem of determining $\boldsymbol{\lambda}_1$ is, however, artificial or circular in that the diagonal elements of $\boldsymbol{\Gamma}^*$ in turn involve $\lambda_{11}^2, \ldots, \lambda_{p1}^2$. The assumption is that the diagonal elements, namely, the communalities h_i^2's are independently known or specifiable. A method of specifying these is mentioned later.

At the second stage of the principal factor method, having determined $\boldsymbol{\lambda}_1$ as above, one seeks to determine $\boldsymbol{\lambda}_2$ so as to maximize $V_2 = \boldsymbol{\lambda}_2'\boldsymbol{\lambda}_2$ subject to a constraint on the residual reduced correlations after removal of the first factor. If

$$\boldsymbol{\Gamma}_1^* = \boldsymbol{\Gamma}^* - \boldsymbol{\lambda}_1\boldsymbol{\lambda}_1' \tag{17}$$

denotes the $p \times p$ matrix of residual reduced correlations after removing f_1, then the constraints are that

$$\boldsymbol{\Gamma}_1^* = [\boldsymbol{\lambda}_2 \cdots \boldsymbol{\lambda}_q] \begin{bmatrix} \boldsymbol{\lambda}_2' \\ \cdot \\ \cdot \\ \cdot \\ \boldsymbol{\lambda}_q' \end{bmatrix}. \tag{18}$$

The present constrained maximization problem, however, is of the same mathematical form as the one at the first stage of the method. Hence the required solution for $\boldsymbol{\lambda}_2$ is proportional to the eigenvector of $\boldsymbol{\Gamma}_1^*$ corresponding to its largest eigenvalue. An eigenanalysis of $\boldsymbol{\Gamma}_1^*$ is, however, not essential since the required solution for $\boldsymbol{\lambda}_2$ may be shown to be equivalent to choosing $\boldsymbol{\lambda}_2$ proportional to the eigenvector associated with the second largest eigenvalue of the original reduced correlation matrix, $\boldsymbol{\Gamma}^*$. Arguments for establishing this computationally convenient result follow.

If $\boldsymbol{\alpha}_k$ is the eigenvector of $\boldsymbol{\Gamma}^*$ corresponding to the eigenvalue γ_k $(k=1, 2, \ldots, p)$, where $\gamma_1 \geq \gamma_2 \geq \cdots$, then

$$\begin{aligned} \boldsymbol{\Gamma}_1^*\boldsymbol{\alpha}_k = (\boldsymbol{\Gamma}^* - \boldsymbol{\lambda}_1\boldsymbol{\lambda}_1')\boldsymbol{\alpha}_k &= (\boldsymbol{\Gamma}^* - \gamma_1\boldsymbol{\alpha}_1\boldsymbol{\alpha}_1')\boldsymbol{\alpha}_k \\ &= \gamma_k\boldsymbol{\alpha}_k - \gamma_1\boldsymbol{\alpha}_1\boldsymbol{\alpha}_1'\boldsymbol{\alpha}_k. \end{aligned} \tag{19}$$

Using the orthonormal property of the set of eigenvectors $\{\boldsymbol{\alpha}_k\}$, it follows from Eq. 19 that $\boldsymbol{\alpha}_1$ is an eigenvector of $\boldsymbol{\Gamma}_1^*$ corresponding to the eigenvalue zero and that, for $k=2, 3, \ldots, p$, γ_k is an eigenvalue of $\boldsymbol{\Gamma}_1^*$ with associated eigenvector $\boldsymbol{\alpha}_k$. In particular, the largest eigenvalue of $\boldsymbol{\Gamma}_1^*$ is γ_2, the second largest eigenvalue of $\boldsymbol{\Gamma}^*$, and the corresponding eigenvector is $\boldsymbol{\alpha}_2$. The required $\boldsymbol{\lambda}_2$ is $\sqrt{\gamma_2}\cdot\boldsymbol{\alpha}_2$.

The remaining stages of the principal factor method now follow in exactly the same manner. Finding $\boldsymbol{\lambda}_3, \ldots, \boldsymbol{\lambda}_q$ so as to maximize the contributions of each corresponding factor to the total communality, subject to constraints on the residual reduced correlations at each stage, turns out to be equivalent to taking $\boldsymbol{\lambda}_j = \sqrt{\gamma_j}\cdot\boldsymbol{\alpha}_j$, for $j=3, \ldots, q$.

The descriptions above have been presented in terms of population characteristics. With data one would use a sample reduced correlation matrix, $\mathbf{R}^*$, in place of $\boldsymbol{\Gamma}^*$ as the input to the eigenanalysis, where

$$\mathbf{R}^* = \begin{pmatrix} \hat{h}_1^2 & r_{12} & \cdots & r_{1p} \\ & \hat{h}_2^2 & \cdots & r_{2p} \\ & & & \cdot \\ & & & \cdot \\ & & & \cdot \\ & & \cdot & \\ & & \cdot & \\ & & \cdot & \\ & & & \hat{h}_p^2 \end{pmatrix}, \tag{20}$$

and $\{\hat{h}_i^2\}$ are communalities estimated from the data, while r_{il} is the correlation coefficient between the ith and lth responses as calculated from the data. Thus, with data, the principal factor solution for $\boldsymbol{\Lambda}$ is

$$\hat{\boldsymbol{\Lambda}} = \mathbf{L} = ((l_{ij})) = [\mathbf{l}_1\mathbf{l}_2\cdots\mathbf{l}_q], \tag{21}$$

where $\mathbf{l}_j = \sqrt{c_j}\mathbf{a}_j$ for $j=1, \ldots, q$, and $c_1 > c_2 > \cdots > c_q > 0$ are the q largest eigenvalues of $\mathbf{R}^*$ with corresponding eigenvectors $\mathbf{a}_1, \ldots, \mathbf{a}_q$. Also, the estimate of δ_i^2 is

$$\hat{\delta}_i^2 = 1 - \sum_{j=1}^{q} l_{ij}^2, \qquad i=1, \ldots, p. \tag{22}$$

The procedure described above is complete except for specification of $\hat{h}_i^2$'s. If $\hat{h}_i^2 = 1$ for $i=1, \ldots, p$, then $\mathbf{R}^* = \mathbf{R}$, the ordinary correlation matrix of the responses, and the principal factor solution is exactly the same as the principal components solution for $\mathbf{R}$. Estimates of h_i^2 whose values are less than 1 include: (i) $\hat{h}_i^2 =$ the highest observed positive correlation of variable y_i with the remaining $(p-1)$ variables $=$ the largest positive element in the ith row (column) of $\mathbf{R}$; (ii) $\hat{h}_i^2 =$ the average (presumed positive) of the observed correlations of y_i with the other variables $= \sum_{\substack{l=1 \\ (l \neq i)}}^{p} r_{il}/(p-1)$; (iii) $\hat{h}_i^2 =$ square of the multiple correlation coefficient of y_i with the other variables $= 1-(1/r^{ii})$, where $((r^{ij})) = \mathbf{R}^{-1}$; and (iv) iterative estimates obtained by starting with an arbitrary set of values for $\hat{h}_i^2$'s to get a principal factor solution, thence using the sums of squares of the factor loadings for each variable in such a solution as the new values of $\hat{h}_i^2$'s, and repeating the process until the sets of successive estimates do not differ greatly. An intuitive basis for the third of these choices is that the squared multiple correlation coefficient measures the proportion of the observed total variability in a specific response that is accounted for by its regression on the remaining $(p-1)$ responses, and hence provides a measure of common or shared variance. A second reason for this choice is that, while 1 is an upper bound on $\hat{h}_i^2$, it can be shown that the squared multiple correlation coefficient involved is a lower bound. Many of the computer programs for performing a principal factor analysis use this choice for $\hat{h}_i^2$ as the standard one. In practice, except for small values of $p\,(\leq 10)$, the different choices of $\hat{h}_i^2$ do not seem in general to lead to noticeably different outcomes.

If some or all of the diagonal elements of $\mathbf{R}^*$ are less than 1, then $\mathbf{R}^*$ need not be positive semidefinite. Hence some of the eigenvalues, $\{c_j\}$, may be negative with the consequence that the vectors of factor loadings, $\mathbf{l}_j$'s, associated with these will be imaginary. In practice, one discards these negative eigenvalues and the associated imaginary vectors of loadings. In fact, since the sum of the eigenvalues of $\mathbf{R}^*$ equals the total communality, the sum of just the positive eigenvalues will exceed the total communality if there are any negative eigenvalues at all. Hence, in extracting the factors, one would not proceed until their number q was as large as the number of positive eigenvalues but, rather, would stop when $\sum_{j=1}^{q} c_j$ was close to $\mathrm{tr}(\mathbf{R}^*)$, the total communality.

Another useful procedure for guiding the choice of a value for q is to compute and study the residual correlation matrix after each factor has been fitted. Although all the numerical computations may be carried out on $\mathbf{R}^*$, from the standpoint of interpretation it is useful to compute the matrix of residual correlation coefficients,

$$\mathbf{R}_j^* = \mathbf{R}^* - \sum_{a=1}^{j} \mathbf{l}_a \mathbf{l}_a',$$

at the jth stage for $j = 1, 2, \ldots$.

Mention has been made that, when $\hat{h}_i^2$, for all i, are taken to be unity, the principal factor method is identical with a principal components analysis of the correlation matrix. In fact, if the communalities $\hat{h}_i^2$'s (or, equivalently, the $\hat{\delta}_i^2$'s) are all essentially equal and q is close to p, the principal factor method as described above and a principal components analysis of $\mathbf{R}$ would both lead to very similar results. The reason is that, if $\hat{\delta}_i^2 = d^2$ for all i, then $\mathbf{R}^* = \mathbf{R} - d^2\mathbf{I}$ and $c_j = b_j - d^2$ is an eigenvalue of $\mathbf{R}^*$ if b_j is an eigenvalue of $\mathbf{R}$. Hence the relationship $\mathbf{R}^*\mathbf{a}_j = c_j\mathbf{a}_j$ is equivalent to $(\mathbf{R} - d^2\mathbf{I})\mathbf{a}_j = (b_j - d^2)\mathbf{a}_j$, or to $\mathbf{R}\mathbf{a}_j = b_j\mathbf{a}_j$, so that the eigenvectors $\{\mathbf{a}_j\}$ of $\mathbf{R}^*$ are also those of $\mathbf{R}$. In practice, however, the $\hat{\delta}_i^2$'s (and $\hat{h}_i^2$'s) are often unequal and $q \ll p$, so that the principal factor method may lead to results that are different from those obtained by a principal components analysis of the correlation matrix. The use of values of $\hat{h}_i^2$ less than 1 has an interesting interpretation in terms of an idea utilized in ridge regression (Theil, 1963; Hoerl & Kennard, 1970). In multiple regression analysis (see Sections 3.3 and 5.2.1), the sum-of-products matrix of the independent variables may be nearly singular in some applications, perhaps because of round-off errors or high intercorrelations amongst the independent variables. The latter cause is referred to as "multicollinearity" in the econometric literature. The near singularity leads not only to numerical difficulties but also to estimates of regression coefficients that have undesirable statistical properties. The idea of ridge regression for handling this problem is to add a constant multiple of the identity matrix to the sum-of-products matrix (i.e., add a positive constant to each of the diagonal elements of the latter) and to utilize the resultant matrix in place of the nearly singular matrix. Thus, in ridge regression, a nearly singular covariance or correlation matrix is adjusted to become "more" positive definite (see also Devlin et al., 1975) by increasing the diagonal elements, whereas decreasing the diagonal elements involved in the principal factor method will have a "deridging" effect. An implication of this, and also of the discussion in the preceding paragraph, is that there is a strong implicit commitment in the principal factor method to a linear model for reducing dimensionality.

Indeed some authors have tried to distinguish between the principal components and principal factor techniques on the grounds of their respective degrees of commitment to a linear model. This, however, does not seem to be a crucial distinction since both techniques have implicit, as well as explicit, linearity considerations underlying them, and both tend to be inadequate in the face of nonlinearity (see Examples 2–4). Perhaps a better distinction to be made is that the factor analysis model (Eqs. 7, 9, 10) is more explicit than the one underlying principal components in assuming a space (linear) of *reduced* dimensionality (i.e., $q \ll p$) for explaining the correlation structure of the original responses.

The Maximum Likelihood Method. This method, originally proposed by Lawley (1940), has received considerable attention from statisticians (see Anderson & Rubin, 1956; Howe, 1955), perhaps because of its usage of the criterion of maximizing a likelihood function, which is a familiar concept and method in statistics.

The assumption in this method is that the observations (viz., the columns of the $p \times n$ matrix $\mathbf{Y}$ of Eq. 8) constitute a random sample from a nonsingular p-variate normal distribution whose covariance matrix $\mathbf{\Sigma}_{yy}$ has a structure specified by Eq. 10. Furthermore, if $p \leq (n-1)$, the sample covariance matrix, $\mathbf{S}$, will be nonsingular with probability 1 and will have a Wishart distribution. Using the Wishart density as a starting point, one obtains the log-likelihood function of $\mathbf{\Lambda}$ and $\mathbf{\Delta}$,

$$\mathscr{L}(\mathbf{\Lambda}, \mathbf{\Delta} \mid \mathbf{S}) = -\frac{n-1}{2}[\ln|\mathbf{\Lambda\Lambda}' + \mathbf{\Delta}| + \text{tr}\{(\mathbf{\Lambda\Lambda}' + \mathbf{\Delta})^{-1}\mathbf{S}\}]. \tag{23}$$

Hence maximizing $\mathscr{L}$ with respect to the elements of $\mathbf{\Lambda}$ and $\mathbf{\Delta}$ is equivalent to minimizing $\ln|\mathbf{\Lambda\Lambda}' + \mathbf{\Delta}| + \text{tr}\{(\mathbf{\Lambda\Lambda}' + \mathbf{\Delta})^{-1}\mathbf{S}\}$, and the resulting values, $\hat{\mathbf{\Lambda}}$ and $\hat{\mathbf{\Delta}}$, are the required maximum likelihood estimates.

The indeterminacy of $\hat{\mathbf{\Lambda}}$ up to rotation is handled in maximum likelihood estimation by imposing the constraint that the matrix

$$\hat{\mathbf{J}} = \hat{\mathbf{\Lambda}}'\hat{\mathbf{\Delta}}^{-1}\hat{\mathbf{\Lambda}} \tag{24}$$

be diagonal. This constraint simplifies the solving of the likelihood equations. The actual equations that need to be solved by iterative methods are

$$\hat{\mathbf{J}}\hat{\mathbf{\Lambda}}' = \hat{\mathbf{\Lambda}}'\hat{\mathbf{\Delta}}^{-1}(\mathbf{S} - \hat{\mathbf{\Delta}}), \qquad \hat{\mathbf{\Delta}} = \text{diag}(\mathbf{S} - \hat{\mathbf{\Lambda}}\hat{\mathbf{\Lambda}}'), \tag{25}$$

where $\hat{\mathbf{J}}$ is defined by Eq. 24, and diag($\mathbf{M}$) denotes a diagonal matrix whose diagonal elements are those of the square matrix $\mathbf{M}$ (for details see Lawley & Maxwell, 1963; Howe, 1955).

Accumulated experience with attempts at solving the likelihood equations (Eq. 25) has indicated that convergence to a solution may be a serious problem. More recently, Jöreskog (1967) and Lawley (1967) have developed numerical approaches for obtaining the maximum likelihood estimates that seem to circumvent this difficulty (see also the expository treatment by Jöreskog & Lawley, 1968). The modifications have two basic features. First, instead of solving the likelihood equations, a direct numerical maximization of the function $\mathscr{L}(\mathbf{\Lambda}, \mathbf{\Delta} \mid \mathbf{S})$ (or, equivalently, a minimization of $-[2/(n-1)]\ \mathscr{L}(\mathbf{\Lambda}, \mathbf{\Delta} \mid \mathbf{S}))$ is attempted. Second, the maximization (or equivalent minimization) is carried out in two parts—for a given $\mathbf{\Delta}$ find a $\mathbf{\Lambda}_{\mathbf{\Delta}}$ that maximizes $\mathscr{L}(\mathbf{\Lambda}, \mathbf{\Delta} \mid \mathbf{S})$, and then determine $\hat{\mathbf{\Delta}}$ as that value of $\mathbf{\Delta}$ which maximizes the function $\mathscr{L}_{\max}(\mathbf{\Delta}) = \mathscr{L}(\mathbf{\Lambda}_{\mathbf{\Delta}}, \mathbf{\Delta} \mid \mathbf{S})$. The required maximum likelihood estimates are $\hat{\mathbf{\Delta}}$ and $\hat{\mathbf{\Lambda}} = \mathbf{\Lambda}_{\hat{\mathbf{\Delta}}}$. The iterative scheme (see Jöreskog, 1967, for details) appears to work well, primarily because the determination of $\mathbf{\Lambda}_{\mathbf{\Delta}}$ for a given $\mathbf{\Delta}$ is quite straightforward in that it simply involves the determination of the q largest eigenvalues and the associated eigenvectors of the matrix $\mathbf{\Delta}^{-1/2} \cdot \mathbf{S} \cdot \mathbf{\Delta}^{-1/2}$. A general iterative numerical optimization technique (e.g., Fletcher & Powell, 1963) is then needed only at the second stage of maximization, viz., the maximization of $\mathscr{L}_{\max}(\mathbf{\Delta})$ with respect to the p diagonal elements of $\mathbf{\Delta}$.

An implication of Eq. 25 is that scaling any observed variable would induce proportional changes in the estimates of the factor loadings for that variable. Independence, in this sense, of the solution of Eq. 25 from the scales of the original variables has the numerical consequence that one may employ the covariance matrix or the correlation matrix of the original variables in seeking the solution. It should be recognized, however, that if one were to use the sample correlation matrix as the starting point, the Wishart density would not provide the basis of the initial likelihood function. The numerical implication of "invariance" of the solution $\hat{\mathbf{\Lambda}}$ of Eq. 25 is, therefore, unrelated to the statistical considerations that underlie the formulation in terms of maximum likelihood estimation. The latter appears to be feasible only in terms of the sample covariance matrix.

A statistical advantage claimed for the maximum likelihood approach is that the asymptotic (viz., n large, $p < n$, and $q \ll p$) properties of such estimates are known and may be used for purposes of statistical inference (see Lawley, 1940; Anderson & Rubin, 1956). In particular, for example, one can obtain a likelihood ratio test for the adequacy of the hypothesized number, q, of common factors. The essential result derived by Lawley (1940) is that the observed value of the statistic

$$(n-1)\left\{\ln\left[\frac{|\hat{\mathbf{\Lambda}}\hat{\mathbf{\Lambda}}' + \hat{\mathbf{\Delta}}|}{|\mathbf{S}|}\right] + \operatorname{tr}[(\hat{\mathbf{\Lambda}}\hat{\mathbf{\Lambda}}' + \hat{\mathbf{\Delta}})^{-1}\mathbf{S}] - p\right\} \tag{26}$$

may be referred to a chi-squared distribution with ν degrees of freedom, where $\hat{\mathbf{\Lambda}}$ and $\hat{\mathbf{\Delta}}$ are the estimates that satisfy Eq. 25, $\mathbf{S}$ is the sample covariance matrix, and

$$\nu = \tfrac{1}{2}[(p-q)^2 - p - q]. \tag{27}$$

[*Note*: For a given p, q has to be $\ll p$ for ν to be positive.] Statistically large observed values of the statistic imply that the number of common factors needed to adequately reproduce the correlations among the original variables is larger than q. Bartlett (1951) has suggested using the multiplicative factor $\{n - p/3 - 2q/3 - 11/6\}$ in place of $(n-1)$ in Eq. 26 for improving the chi-squared approximation.

Table 1 provides a summary comparison of features of the principal factor and maximum likelihood methods.

Table 1

Feature	Principal Factor Method	Maximum Likelihood Method
1. Estimates of communalities	Required	Not required
2. Dimensionality of common factors space	Inferable from manner of computing	Assumed for obtaining a solution but then may be statistically tested for adequacy
3. Distributional assumptions	None specific	Multivariate normal
4. Formal statistical inference status	Not much is known	Large-sample theory is available
5. Iteration for obtaining the solution	Optional (i.e., not required unless one chooses to estimate comunalities iteratively)	Necessary
6. Convergence of iterative procedure	Good	May be poor for solving Eq. 25, but modified method seems good
7. Scale "invariance"	No	Yes

A useful graphical technique, associated with both methods of factor analysis, is to represent the original variables in terms of their factor loadings in a space that corresponds to the common factors. Thus, using pairs (and/or triplets) of axes, one obtains p points whose coordinates are factor loadings with respect to pairs (and/or triplets) of the common factors. Such plots can often aid in interpreting the nature of the factors, as well as in suggesting "rotations" to more meaningful sets of coordinates for the factors.

Although factor analysis has been used most extensively as a tool in psychology and the social sciences, applications have been made to other fields as well. Seal (1964) summarizes various biological applications of factor analysis, and Imbrie (1963), Imbrie & Van Andel (1964), and Imbrie & Kipp (1971) have used it in analyzing certain geological problems.

Some applications of factor analysis, especially in the social sciences, raise questions concerning its usefulness for achieving parsimony of description or for incisively understanding a complex of observed variables in terms of a few underlying variables.Often in questionnaire survey data, for example, built-in or a priori groupings of the initial variables are the ones that are uncovered by using factor analysis. Even in such examples, however, the technique is perhaps useful in that it provides a more quantitative understanding of the qualitative prior groupings.

A different issue related to the usefulness of the method is its inadequacy in the face of nonlinearity of the underlying relationships. Some fairly recent work in the literature (McDonald, 1962, 1967; Carroll, 1969) is directed toward nonlinear factor analysis methodology. Other nonlinear techniques are considered next in this chapter.

2.3. NONMETRIC METHODS FOR NONLINEAR REDUCTION OF DIMENSIONALITY

A class of procedures, collectively designated as *multidimensional scaling techniques*, has been developed in connection with the following problem: given a set of observed measures of similarity or dissimilarity between every pair of n objects, find a representation of the objects as points in Euclidean space such that the interpoint distances in some sense "match" the observed similarities or dissimilarities. Some examples of measures of similarity are (i) confusion probabilities or the proportion of times one stimulus is identified as another among n stimuli, (ii) the absolute value of a correlation coefficient, and (iii) any index of profile similarity.

Several approaches have been proposed (see Coombs, 1964, for a general discussion) to the problem of multidimensional scaling, but for present purposes only the technique developed by Shepard (1962a, b) and further refined by Kruskal (1964a, b) is considered. A central feature of the Shepard-Kruskal approach is the specification of monotonicity as the sense in which interpoint distances are to match the observed dissimilarities among the objects; i.e., the larger the specified dissimilarity between two objects, the larger should the interpoint distance be in the Euclidean representation of these objects. Kruskal (1964a, b) not only developed efficient algorithms for using the method but also proposed an explicit measure for judging the degree of conformity to monotonicity in any solution. Furthermore, as an integral part of their approach, Shepard and Kruskal obtain a graphical display of the data-determined monotone relationship between dissimilarity and distance (see details below).

Another important characteristic of the approach, demonstrated empirically by Shepard (1962a), is that one could start just from the nonmetric rank order information about the dissimilarities and still obtain quite "tightly determined" configurations. The technique exploits nonmetric information, when enough of it is available, to derive metric representations of the data (see also Abelson & Tukey, 1959). In this respect it is an interesting example of a data-analytic technique with a counterobjective to that underlying the practice of replacing metric observations by their ranks as a prelude to employing some nonparametric statistical procedures.

As motivation for the concepts and procedures involved in the Shepard-Kruskal approach, consider the case wherein one has four $(=n)$ objects and six observed values of dissimilarity for the six possible pairs of the objects. If δ_{ij} denotes the dissimilarity value for the pair of objects i and j, for $i, j=1, 2, 3, 4$, then suppose, for example, that the following rank ordering among the six observed dissimilarity values holds:

$$\delta_{23}<\delta_{12}<\delta_{34}<\delta_{13}<\delta_{24}<\delta_{14}. \tag{28}$$

In other words, the second and third objects are judged to be least dissimilar (or most similar), the first and second objects next least dissimilar, and so on, with objects 1 and 4 ranked as most dissimilar (or least similar). Suppose that the objects are represented as points in a Euclidean space of a specified dimensionality, and let $\mathbf{y}_i$ denote the column vector of coordinates of the point corresponding to the ith object, $i=1, \ldots, 4$. Then the familiar unweighted Euclidean distance between the points representing objects i and j is

$$d_{ij}=[(\mathbf{y}_i-\mathbf{y}_j)'(\mathbf{y}_i-\mathbf{y}_j)]^{1/2}, \tag{29}$$

$i<j=2, 3, 4$. The monotonicity constraint, which is central in the approach, is said to be met perfectly in this simple example if, corresponding to the ordering of the observed dissimilarities shown in Eq. 28, the d_{ij}'s calculated by using Eq. 29 turn out to satisfy the following:

$$d_{23}\leq d_{12}\leq d_{34}\leq d_{13}\leq d_{24}\leq d_{14}. \tag{30}$$

In other words, the order relationship among the interpoint distances in the Euclidean representation of the objects is in exact concordance with the order relationship among the observed dissimilarities. Such a perfect match may, of course, not hold in a particular Euclidean representation, and one then needs both a measure to evaluate the closeness of match and a method of determining a configuration so as to achieve as close a match as possible.

A graphical representation that facilitates understanding the explicit measure of monotonicity and the mode of analysis proposed by Kruskal (1964a, b) is a scatter plot of points whose coordinates are (d_{ij}, δ_{ij}). Thus, in the above simple example, one can obtain a plot of six points as shown, for instance, in Figure 2*a* by the crosses, which are labeled by the pair of object numbers to which each of them corresponds. Corresponding to the perfect monotonicity implied by Eqs. 28 and 30, the configuration of the crosses is such that the line segments joining the points form a chain in which, as one moves upward, one moves always to the right as well.

Complete conformity to monotonicity is always achievable by using a representation in a space of sufficiently high ($\geq n-1$ with n objects) dimensionality. The primary interest, therefore, is to find a low-dimensional representation in which conformity to monotonicity is achieved to a reasonable degree if not perfectly.

Thus suppose that, in the same example, one has a Euclidean representation of the four objects in which the ordered interpoint distances turn out to be

$$d_{23} < d_{34} < d_{12} < d_{13} < d_{14} < d_{24}. \tag{31}$$

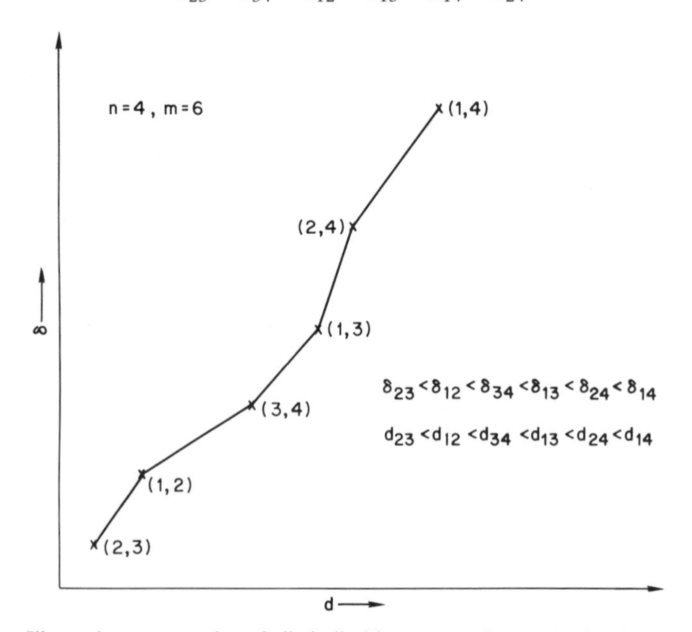

Fig. 2*a*. Illustrative scatter plot of dissimilarities versus distances wherein monotonicity constraint is satisfied.

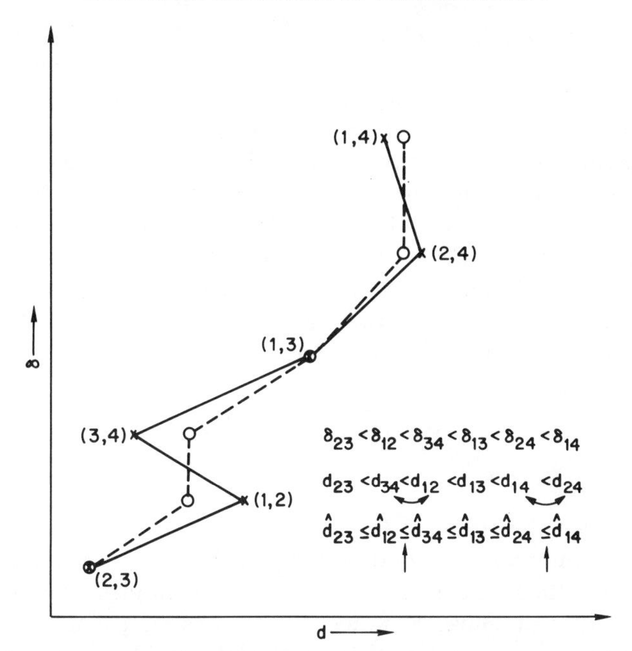

Fig. 2*b*. Illustrative scatter plot of dissimilarities versus distances wherein monotonicity constraint is not satisfied.

The monotonicity constraint is now violated in that the order relationship between the interpoint distances between objects 3 and 4 and between objects 1 and 2 is not the same as the order relationship between the corresponding observed dissimilarities as specified by Eq. 28. There is also a violation of monotonicity by the interpoint distances between objects 1 and 4 and between objects 2 and 4. The scatter plot of (d_{ij}, δ_{ij}) corresponding to this situation is shown by the $\times$'s in Figure 2*b*, and the chain of lines joining the points is now observed to zigzag instead of always moving to the right as one moves upward. In this situation, one may wish to "fit" a set of values, $\hat{d}_{ij}$'s, such that the fitted values will indeed satisfy the monotonicity constraint so that

$$\hat{d}_{23} \leq \hat{d}_{12} \leq \hat{d}_{34} \leq \hat{d}_{13} \leq \hat{d}_{24} \leq \hat{d}_{14}, \tag{32}$$

corresponding to the order relationship in Eq. 28. A satisfactory set of fitted values in this example would be the abscissa values of the "fitted" points which are shown as O's and joined by dashed line segments in Figure 2*b*. Notice that only the $\hat{d}$ values for distances that did not conform to monotonicity are different from the corresponding d values. In fact, in the example the $\hat{d}$ value for both d_{12} and d_{34}, which, for instance, violate monotonicity, is just the average of d_{12} and d_{34}, i.e.,

$\hat{d}_{12}=\hat{d}_{34}=(d_{12}+d_{34})/2$; and, similarly, $\hat{d}_{24}=\hat{d}_{14}=(d_{14}+d_{24})/2$. Apart from conforming to monotonicity, however, these fitted values may not, in fact, be distances in the sense that there is a configuration of points in Euclidean space whose interpoint distances are these values.

One measure for assessing the fit (viz., the conformity to monotonicity) of any proposed configuration is the sum of squares of deviations, $\sum_{i<j}(d_{ij}-\hat{d}_{ij})^2$. This measure of goodness of fit, although invariant under shifts (translations), reflections, and rotations (orthogonal transformations) of the coordinates in the Euclidean representation of the objects, is not invariant under uniform shrinking or dilation. Hence Kruskal (1964a) proposed the following normalized measure of goodness of fit:

$$S=\left[\frac{\sum_{i<j}(d_{ij}-\hat{d}_{ij})^2}{\sum_{i<j}d_{ij}^2}\right]^{1/2}, \tag{33}$$

which he called the *stress.*

The stress may now be used as the basis for a systematic method of obtaining the fitted values. Given a set of d_{ij}'s, in fact, one may choose the $\hat{d}_{ij}$'s so as to minimize S subject to the constraint that they are to be monotone nondecreasing with the observed dissimilarity values, δ_{ij}'s, This minimization problem is equivalent to so-called *monotone least squares* or *monotone regression* and has been considered conceptually and algorithmically in other contexts of statistical applications (see Bartholomew, 1959; Miles, 1959; Barton & Mallows, 1961; van Eeden, 1957a, b). To avoid further notation, it is assumed that the fitted values, $\hat{d}_{ij}$'s, are in fact always obtained by this process of minimization, and the stress of a given configuration representing the initial objects in a Euclidean space is the value of S given by Eq. 33, using such fitted values. This value of S, of course, depends on the given configuration, and one may wish to make this relationship clearer by denoting it as $S(\mathbf{y}_1, \mathbf{y}_2, \ldots)$, where $\mathbf{y}_i$ is the vector of coordinates of the point corresponding to the ith object.

The next step is to determine the "best" configuration in a Euclidean space of specified dimensions. Such a configuration is one in the space of specified dimensionality whose stress is a minimum among all configurations in that space, i.e., one wishes to determine $(\mathbf{y}_1^*, \mathbf{y}_2^*, \ldots, \mathbf{y}_n^*)$ so that

$$S(\mathbf{y}_1^*, \mathbf{y}_2^*, \ldots, \mathbf{y}_n^*)=\min_{\{\mathbf{y}_1,\ldots,\mathbf{y}_n\}} S(\mathbf{y}_1, \mathbf{y}_2, \ldots, \mathbf{y}_n).$$

Viewed as a trial and error process, what is involved is to start with a trial configuration, and then if $\hat{d}_{ij}<d_{ij}$ to move $\mathbf{y}_i$ and $\mathbf{y}_j$ closer, or if $\hat{d}_{ij}>d_{ij}$ to move $\mathbf{y}_i$ and $\mathbf{y}_j$ apart, so that in either case one is attempting to make d_{ij} resemble $\hat{d}_{ij}$ more closely. A systematic approach to this problem is provided by considering $S(\mathbf{y}_1, \ldots, \mathbf{y}_n)$ as a function of the coordinates of all n points (i.e., a function of $n \times t$ variables if one is using a space of t dimensions for the representation), and then using a general numerical technique of function optimization, such as steepest descent, for determining the location of the minimum value of $S(\mathbf{y}_1, \ldots, \mathbf{y}_n)$.

Next, there is the issue of the choice of the dimensionality for the Euclidean representation. If $S_0(t) = S(\mathbf{y}_1^*, \mathbf{y}_2^*, \ldots, \mathbf{y}_n^*)$ denotes the minimum value of the stress associated with the minimum stress configuration in a t-dimensional space, Kruskal (1964a) suggests basing the choice of q, the minimum adequate value of t, on a study of a plot of $S_0(t)$ versus t. As t increases, $S_0(t)$ will decrease and, in fact, will be 0 for $t \geq (n-1)$. As general though not rigid benchmarks, Kruskal (1964a) proposes that a value of $S_0(t)$ of 20% be interpreted as suggesting a poor fit, 10% a fair fit, 5% a good one, and $2\frac{1}{2}$% an excellent one, with 0% being a perfect fit. In addition to these general guidelines, one may decide on a value for q by looking for an "elbow" in the plot of $S_0(t)$ versus t (see the discussion in Example 5).

The entire procedure can now be summarized in terms of the following steps:

1. For n objects, obtain the initial information, which is the rank ordering of the $m = n(n-1)/2$ dissimilarity values, δ_{ij}'s, among every pair of the n objects.

2. Given the m dissimilarity values, with the ordering

$$\delta_{i_1j_1} < \delta_{i_2j_2} < \cdots < \delta_{i_mj_m}, \tag{34}$$

and using some initial trial configuration of points $\mathbf{y}_{i0}$ ($i = 1, 2, \ldots, n$), in $t(\geq 1)$ dimensions, determine the interpoint distances d_{ij}'s (see Eq. 29) and fit $\hat{d}_{ij}$'s so that

$$\hat{d}_{i_1j_1} \leq \hat{d}_{i_2j_2} \leq \cdots \leq \hat{d}_{i_mj_m}. \tag{35}$$

For a given configuration, the $\hat{d}_{ij}$'s are chosen so as to minimize the stress S (see Eq. 33) subject to the monotonicity constraint, Eq. 35. The algorithm required here is that of so-called monotone regression.

3. Next, using these $\hat{d}_{ij}$'s and considering S as a function of the $n \times t$ coordinates of the n points in the representation, determine an improved configuration, $\{\mathbf{y}_{i1}\}$, and thence the new d_{ij}'s, $\hat{d}_{ij}$'s, etc., until the best configuration in t dimensions is found as the one whose stress is $S_0(t) = \min\{S\}$, where the minimum is over all configurations in t dimensions. Steepest descent or some other general function minimization algorithm may be used for this step.

4. Finally, plot $S_0(t)$ versus t and choose q, the number of dimensions, as the minimum "adequate" value of t from the indications in such a plot.

For simplicity of exposition, the above discussion has assumed that the initially observed dissimilarity values are symmetric ($\delta_{ij} = \delta_{ji}$), that there are no ties among them, and that they are available for all possible pairs of the objects. Kruskal (1964a, b) suggests methods for handling asymmetries, ties, and missing observations, and also describes the details of the algorithms developed and implemented by him for using the technique.

Neither the final "best fitting" configuration nor the configuration at any stage is unique in that any similarity transformation (i.e., translation, rotation, reflection, or dilation) of the configuration will also have the same value of stress. In particular, if one so desires, one can rotate to the principal components axes of the configuration (see Section 2.2.1) and look at the projections of the points in the two- and three-dimensional spaces of pairs and triplets of these principal components axes.

The above ideas and methods of multidimensional scaling are directly relevant to reduction of dimensionality for a body of multiresponse data. Indeed, if n observed points in p-space are located close to a q-dimensional linear subspace, the use of the interpoint Euclidean distances of the points in the p-space as the initial measures of dissimilarity in multidimensional scaling could lead to a "solution" in q-space, in correspondence with the results of a principal components analysis of the original covariance matrix.

However, if the n points in p-space are located close to certain kinds of *curved* q-dimensional subspaces, multidimensional scaling may produce a solution in q-space which would not necessarily be indicated by the linear principal components analysis or usual factor analysis. The point is that multidimensional scaling attempts to preserve the monotone relation of distances, and, if the distances along the curved q-dimensional subspace are reasonably monotone with the Euclidean distances, the procedure will recognize the lower-dimensional curved space. For instance, in the oversimplified example of points lying on a semicircle, since the interpoint Euclidean distances (viz., chord lengths) are a strictly monotone function of distances measured along the curve (viz., arc lengths), multidimensional scaling will recover the spacing of the points along the unidimensional curve and a single dimension will be indicated as providing a perfect fit.

Example 2. This example derives from Ekman (1954), and the data, used by Shepard (1962b), consisted of similarity ratings by 31 subjects of every pair among 14 color stimuli, which varied primarily in hue. Thus $n = 14$ and $m = 91$ here. The subjective similarity rating of each pair by every subject was on a five-point scale, and the mean ratings from all 31 subjects were scaled to go from 0 ("no similarity at all") to 1 ("identical"). A 14×14 matrix of such mean similarity ratings was obtained and treated by Ekman (1954) as a correlation matrix for purposes of a factor analysis, which led to a five-factor description. The five factors were identified as violet, blue, green, yellow, and red. On the other hand, as mentioned by Shepard (1962b), intuition and the familiar concept of the "color circle" for representing colors differing in hue might suggest the reasonableness of a two-factor (or perhaps even a one-factor) solution. Even if experimentally unintended variations in "brightness" and "saturation" were involved in the subjective ratings, one would still expect three and not five factors.

Exhibit 2*a*, taken from Shepard (1962b), shows the two-dimensional solution obtained by a multidimensional scaling algorithm. [*Note*: The axes shown in the figure were obtained by rotation of the ones in the solution to principal axes; however, with the essentially circular configuration involved here this makes hardly any difference.] The multidimensional scaling solution, of course, consists merely of the coordinate representation of the 14 points, and the smooth line was drawn through the points by Shepard to emphasize the similarity of the configuration to the color circle.

Exhibit 2*b* is a scatter plot of the original measures of similarity against the interpoint distances as computed from the two-dimensional solution shown in Exhibit 2*a*. The monotone relation between similarity and interpoint distance seen in this plot is, of course, a constraint of the multidimensional scaling procedure. The greater the observed similarity between two stimuli, the smaller is the distance between the two points representing the stimuli. The plot provides a graphical display of the data-determined monotone relationship involved, and in the present

Exhibit 2a. Multidimensional scaling solution for 14 colors (Ekman, 1954; Shepard, 1962b)

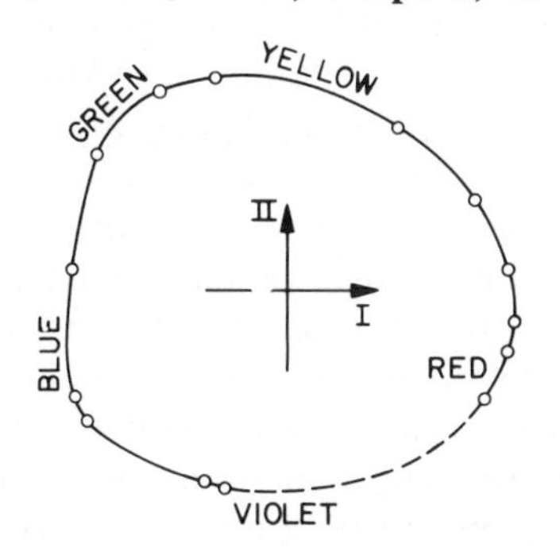

Exhibit 2b. Scatter plot of similarity versus distance for the example of Exhibit 2a (Shepard, 1962b)

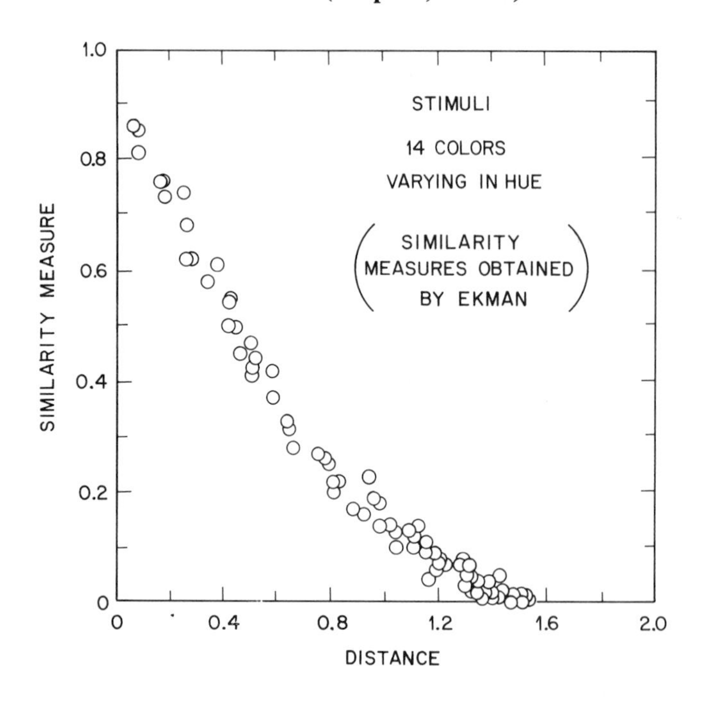

example it appears to be a relatively smooth, nonlinear (perhaps quadratic) relationship.

In this example, multidimensional scaling has produced both an intuitively appealing and a scientifically adequate parsimonious representation. It is perhaps reasonable to inquire about the details of the factor analysis solution and to try to understand the nature of and the reasons for the differences between the two solutions. In particular, after fitting the first two factors by the principal factor method, only 64% of the total variance (see Eq. 13) had been accounted for. Moreover, the residual correlations at that stage, i.e., the off-diagonal elements of $\mathbf{R}_2^* = (\mathbf{R}^* - \mathbf{l}_1\mathbf{l}_1' - \mathbf{l}_2\mathbf{l}_2')$, where $\mathbf{R}^*$, $\mathbf{l}_1$, and $\mathbf{l}_2$ are defined by Eqs. 20 and 21, were still reasonably large. This implies that the original correlations (which in this example were, in fact, similarities) were not adequately reproducible from the two-factor solution. Shepard (1962b), however, fitted a quadratic to the plot shown in Exhibit 2*b*, obtained fitted values for similarities from such a quadratic, and demonstrated that such fitted values were adequate reproductions of (i.e., were sufficiently close in value to) the original measures of similarity. This is not surprising in view of the indication of "tightness" of the points about a quadratic which can be visualized in Exhibit 2*b*.

The main reason for the difference in the dimensionalities suggested by the two methods is perhaps the inadequacy of the inherent linearity in the factor analysis approach to handle nonlinear reductions of dimensionality. In the present example, there may also be an effect on the factor analysis solution due to the use of similarity measures (with a range of 0 to 1) as inputs instead of the usual correlation coefficients (with a range of -1 to 1).

A modification of the scaling approach, due to Shepard & Carroll (1966), is directed toward improving the recognition of near singularities of a nonlinear nature among multidimensional observations. This modification focuses attention mainly on retaining the monotone relationship between interpoint distances and similarities only for nearby points rather than for all the points. The idea is illustrated by the next example, taken from Shepard & Carroll (1966).

Example 3. The data are from Boynton & Gordon (1965) and were used by Shepard & Carroll (1966) for illustrating the modified multidimensional scaling approach. The general concern and nature of the experiment that gave rise to the data are somewhat similar to those in the Ekman experiment described in Example 2, although the experimental detail and the nature of the data are different here. Specifically, 23 spectral colors differing only in their wavelengths were projected in random sequence several times to a group of observers. For each color the relative frequencies with which the observers denoted it as blue, green, yellow, or red were noted, thus giving a four-dimensional response associated with each color. In this example, $n=23$ and $p=4$. Exhibit 3*a* shows a pictorial representation of the data. The 23 colors (observations) are labeled A through W, and the four-dimensional vector of observations for each color is shown in a profile format. [*Note*: The four relative frequencies for any color are not required to add up to 100%, and they do not.]

A 4×4 correlation matrix may be calculated from the 23 observations; and, when a principal components analysis of the correlation matrix is performed, three of the four eigenvalues turn out to be relatively important, whereas the smallest eigenvalue is comparatively small and negligible. Thus, using a linear technique would lead one to conclude that a linear space of reduced dimensionality, $q=3$, would be feasible and might be adequate in the present problem. Shepard & Carroll (1966) show a representation of the 23 stimuli in the space of the first three principal components, and Exhibit 3*b* is a reproduction of their ingenious two-dimensional display of the three-dimensional representation. Two of the

Exhibit 3a. Profiles of relative frequencies of identifications of each of 23 spectral colors as blue, green, yellow, or red (Boynton and Gordon, 1965; Shepard and Carroll, 1966)

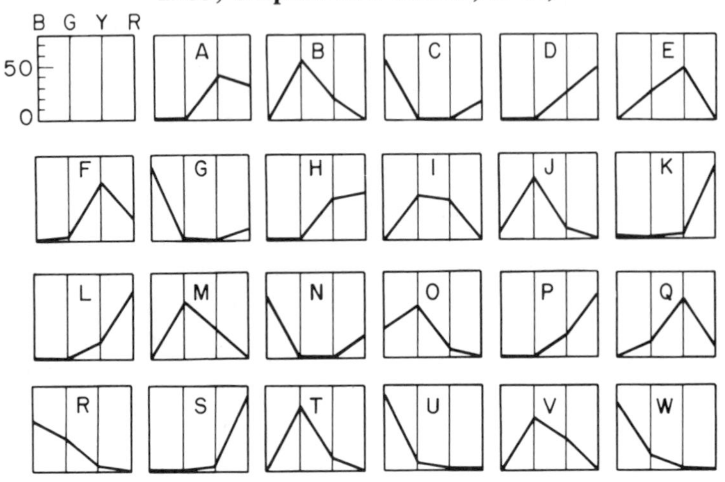

axes are on the plane of the picture, while the third axis is to be visualized as emanating out toward the viewer. The 23 points are labeled A through W to correspond with the stimuli, and the size of the circle around a point corresponds to its distance away from the picture plane. Thus F and Q are about the closest to the viewer, while S, K, and J are among the farthest away.

From Exhibit 3*b* it is clear that, although $q=3$, the 23 points are not scattered throughout three-dimensional space but, rather, appear to lie on a reasonably smooth one-dimensional curve that winds through the three-dimensional space. To emphasize this feature, the points in Exhibit 3*b* are also labeled 1 through 23 (shown alongside their original identifications by the letters A through W) to correspond with their positions on the one-dimensional curve. It is, of course, known that a single dimension (viz., wavelength) underlies the data; in fact, in the experimental setup the variation in wavelength of the 23 stimuli was accomplished by turning a single knob to different settings. (The numbering 1 through 23, in fact, corresponds with the known ordering of the stimuli on the single dimension of wavelength!) Hence one might ask whether it is possible to obtain an adequate one-dimensional representation of the data.

If $\mathbf{y}_i'=(y_{i1},\ldots,y_{i4})$ denotes the four-dimensional observation corresponding to the ith stimulus ($i=1,\ldots,23$), one could use the so-called city-block distance between the ith and jth observations, $\sum_{l=1}^{4}|y_{il}-y_{jl}|$, for $i<j=1,\ldots,23$, as a measure of dissimilarity (δ_{ij}) between stimuli i

and j. Thus 253 ($=m$) dissimilarities may be obtained and used as input to multidimensional scaling. Shepard & Carroll (1966) performed such an analysis and found that the minimum stress in one-dimension, $S_0(1)$, was not adequately small but that $S_0(2)$ was sufficiently and markedly smaller. The two-dimensional solution obtained by Shepard & Carroll (1966) is shown in Exhibit 3*c*, with the points joined by a smooth line. Except for being confined to a two-dimensional space, the curve in Exhibit 3*c* is qualitatively (including the location of bends) the same as the one which manifested itself in the principal components representation of Exhibit 3*b*. The tendency of both curves to close the loop is similar to the color circle concept and is explainable by the phenomenon that violet [stimulus N (or 1) with the lowest wavelength] is judged to contain some

Exhibit 3b. Representation of three-dimensional principal factor solution for the data of Exhibit 3a (Shepard and Carroll, 1966)

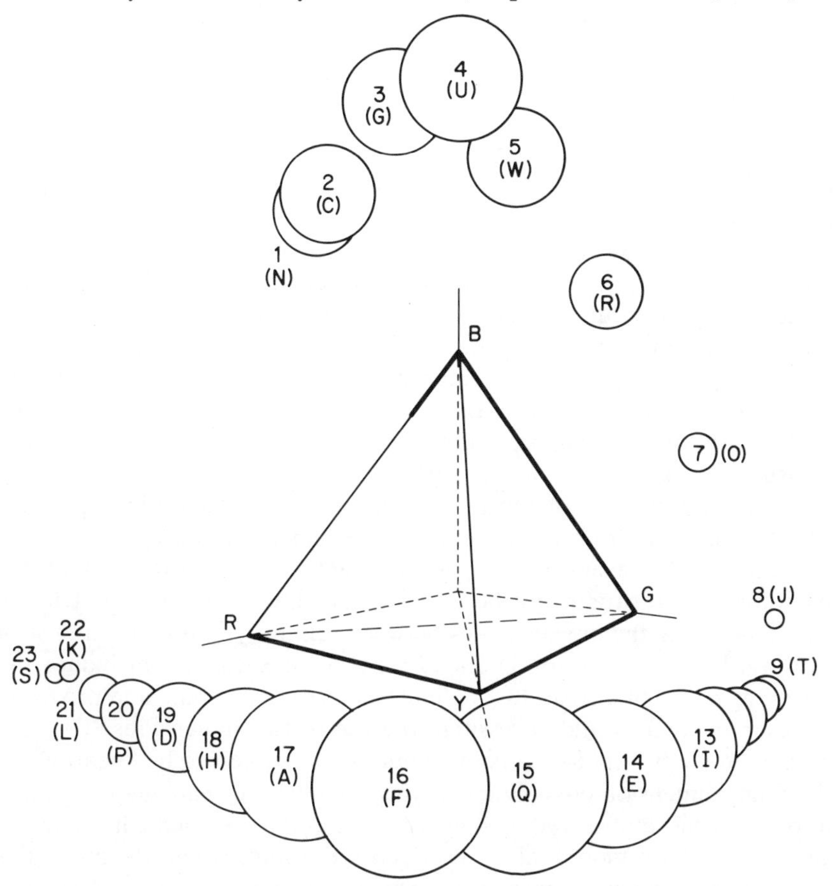

Exhibit 3c. Multidimensional scaling solution for the data of Exhibit 3a (Shepard and Carroll, 1966).

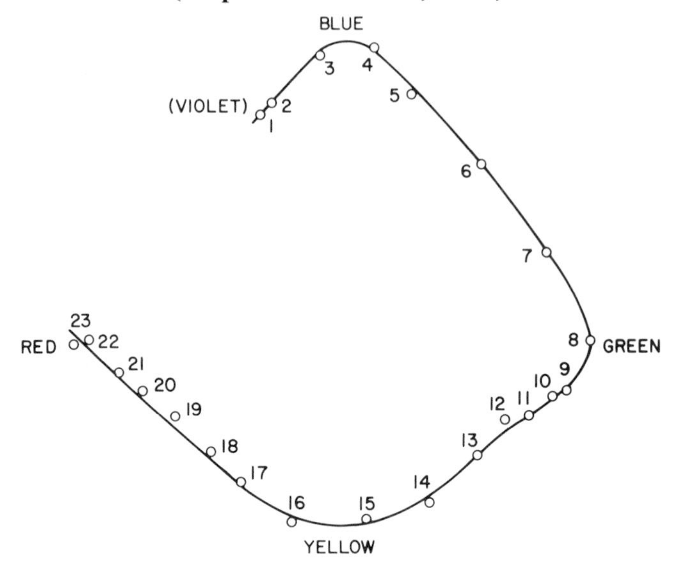

red [stimulus S (or 23) with the highest wavelength] along with a dominance by blue.

Exhibit 3*d* shows the scatter plot of interpoint Euclidean distances in the solution against the original dissimilarities (viz., the city-block distances). The near linearity of this configuration suggests that the interpoint Euclidean distances in the two-dimensional representation, determined as the output of the multidimensional scaling algorithm, are essentially linearly related to the city-block distances (calculated in the initial four-dimensional space of observations) that constituted the input measures of dissimilarity to the algorithm.

Next, since the use of the regular multidimensional scaling approach still does not lead to recovering the single dimension known to underlie the data in this example, Shepard & Carroll (1966) suggest that the monotonicity constraint not be imposed globally. The idea is not to try accommodating the dissimilarities between relatively remote profiles such as those for stimuli N (or 1) and S (or 23), since this might induce the "bending over" of a basically unidimensional phenomenon, and provision for such bending would necessitate the use of two dimensions. To focus on monotonicity only for pairs of stimuli that are likely to be "nearby" on the single dimension possibly underlying the data, one can ignore all pairs of objects whose observed dissimilarities exceed a specified cut-off value, and then require monotonicity between the distances and dissimilarities only for the remaining pairs of objects whose dissimilarities are smaller than the cut-off value. Using such a procedure and requiring monotonicity only for the pairs of stimuli with the smallest 100 dissimilarities (i.e., ignoring the 153 larger dissimilarity values) led Shepard & Carroll (1966) to the one-dimensional solution shown along the bottom of Exhibit 3*e*.

The configuration, including the spacing, is essentially the one that would be obtained by "unbending" the curve of Exhibit 3*c*. The ordering of the stimuli from 1 through 23 in Exhibit 3*e* does correspond to increasing wavelength. Their spacing, however, is not the same as it is on wavelength, as is evident from the scale at the top of Exhibit 3*e*, which shows the wavelengths corresponding to the 23 stimuli. Also shown in the figure are plots of values of each of the four original responses (B, G, Y, and R) for the 23 stimuli, against the spacings as determined in the one-dimensional solution. The observed values of the original responses are thus seen to be nonlinear functions of the single underlying dimension. Shepard & Carroll (1966) noticed the interesting fact that these curves are more regular and symmetrical than those obtained by Boynton & Gordon (1965), showing the responses plotted directly against wavelength. The experiment clearly involves the psychological perception of colors, and for this reason the data may not be reflecting only the experimentally controlled physical dimension of wavelength.

Exhibit 3d. Scatter plot of dissimilarity versus distance associated with Exhibit 3c (Shepard and Carroll, 1966)

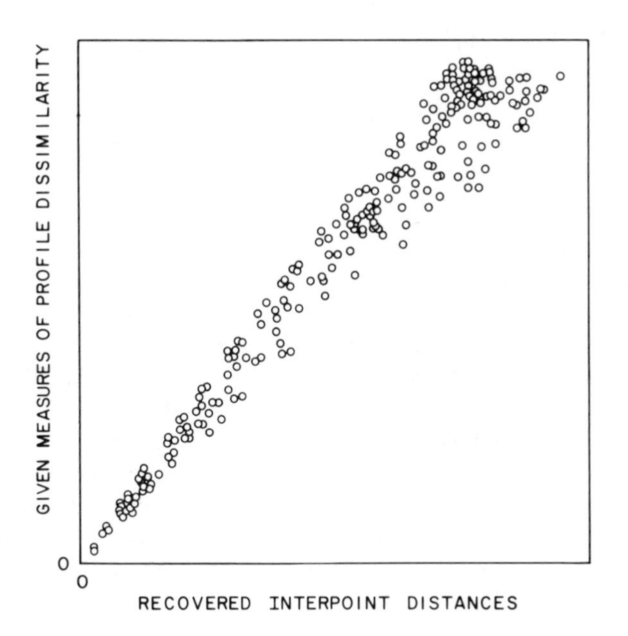

Exhibit 3e. Multidimensional scaling solution with local monotonicity constraint for the data of Exhibit 3a (Shepard and Carroll, 1966)

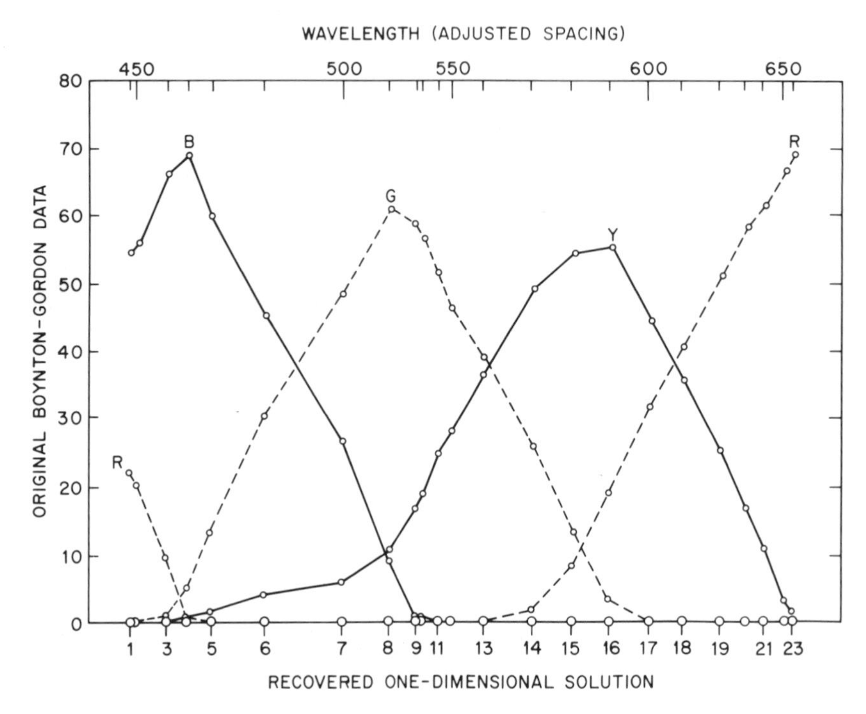

Exhibit 3*f* shows the scatter plot of interpoint Euclidean distances versus the initial dissimilarities for the one-dimensional solution shown in Exhibit 3*e*, and the departure from monotonicity at the top end may be seen clearly. The interpoint distances and dissimilarities, for stimuli such as *N* and *S* whose profiles (initial data) are very different, are indeed coordinates of the points at the top end of Exhibit 3*f*.

An important difficulty with the above simple modification of confining monotonicity to the smallest dissimilarities and the corresponding recovered distances is that, when the underlying dimensionality of the curved manifold is larger than 1, the procedure may not lead to detecting this. Hence the modified multidimensional scaling procedure may be adequate when $q=1$ but not when $q>1$. As an example, Shepard & Carroll (1966) mention the case in which the points lie on the surface of a fish bowl or a sphere with a hole. Here $p=3$ and $q=2$, and an appropriate solution would be a representation of the points on a two-dimensional disk obtained by pulling out the sphere at the hole and flattening out into a disk. However, in such a solution, points at the rim of the hole which were close together in three-dimensional space end up being far apart on the two-dimensional disk; i.e., points with small initial dissimilarities end up with large interpoint distances in the recovered configuration. An alternative modification of multidimensional scaling to handle this difficulty might be to impose regional monotonicity; in other words, monotonicity might be required separately within a region surrounding each point (see Bennett, 1965).

Rather than pursuing such a modification, however, Shepard & Carroll (1966) suggest a different procedure that involves maximizing an index of continuity, so as to find a representation of the original p-dimensional points in terms of $q(<p)$ new coordinates that are "smoothly" related to the old ones. Specifically, if $\mathbf{y}_1, \mathbf{y}_2, \ldots, \mathbf{y}_n$ are points in an initial p-dimensional representation of n objects, they suggest finding a configuration $\mathbf{x}_1, \mathbf{x}_2, \ldots, \mathbf{x}_n$ in q-space so as to minimize an index of the form

$$\kappa = \frac{\sum\sum_{i<j=1}^{n} (d_{ij}^2/D_{ij}^2)w_{ij}}{\text{Normalizing factor}}.$$

Here d_{ij} is the Euclidean distance between $\mathbf{y}_i$ and $\mathbf{y}_j$, while D_{ij} is the Euclidean distance between $\mathbf{x}_i$ and $\mathbf{x}_j$, and w_{ij}'s are weights that decrease

Exhibit 3f. Scatter plot of dissimilarity versus distance associated with the multidimensional scaling solution in Exhibit 3e (Shepard and Carroll, 1966)

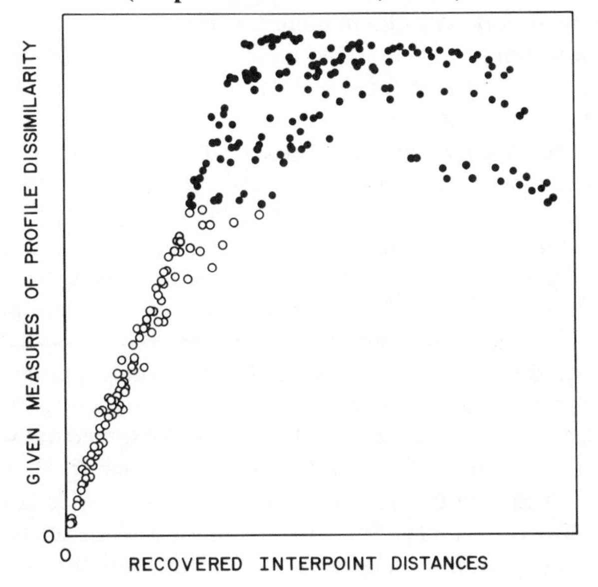

as the distance between the corresponding points in the x-space increases. The numerator of κ is a multivariate generalization of the mean square successive difference whose ratio to the variance is a statistic which has been used as an inverse measure of trend in univariate time series. The smaller the value of κ, the "smoother" the relationship between $\mathbf{y}$ and $\mathbf{x}$ is considered to be.

Using the desiderata of invariance of the ratios of the weights under translations, and uniform shrinking or dilation of the x-space, Shepard & Carroll (1966) recommend the choice $w_{ij} = 1/D_{ij}^2$. Similarly, arguing that it would be desirable for the unconstrained minimum of the index to be attained when D_{ij}^2 is proportional to d_{ij}^2 for every i and j (i.e., the configurations in x-space and y-space match except for a similarity transformation), they suggest using the normalizing factor, $[\sum\sum_{i<j} 1/D_{ij}^2]^2$, in the denominator of κ.

Thus, given the initial configuration $\mathbf{y}_1, \ldots, \mathbf{y}_n$ in p-space, the approach suggested by Shepard & Carroll (1966) is to choose the nq coordinates involved in $\mathbf{x}_1, \ldots, \mathbf{x}_n$ so as to minimize

$$\kappa = \frac{\sum\sum_{i<j=1}^{n} (d_{ij}^2/D_{ij}^4)}{\left\{\sum\sum_{i<j=1}^{n} (1/D_{ij}^2)\right\}^2}. \tag{36}$$

Starting from a trial configuration in q-space, one could iterate to the desired configuration, with the minimum value of κ, by using a numerical optimization technique, such as the method of steepest descent. Also, as was the case in multidimensional scaling, one could repeat the process for a series of values of q and choose the smallest "adequate" value of q by studying the achieved values of κ for the different values of q.

In contradistinction to multidimensional scaling, the above procedure assumes the initial format of the data to be a Euclidean representation (i.e., n points in p-space) and not to consist only of rank order information about the pairwise dissimilarities. Also, little is known about the dependence of the final solution on the use of (i) other measures of distance besides the Euclidean measure in the x- or y-space, (ii) other weights, w_{ij}, and (iii) other normalizing factors. The experience with this approach, using Monte Carlo or real data, is too limited for specification of a yardstick for assessing the smallness of an observed value of κ. The unconstrained minimum of κ (corresponding to which there need not, of course, be a Euclidean configuration in x-space) is easy to compute, and in their published examples Shepard & Carroll (1966) seem to use this value for judging how small the κ is for the optimum configuration determined by the procedure. Despite these limitations, some interesting examples of the application of the procedure are discussed by Shepard & Carroll (1966), and the following example is taken from their work.

Example 4. The artificially generated data consisted of 62 points on the surface of a sphere—12 points on each of five equally spaced parallels, and the two poles. Hence $n = 62$, $p = 3$, and it is known here that $q = 2$. Exhibit 4a shows the data; Exhibit 4b, the solution obtained in two dimensions by minimizing κ. The solution consists of two hemispheres in three-dimensional space opened out on a hinge at the equator and then flattened out into a common plane. The equatorial circle has been distorted into an S-shaped curve. The reader is reminded, however, that the computer output in this solution (exactly as in the uses of multidimensional scaling) consists *only* of the coordinates of the points corresponding to the n objects, and the lines are drawn in from extraneous knowledge of some structure among the objects.

Exhibit 4a. Artificial data consisting of 62 points on the surface of a sphere (Shepard and Carroll, 1966)

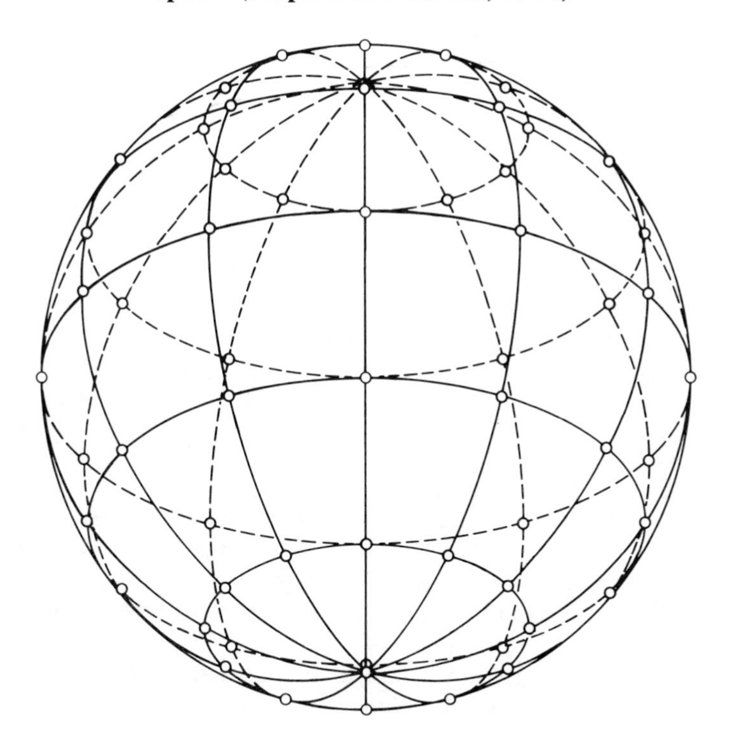

Exhibit 4b. Two-dimensional parametric mapping of data of Exhibit 4a (Shepard and Carroll, 1966)

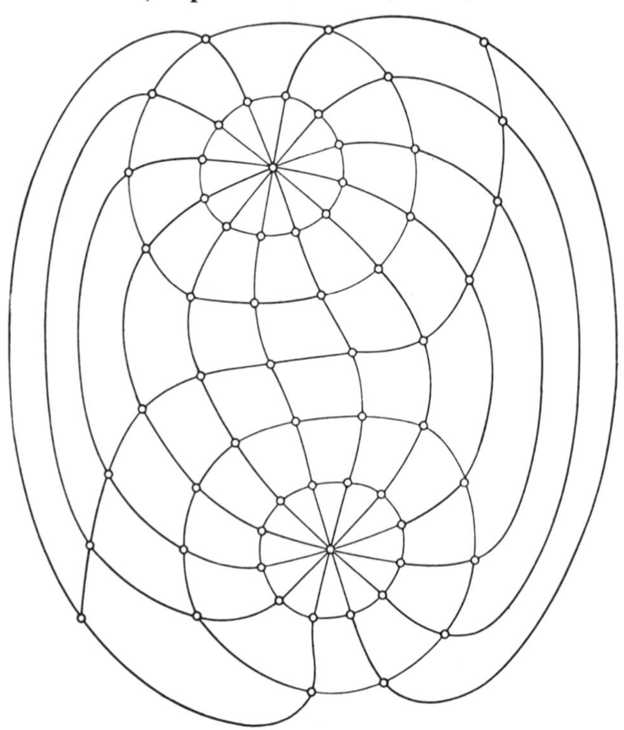

All of the procedures described in the present subsection of the book depend on the use of an index of achievement (viz., minimum stress or minimum κ) as an informal basis for assessing the comparative adequacy of the successive dimensions employed. This index is to be used with appropriate judgment relative to meaningfulness of interpretation in the subject matter area. A firm commitment to benchmarks for comparing achieved values of the index, without regard for issues such as interpretability, is neither necessary nor recommended for using these tools of data analysis.

Example 5. The data, from an experiment of Rothkopf (1957) (see also Shepard, 1963; Kruskal, 1964a), were obtained from 598 subjects who were asked to judge whether or not pairs of successively presented Morse code signals were the same. Thirty-six signals were employed: 26 for the letters of the alphabet and 10 for the digits 0 through 9. Exhibit 5*a* is a matrix of the percentages of times that a signal corresponding to the row label was identified as being the same as the signal corresponding

Exhibit 5a. Data matrix of percentages of confusions between pairs of Morse code signals (Rothkopf, 1957; Shepard, 1963)

	A	B	C	D	E	F	G	H	I	J	K	L	M	N	O	P	Q	R	S	T	U	V	W	X	Y	Z	1	2	3	4	5	6	7	8	9	Ø	
A	92	04	06	13	03	14	10	13	46	05	22	03	25	34	06	06	09	35	23	06	37	13	17	12	07	03	02	07	05	05	08	06	05	06	02	03	A
B	05	84	37	31	05	28	17	21	05	19	34	40	06	10	12	22	25	16	18	02	18	34	08	84	30	42	12	17	14	40	32	74	43	17	04	04	B
C	04	38	87	17	04	29	13	07	11	19	24	35	14	03	09	51	34	24	14	06	06	11	14	32	82	38	13	15	31	14	10	30	28	24	18	12	C
D	08	62	17	88	07	23	40	36	09	13	81	56	08	07	09	27	09	45	29	06	17	20	27	40	15	33	03	09	06	11	09	19	08	10	05	06	D
E	06	13	14	06	97	02	04	04	17	01	05	06	04	04	05	01	05	10	07	67	03	03	02	05	06	05	04	03	05	03	05	02	04	02	03	03	E
F	04	51	33	19	02	90	10	29	05	33	16	50	07	06	10	42	12	35	14	02	21	27	25	19	27	13	08	16	47	25	26	24	21	05	05	05	F
G	09	18	27	38	01	14	90	06	05	22	33	16	14	13	62	52	23	21	05	03	15	14	32	21	23	39	15	14	05	10	04	10	17	23	20	11	G
H	03	45	23	25	09	32	08	87	10	10	09	29	05	08	08	14	08	17	37	04	36	59	09	33	14	11	03	09	15	43	70	35	17	04	03	03	H
I	64	07	07	13	10	08	06	12	93	03	05	16	13	30	07	03	05	19	35	16	10	05	08	02	05	07	02	05	08	09	06	08	05	02	04	05	I
J	07	09	38	09	02	24	18	05	04	85	22	31	08	03	21	63	47	11	02	07	09	09	09	22	32	28	67	66	33	15	07	11	28	29	26	23	J
K	05	24	38	73	01	17	25	11	05	27	91	33	10	12	31	14	31	22	02	02	23	17	33	63	16	18	05	09	17	08	08	18	14	13	05	06	K
L	02	69	43	45	10	24	12	26	09	30	27	86	06	02	09	37	36	28	12	05	16	19	20	31	25	59	12	13	17	15	26	29	36	16	07	03	L
M	24	12	05	14	07	17	29	08	08	11	23	08	96	62	11	10	15	20	07	09	13	04	21	09	18	08	05	07	06	06	05	07	11	07	10	04	M
N	31	04	13	30	08	12	10	16	13	03	16	08	59	93	05	09	05	28	12	10	16	04	12	04	06	11	05	02	03	04	04	06	02	02	10	02	N
O	07	07	20	06	05	09	76	07	02	39	26	10	04	08	86	37	35	10	03	04	11	14	25	35	27	27	19	17	07	07	06	18	14	11	20	12	O
P	05	22	33	12	05	36	22	12	03	78	14	46	05	06	21	83	43	23	09	04	12	19	19	19	41	30	34	44	24	11	15	17	24	23	25	13	P
Q	08	20	38	11	04	15	10	05	02	27	23	26	07	06	22	51	91	11	02	03	06	14	12	37	50	63	34	32	17	12	09	27	40	58	37	24	Q
R	13	14	16	23	05	34	26	15	07	12	21	33	14	12	12	29	08	87	16	02	23	23	62	14	12	13	07	10	13	04	07	12	07	09	01	02	R
S	17	24	05	30	11	26	05	59	16	03	13	10	05	17	06	06	03	18	96	09	56	24	12	10	06	07	08	02	02	15	28	09	05	05	05	02	S
T	13	10	01	05	46	03	06	06	14	06	14	07	06	05	06	11	04	04	07	96	08	05	04	02	02	06	05	05	03	03	03	08	07	06	14	06	T
U	14	29	12	32	04	32	11	34	21	07	44	32	11	13	06	20	12	40	51	06	93	57	34	17	09	11	06	06	16	34	10	09	09	07	04	03	U
V	05	17	24	16	09	29	06	39	05	11	26	43	04	01	09	17	10	17	11	06	32	92	17	57	35	10	10	14	28	79	44	36	25	10	01	05	V
W	09	21	30	22	09	36	25	15	04	25	29	18	15	06	26	20	25	61	12	04	19	20	86	22	25	22	10	22	19	16	05	09	11	06	03	07	W
X	07	64	45	19	03	28	11	06	01	35	50	42	10	08	24	32	61	10	12	03	12	17	21	91	48	26	12	20	24	27	16	57	29	16	17	06	X
Y	09	23	62	15	04	26	22	09	01	30	12	14	05	06	14	30	52	05	07	04	06	13	21	44	86	23	26	44	40	15	11	26	22	33	23	16	Y
Z	03	46	45	18	02	22	17	10	07	23	21	51	11	02	15	59	72	14	04	03	09	11	12	36	42	87	16	21	27	09	10	25	66	47	15	15	Z
1	02	05	10	03	03	05	13	04	02	29	05	14	09	07	14	30	28	09	04	02	03	12	14	17	19	22	84	63	13	08	10	08	19	32	57	55	1
2	07	14	22	05	04	20	13	03	25	26	09	14	02	03	17	37	28	06	05	03	06	10	11	17	30	13	62	89	54	20	05	14	20	21	16	11	2
3	03	08	21	05	04	32	06	12	02	23	06	13	05	02	05	37	19	09	07	06	04	16	06	22	25	12	18	64	86	31	23	41	16	17	08	10	3
4	06	19	19	12	08	25	14	16	07	21	13	19	03	03	02	17	29	11	09	03	17	55	08	37	24	03	05	26	44	89	42	44	32	10	03	03	4
5	08	45	15	14	02	45	04	67	07	14	04	41	02	00	04	13	07	09	27	02	14	45	07	45	10	10	14	10	30	69	90	42	24	10	06	05	5
6	07	80	30	17	04	23	04	14	02	11	11	27	06	02	07	16	30	11	14	03	12	30	09	58	38	39	15	14	26	24	17	88	69	14	05	14	6
7	06	33	22	14	05	25	06	04	06	24	13	32	07	06	07	36	39	12	06	02	03	13	09	30	30	50	22	29	18	15	12	61	85	70	20	13	7
8	03	23	40	06	03	15	15	06	02	33	10	14	03	06	14	12	45	02	06	04	06	07	05	24	35	50	42	29	16	16	09	30	60	89	61	26	8
9	03	14	23	03	01	06	14	05	02	30	06	07	16	11	10	31	32	05	06	07	06	03	08	11	21	24	57	39	09	12	04	11	42	56	91	78	9
Ø	09	03	11	02	05	07	14	04	05	30	08	03	02	03	25	21	29	02	03	04	05	03	02	12	15	20	50	26	09	11	05	22	17	52	81	94	Ø
	A	B	C	D	E	F	G	H	I	J	K	L	M	N	O	P	Q	R	S	T	U	V	W	X	Y	Z	1	2	3	4	5	6	7	8	9	Ø	

to the column label. These percentages may be considered as measures of similarity between the pairs of Morse code signals. [*Note*: Although large, the diagonal values in Exhibit 5*a* are not 100% (similarity of a signal to itself is not perfect), and also the matrix is not symmetric, so that $\delta_{ij} \neq \delta_{ji}$.] To use multidimensional scaling in a direct manner, the averages of each pair of symmetrically situated off-diagonal elements of this matrix may serve as input measures of similarity between the corresponding pair in the 36 signals.

Exhibit 5*b*, taken from Kruskal (1964a), shows the minimum stress achieved by the multidimensional scaling solution plotted against the number of dimensions employed for that solution. Using the benchmarks recommended by Kruskal (1964a) and mentioned earlier, as well as the rule of choosing the dimensionality by looking for an "elbow" in a plot such as Exhibit 5*b*, one might feel that a choice of $p=2$ in this example would, from the point of view of the index of achievement, be between fair and poor, and if not the value of 5 for p one should at least consider the value 3.

However, the two-dimensional solution obtained and interpreted by Shepard (1963) is shown in Exhibit 5*c*. The vertical axis in this solution is seen to correspond to the number of components in the Morse code symbol (i.e., the total number of dots and dashes), while the horizontal axis characterizes the composition of the symbol (i.e., the ratio of number of dots to number of dashes). Corresponding to the large value of stress

Exhibit 5b. *Plot of stress versus number of dimensions for the example in Exhibit 5a (Kruskal, 1964a; Shepard, 1963)*

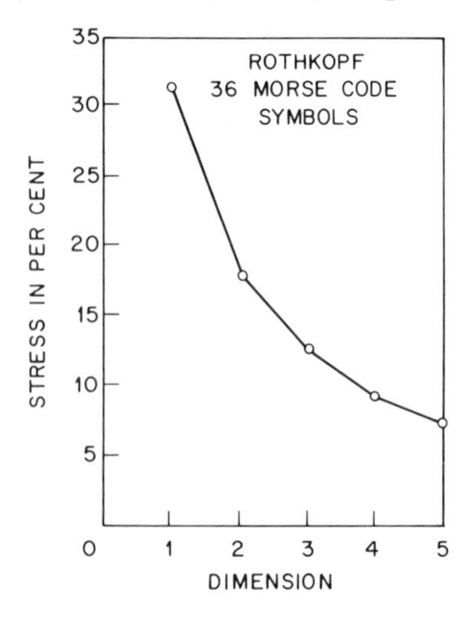

Exhibit 5c. Multidimensional scaling solution for the Morse code signals (Shepard, 1963)

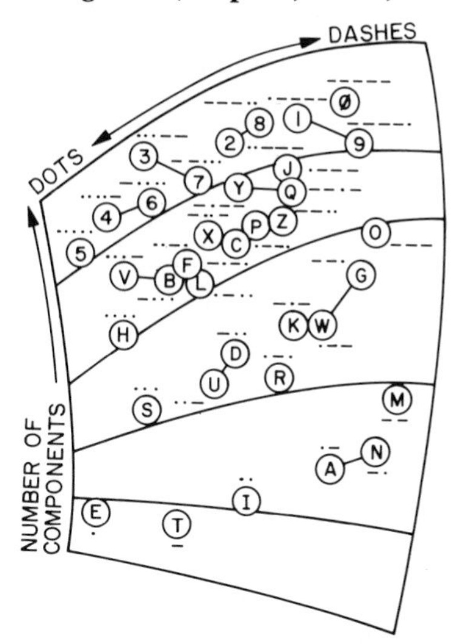

($\simeq$18%) for this solution, one observes that the distances between the representations of the signals do not closely match the observed similarities in Exhibit 5*a* [e.g., compare the distance between the pair (*B*, *X*) with those between (*B*, *F*) and (*B*, *L*)]. However, as Kruskal (1964a) points out, the fact that Shepard could not extract additional interpretable structure by going to three dimensions suggests that $p = 2$ would be a better choice than $p = 3$ for this problem regardless of any contraindication from Exhibit 5*b*. Interpretability and simplicity are important in data analysis, and any rigid inference of optimal dimensionality, in the light of the observed values of a numerical index of goodness of fit, may not be productive.

In this example, as well as the previous ones, of the use of multidimensional scaling, the initial dissimilarity (or similarity) values were averages across several subjects, and no provision was made for differences among subjects. More recently Carroll & Chang (1970) have proposed a scheme for multidimensional scaling that allows for individual differences. In this approach the coordinates recovered by multidimensional scaling are assumed to be the same for all individuals, but the weights assigned to the different coordinates are not assumed to be identical for all individuals.

The ideas and procedures involved in the above scaling types of approaches are imaginative and insightful. The procedures, however, have limitations in practical application in that they involve extensive iteration on $n(n-1)/2$ quantities—the interpoint similarities or distances. The currently available computer programs, for instance, can effectively and economically handle up to about $75\,(=n)$ objects only. Also, the solution space produced by these procedures does not have an analytic description that is simply interpretable in terms of the original response space. This becomes especially important when the solution space is of dimensionality greater than three and graphical representations are no longer available. Also, when one starts with a metric representation and uses some measure of distance in the initial space as the input measure of dissimilarity (see Example 3), questions arise concerning the nature of the dependence of the recovered configuration on the initial choice of distance function.

2.4. NONLINEAR SINGULARITIES AND GENERALIZED PRINCIPAL COMPONENTS ANALYSIS

2.4.1 Nonlinear Singularities

The problem of reduction in dimensionality concerns the recognition of lower-dimensional, possibly nonlinear, subspaces near which the multiresponse observations may, statistically speaking, lie. This is, of course, not a well-defined concept, in a sense very similar to the indefiniteness involved in the notion of "fitting a curve" to a scatter plot of y against x.

One major source of difficulty in the problem is the fact that in the analysis of high-dimensional data there are not available the informal, mainly graphical, internal comparisons procedures, such as scatter plots, that guide so much single-response, and some two-response, data analysis.

So far as near-linear singularities in a body of data are concerned, these may be statistically indicated by principal components analysis (but see Section 6.6 for a discussion of possible effects of outliers). However, nonlinear singularities will not necessarily be indicated by principal components, and one may not be able to infer their existence even from various obvious two-dimensional scatter plot representations of the data.

Example 6. The C-shaped configuration shown in Exhibit 6*a* is an oversimplified example of data configurations that may not be revealed by classical linear principal components analysis. Clearly, the three-dimensional analogue of this, namely, a cup-shaped configuration of data, may not be revealed even by a combination of principal components analysis in 3-space and two-dimensional marginal scatter plots. Example 7 discusses such an example.

Exhibit 6*b* is a plot of 50 computer-generated random bivariate normal

Exhibit 6a. Example of a curved configuration of data

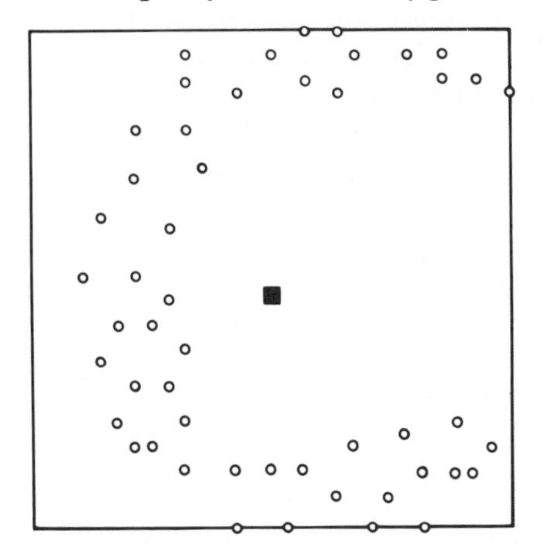

samples with an underlying positive correlation coefficient. This is, therefore, an example of typical normal distribution scatter when $p=2$.

The filled-in squares in Exhibits 6*a* and 6*b* are the centers of gravity (means) of the data.

Exhibit 6b. Simulated sample of bivariate normal scatter

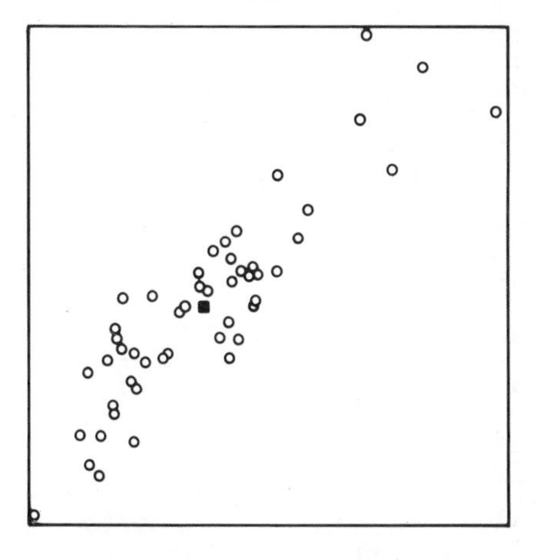

One elementary technique for detecting the existence of curved configurations, such as the one considered in Example 6, involves the computation of the squared generalized distances (using the inverse of the sample covariance matrix) of each point from the center of gravity, viz., $d_i^2 = (\mathbf{y}_i - \bar{\mathbf{y}})'\mathbf{S}^{-1}(\mathbf{y}_i - \bar{\mathbf{y}})$ for $i=1, \ldots, n$. For a typical multivariate normal scatter (see Exhibit 6*b*), either throughout p dimensions or mainly on a linear subspace, these distances will have approximately a chi-squared or gamma distribution. Hence, on an appropriately selected gamma probability plot (Wilk, Gnanadesikan, & Huyett, 1962a; see also the discussion of probability plots in Section 6.2), using shape parameter values in the neighborhood of $p/2$, they will tend to show as a linear configuration oriented toward the origin. For curved singularity, such as the illustration, however, it is clear that there will be a deficiency of small distances. This will show on the gamma plot by orientation of the configuration toward a nonzero intercept at the "small" end. A histogram of the observed distances would also indicate the sparseness of small ones, but the probability plot may provide additional insights. The following simulated three-dimensional example illustrates some of these ideas and procedures.

Example 7. The 61 triads shown in Exhibit 7*a* were obtained by appending a standard normal deviate to each of the coordinates of points on the surface of a specified paraboloid. Exhibits 7*b*, 7*c*, and 7*d*, show the

Exhibit 7a. Simulated data with observations scattered off the surface of a paraboloid

-2.732	6.557	25.507	-3.452	2.948	25.591
-5.264	5.253	24.200	-7.261	6.959	26.789
-5.103	5.986	26.446	-2.370	3.617	25.510
-3.335	5.888	23.947	-4.181	4.530	29.118
-5.420	5.607	25.321	-2.360	3.916	24.879
-3.261	7.697	27.479	-5.297	5.802	29.073
-4.607	6.651	26.518	-1.585	2.524	26.954
-4.236	4.220	24.416	-3.267	4.402	28.899
-4.947	5.363	26.918	-1.187	3.257	26.100
-2.189	5.881	26.282	-2.095	6.931	27.269
-2.913	5.953	26.962	-4.800	3.339	27.011
-4.838	5.909	25.196	-5.602	5.322	28.759
-3.448	5.610	27.489	-1.478	1.644	26.057
-0.990	5.391	25.667	-5.151	4.481	27.583
-6.116	6.326	30.189	-0.694	3.408	24.997
-2.715	4.645	25.613	-5.687	4.766	29.640
-5.849	6.876	26.070	-1.733	3.932	26.198
0.162	5.521	25.027	-6.154	4.932	29.631
-5.360	5.494	28.675	-3.823	3.784	25.123
-1.740	4.070	27.311	-2.588	4.923	28.343
-2.975	6.716	27.999	-3.237	3.648	26.249
-4.220	3.853	26.396	-5.740	4.537	30.277
-6.306	4.573	25.715	-0.709	1.542	27.240
-1.972	5.615	24.900	-6.568	5.335	29.631
-4.497	5.314	27.978	-1.669	1.501	25.413
-2.005	3.352	24.599	-7.690	4.578	30.863
-3.809	5.421	28.794	0.837	1.271	25.303
-2.081	3.795	25.542	-5.832	7.020	28.915
-4.907	7.120	27.449	-0.405	3.669	27.587
-0.742	2.800	26.394	-3.019	3.752	29.665
-2.750	2.233	27.669			

Exhibits 7b–d. Scatter plots of bivariate subsets of the data of Exhibit 7a

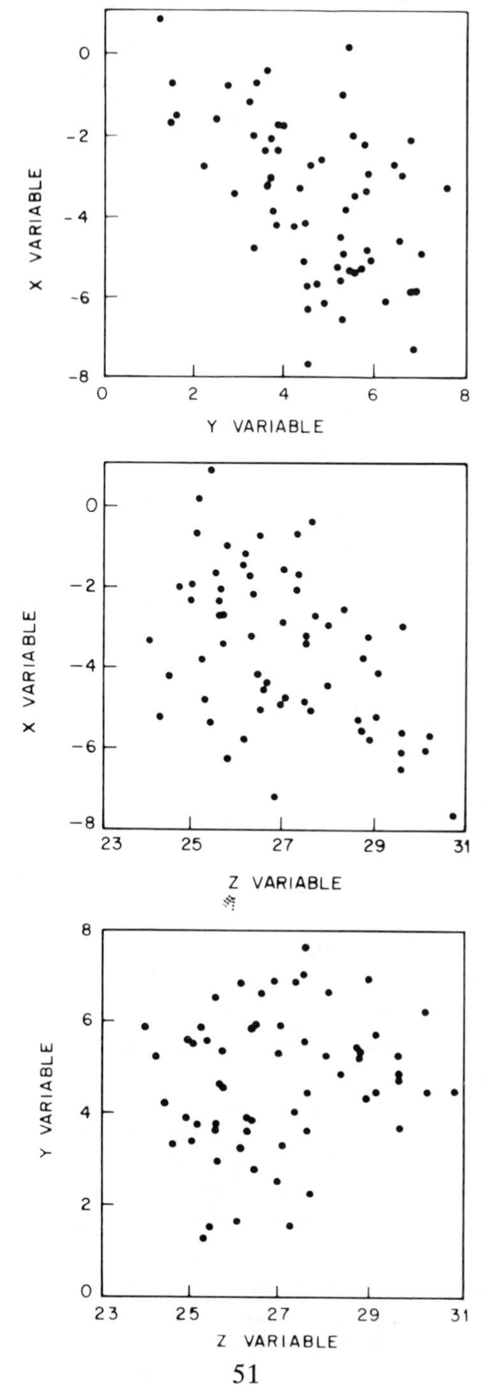

Exhibit 7e. Gamma probability plot (with shape parameter = 3/2) of generalized squared distances for data of Exhibit 7a

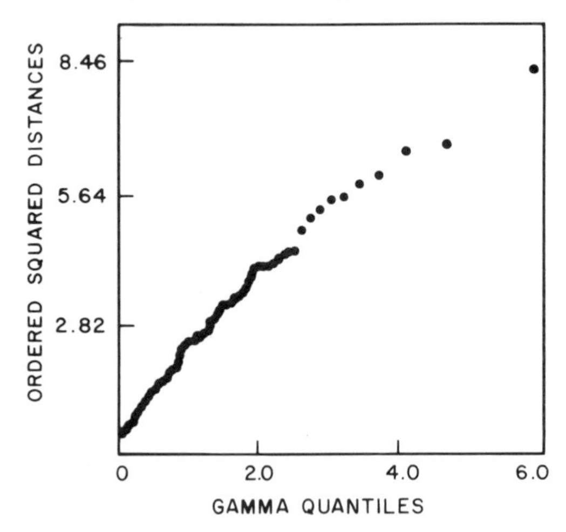

three possible two-dimensional scatter plots of these data with respect to pairs of the original three coordinates. None of these data displays is suggestive of the observations in three dimensions lying "near" a curved subspace.

The inverse of the sample covariance matrix, **S**, of these observations was employed to compute the generalized squared distance of each of the 61 points from their centroid. (See Section 4.2.1 for a discussion of distance measures.) A gamma probability plot of the ordered squared distances is shown in Exhibit 7*e*. The value of the shape parameter used for this plot was $\eta = 3/2$, and hence it is essentially a probability plot for a chi-squared distribution with 3 degrees of freedom. The configuration, which has a nonzero intercept, shows clearly the "deficiency" of small values, indicating a "hole" in the data. Furthermore, the slight tendency of the configuration to "bend over" at the top right-hand corner suggests that the data may be near a dish or a shallow paraboloid rather than a deep paraboloid. The nature of the indicated peculiarity may be investigated further by using the methods discussed below.

The simple-minded idea illustrated in Example 7 will, of course, not be indicative when the nonlinear surface near which the data lie has several bends, such as a sinusoidal shape. Furthermore, if a peculiarity is indicated, the probability plot does not tell very much about its nature.

A method proposed by Gnanadesikan & Wilk (1966, 1968) and considered independently by Van de Geer (1968) is useful for analyzing multidimensional linear or nonlinear singularities. The technique is a generalization of classical linear principal components analysis.

2.4.2. Generalized Principal Components Analysis

If one has nearly linear singularity of the data, what one wishes to do is to determine the *linear* coordinate system that is most nearly in concordance with the data configuration. Then the expression of the data in the new coordinate system is simpler, in that the effective description can be given by the use of fewer coordinates. This is accomplished by linear principal components analysis.

If one has nearly nonlinear singularity of the data, what one wishes to do is to determine the *nonlinear* coordinate system that is most nearly in agreement with the data configuration, just as in the linear case. Given a class of possible nonlinear coordinates, one needs to select that one along which the data variance is maximum, and then obtain another, uncorrelated with the first, along which the variance is next largest, and so on. For any class of coordinates that consist of an unspecified linear combination of arbitrary *functions* of the original coordinates, the solution to this problem is simply an enlarged eigenvalue-eigenvector problem. The essential concepts and computations may be illustrated by considering the bivariate ($p=2$) case.

Suppose that $\mathbf{y}'=(y_1, y_2)$ denotes the original bivariate response, and for illustrative purposes suppose that one is seeking a quadratic coordinate system. Thus, as a first step, one wishes to find

$$z = ay_1+by_2+cy_1y_2+dy_1^2+ey_2^2 \tag{37}$$

such that the variance of z is maximum among all such quadratic functions of y_1 and y_2.

Denoting y_1y_2 as y_3, y_1^2 as y_4, and y_2^2 as y_5, one can consider

$$\mathbf{y}^{*\prime}=(y_1, y_2, y_3, y_4, y_5) \tag{38}$$

as a five-dimensional vector of responses, two of which are just the original variables and the remaining three are functions (squares and cross product) derived from them. If

$$\mathbf{a}^{*\prime}=(a, b, c, d, e) \tag{39}$$

denotes the vector of coefficients used in Eq. 37 in defining z, the first stage of a *quadratic principal components* analysis may be formulated as the problem of determining $\mathbf{a}^*$ so that the variance of $z(=\mathbf{a}^{*\prime}\mathbf{y}^*)$ is maximum subject to a normalizing constraint, such as $\mathbf{a}^{*\prime}\mathbf{a}^*=1$, exactly as in linear principal components analysis.

On the basis of a sample of n observations on the initial response variables y_1 and y_2, one can generate n observations on $\mathbf{y}^*$ and thence obtain the "sample" mean vector and covariance matrix:

$$\bar{\mathbf{y}}^{*\prime} = (\bar{y}_1, \bar{y}_2, \bar{y}_3, \bar{y}_4, \bar{y}_5),$$

$$\mathbf{S}^* = \frac{1}{n-1} \sum_{i=1}^{n} (\mathbf{y}_i^* - \bar{\mathbf{y}}^*)(\mathbf{y}_i^* - \bar{\mathbf{y}}^*)'.$$

For a given $\mathbf{a}^*$ the observed variance of z will be $\mathbf{a}^{*\prime}\mathbf{S}^*\mathbf{a}^*$, and the first stage of the quadratic principal components analysis will then result in choosing $\mathbf{a}^* = \mathbf{a}_1^*$, where $\mathbf{a}_1^*$ is the eigenvector corresponding to the largest eigenvalue, c_1^*, of $\mathbf{S}^*$. Furthermore, if at the next stage one wishes a second quadratic function that is uncorrelated (in the sample) with the first and has maximum variance subject to being thus uncorrelated, then, exactly as in linear principal components, one will choose the set of coefficients for the second quadratic function as the eigenvector, $\mathbf{a}_2^*$, of $\mathbf{S}^*$ corresponding to its second largest eigenvalue, c_2^*. The process can be repeated to determine additional quadratic functions with decreasing variances at each stage and with zero correlations with each of the quadratic functions determined at earlier stages. Thus, if $c_1^* \geq c_2^* \geq \cdots \geq c_5^* \geq 0$ are the eigenvalues of $\mathbf{S}^*$ with corresponding eigenvectors $\mathbf{a}_1^*$, $\mathbf{a}_2^*$, $\mathbf{a}_3^*$, $\mathbf{a}_4^*$, and $\mathbf{a}_5^*$, one can derive five quadratic functions of the two original variables from the relationships

$$z_i = \mathbf{a}_i^{*\prime}\mathbf{y}^*, \qquad i = 1, 2, \ldots, 5. \tag{40}$$

As a method for the reduction of dimensionality, the interest will lie in the functions defined by "smallest" eigenvalues. For instance, if the bivariate observations lie on a quadratic curve, one will expect $c_5^* = 0$ and z_5 will be a constant for all observations, $\mathbf{y}_i^*$.

More generally, given n p-dimensional observations, if one wishes to perform a quadratic principal components analysis, one will augment the original p variables by $p + [p(p-1)/2]$ derived variables (viz., all squared and cross-product terms) and carry out a regular principal components analysis of the

$$\left[2p + \frac{p(p-1)}{2}\right] \times \left[2p + \frac{p(p-1}{2}\right]$$

covariance matrix of the enlarged set of variables.

The generalization is immediate to cubic and higher-order polynomial principal components, as well as to any situation in which the system of derived coordinates sought consists of (unknown) linear combinations of (specified, possibly nonlinear) functions of the original variables. Thus a typical member of the class of derived coordinates that can be handled by the above generalization of principal components is

$$z = \sum_{j=1}^{k} a_j f_j(y_1, y_2, \ldots, y_p), \tag{41}$$

where the a_j's are to be determined, and the f_j's are completely specified, but otherwise arbitrary, functions of the original variables.

The above generalization, formulated in terms of properties of variances of and correlations among the derived coordinates, is an obvious extension to the nonlinear case of Hotelling's approach to linear principal components analysis. In particular, linear principal components may be viewed as a special case of polynomial principal components. In a linear principal components analysis, one starts with p correlated coordinates and derives a set of p uncorrelated (in the sample) coordinates that are linear functions of the original variables. In a quadratic principal components analysis, however, one starts with p correlated coordinates and obtains a set of $2p+[p(p-1)/2]$ uncorrelated quadratic coordinates that are quadratic functions of the original variables. As a method for nonlinear reduction of dimensionality, in using quadratic (or otherwise generalized) principal components analysis, the interest will lie in the functions defined by the "smallest" eigenvalues.

Example 8. The data consisted of 41 points lying on the parabola $Y_2=2+4Y_1+4Y_1^2$, with $Y_1=-1.5(0.05)0.5$. A quadratic principal components analysis was applied to this data, and the dimension of the enlarged eigenvalue problem was, therefore, $2p+[p(p-1)/2]=5$, since $p=2$. The resulting eigenvalues and eigenvectors are shown in Exhibit 8*a*. The largest eigenvalue is seen to be over 2000, while the smallest, which is "known" to be 0, is computed as 9×10^{-6}.

Each eigenvector provides a nonlinear (quadratic in this case) coordinate in the original space, and Exhibit 8*b* shows the coordinates determined by the eigenvectors corresponding to the smallest and largest of the five eigenvalues. The parabolic coordinate is from the smallest eigenvalue, and the flat elliptic coordinate is from the largest. In the absence of statistical errors in the data, one of the parabolas (viz., the middle one) passes exactly through the points.

Exhibit 8a. Eigenanalysis for example of quadratic principal components analysis

CHARACTERISTIC ROOTS

2163.634	219.915	2.258	2.223	0.000009

CHARACTERISTIC VECTORS

-.002513	.246077	.508757	-.442445	.696310
.169321	.011882	.758811	.604229	-.174076
-.094253	.932909	-.212548	.274991	.0000004
.044843	-.243106	.319056	.593499	.696311
.980015	.099425	-.135640	-.106239	.0000003

Exhibit 8b. Data and coordinates from the largest and smallest eigenvalues

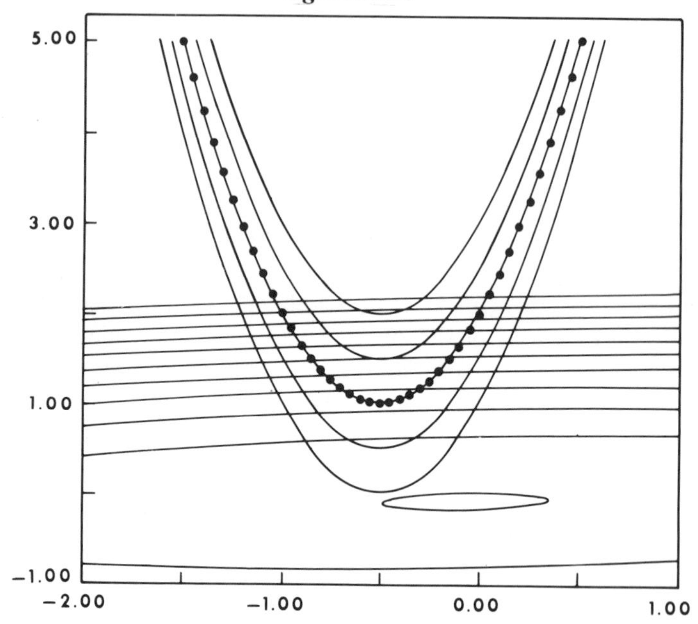

Example 9. To illustrate the effect of "noise" on the technique, the data of Example 8 were modified by adding random normal components (with mean 0 and variance $\frac{1}{16}$) to each of the two coordinates of every point. The results of an eigenanalysis, comparable to the one in Example 8, are shown in Exhibit 9*a*. It is seen that, although the largest eigenvalue is still about 2000, the smallest one is now 1.081. The first two eigenvalues, and especially the corresponding eigenvectors, are, in fact, quite similar to the ones in Exhibit 8*a*. The indications from the remaining

Exhibit 9a. Eigenanalysis for example of quadratic principal components analysis

CHARACTERISTIC ROOTS

1969.563	285.938	5.717	2.034	1.081

CHARACTERISTIC VECTORS

-.010915	.260105	-.269383	.610264	.698023
.172863	.012363	.189273	.761768	-.594853
-.106665	.930317	.301394	-.146763	-.103707
.062983	-.230187	.892766	.062053	.377047
.977064	.117118	-.061142	-.147978	.077413

Exhibit 9b. Data and coordinates from the largest and smallest eigenvalues

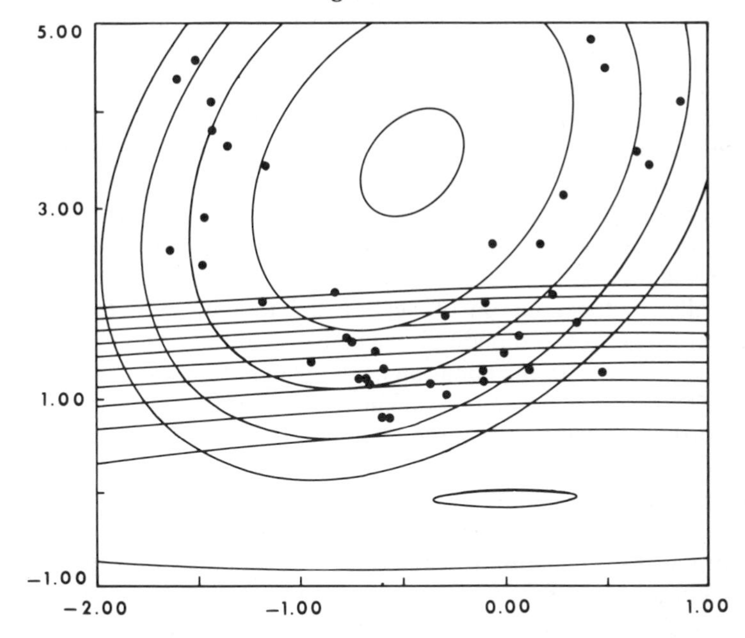

three eigenvalues and eigenvectors, however, are different because of the added noise in the data of this example. In particular, in the eigenvector that corresponds to the smallest eigenvalue it is seen that the elements which provide the coefficients for $Y_1 Y_2$ and Y_2^2 are no longer negligible and only the coefficient of Y_1 appears to be essentially the same as it was for the noiseless data.

Exhibit 9*b* shows the data and the quadratic coordinates defined by the eigenvectors corresponding to the smallest and largest eigenvalues. The smallest eigenvalue now leads to an elliptical coordinate system because of the influence of the statistical errors. This distortion of the parabolic coordinate into an elliptical one, however, does not seem to be unreasonable relative to the configuration of the data.

The illustration of the technique of generalized principal components in Examples 8 and 9 is trivial since $p=2$ for both cases. With $p=3$ more interesting possibilities begin to arise since the points may then lie on $(q=)$ one- or two-dimensional curved subspaces. Example 10 is a case with $p=3$ and $q=2$. The purpose of Examples 8 and 9 is to illustrate the basic concept of nonlinear coordinate transformations and the analytical and algorithmic aspects of the method, which, of course, are valid for any value of p. The method is not crucial for two- or three-dimensional problems which lend themselves to graphical representation and study.

The above development of generalized principal components analysis in terms of minimizing (or maximizing) variances, and of requiring the different coordinates to be uncorrelated in the sample, is a statistical extension of Hotelling's (1933) approach for the linear case. However, to see the problem of nonlinear singularities in a broader context of nonlinear coordinate systems, one can formulate the question in function-fitting terms. Thus, in the linear case, the eigenvector corresponding to the smallest eigenvalue essentially determines a plane of closest fit, where closeness is measured by the sum of squares of perpendicular distances. Specifically, in the notation of Section 2.2.1, if $\mathbf{a}_1', \ldots, \mathbf{a}_p'$ denote the eigenvectors of $\mathbf{S}$ corresponding, respectively, to the ordered eigenvalues $c_1 \geq c_2 \geq \cdots \geq c_p > 0$, the equation of the plane of closest fit to the data is $\mathbf{a}_p'\mathbf{y} = \mathbf{a}_p'\bar{\mathbf{y}}$, where $\bar{\mathbf{y}}$ is the sample mean vector (see Eq. 1). Also, among all planes orthogonal to the first, the equation of the plane of next closest fit to the data is $\mathbf{a}_{p-1}'\mathbf{y} = \mathbf{a}_{p-1}'\bar{\mathbf{y}}$, and so on. This indeed was Karl Pearson's (1901) formulation leading to the linear principal components.

In the linear case the statistical approach through variance minimization and mutual uncorrelatedness turns out to be identical with the approach of fitting mutually orthogonal planes by minimizing the sum of squares of perpendicular distances from the data to the planes. This equivalence of the two approaches does not, however, carry over to the general nonlinear case with "noisy" data. Developing algorithms for fitting even specific types (e.g., quadratic polynomials) of nonlinear functions to data by minimizing the sum of squares of "perpendicular" distances would be quite useful.

One value of such an algorithm would be the intuitive appeal of its criterion in function-fitting situations involving variables all of which may be subject to random errors, a circumstance which is not rare and in which the use of the usual least squares criterion may be questionable. Another feature of the function-fitting approach would make it particularly appealing as a statistical tool for the reduction of dimensionality. Unlike the eigenvector algorithm involved in the variance-minimization approach, the function-fitting algorithm is sensitive only to the scales of the original variables and not to the scales of additional nonlinear functions of them. For instance, in the bivariate quadratic case discussed earlier, the solution of the enlarged eigenvector problem is sensitive not only to the scales of Y_1 and Y_2 but also to those of Y_1^2, Y_2^2, and Y_1Y_2, which are necessarily noncommensurable with the original coordinates. However, the criterion of minimizing the sum of squares of perpendicular deviations, although dependent on the scales of Y_1 and Y_2, does not depend in any way on the scales of the quadratic terms Y_1^2, Y_2^2, and $Y_1 Y_2$. This scale-resistant nature would be particularly desirable in many practical applications. One way of handling this issue, while still retaining the variance-minimization approach, is to carry out the eigenanalysis on the enlarged correlation matrix rather than on the corresponding enlarged covariance matrix, an idea discussed earlier (see pp. 11–12) in the context of linear principal components analysis.

A main advantage of the variance-minimization approach is the computational simplicity of the algorithm involved—merely an eigenanalysis. In the absence of a general algorithm for function fitting based on minimizing perpendicular deviations, one is limited to this approach anyway. In the absence of noise in the data, as in Examples 8 and 10, the issue of differences between the two approaches does not arise. When the variance-minimization approach is used for generalized principal components analysis, the equation of the nonlinear subspace to which the observations may possibly be confined will be defined by means of a procedure analogous to the one employed in the linear case. Specifically, for instance, in the bivariate case considered above, if the observations lie on a quadratic curve (as in Example 8), the variance-minimization approach will be expected to lead to $c_5^* = 0$ and the equation of the quadratic curve will be $(z_5 =)\, \mathbf{a}_5^{*\prime}\mathbf{y}^* = \mathbf{a}_5^{*\prime}\bar{\mathbf{y}}^*$, where $\mathbf{y}^*$, $\bar{\mathbf{y}}^*$, c_5^*, and $\mathbf{a}_5^*$ were defined earlier. In fact, the "middle" one of the five parabolas in Exhibit 8*b* and the "middle" one of the five ellipses in Exhibit 9*b* have equations that are specified in this manner. The same idea is also used in obtaining the equation of the quadratic surface involved in the next example.

Example 10. The data, shown in Exhibit 10*a*, consist of 19 points lying on the surface of the sphere, $X^2 + Y^2 + Z^2 = 25$. Thus, in this example, $p = 3$ and $q = 2$.

A quadratic principal components analysis, involving an eigenanalysis of dimension nine, yields the eigenvalues and eigenvectors shown in Exhibit 10*b*. The largest eigenvalue is seen to be about 96, while the smallest, which in the absence of "noise" in the data is 0, was computed to be less in value than 10^{-7} and hence is shown as being 0. Also shown at the bottom of Exhibit 10*b* is the equation of the sphere on which the data lie as determined by the eigenvector for the zero (smallest) eigenvalue. With the error-free data, the original sphere is recovered. Since $p = 3$, it is possible to represent the data of this example and the fitted sphere in a stereoscopic three-dimensional display, which can be obtained by using current capabilities in computer software and graphical aids.

In utilizing polynomial principal components, such as the quadratic one

Exhibit 10a. Artificial data—19 points on the surface of a sphere

X	Y	Z
-5.0	0.0	0.0
-4.0	1.0	±2.828
-3.0	0.5	±3.969
-2.0	4.0	±2.236
-1.0	0.0	±4.899
0.0	3.0	±4.000
1.0	2.0	±4.472
2.0	4.0	±2.236
3.0	3.3	±2.261
4.0	2.4	±1.800

illustrated in Examples 8–10, the expectation is that these will respond to local nonlinearities. The use of quadratic analysis may produce a significant improvement in sensitivity to local nonlinearities, and the hope is that one will not need polynomials of very high degree for accommodating most nonlinearities met in practice.

The choice of the degree of polynomials to be used also has certain implications for the number, n, of observations that will be required. When $p=1$, of course, one needs at least two observations to obtain a nonzero estimate of variance; and similarly, when $p=2$, one needs $n \geq 3$ observations for obtaining a nonsingular estimate of the covariance matrix. In the same vein, for a problem in p-space, a quadratic principal components analysis will involve the eigenvector solution for a

$$\left[2p+\frac{p(p-1)}{2}\right]\times\left[2p+\frac{p(p-1)}{2}\right]$$

matrix, so that a nontrivial solution can be obtained only if $n > 2p+[p(p-1)/2]$; thus, for $p=8$, n must exceed 44. Using a cubic coordinate system with $p=5$ will require $n>55$. In practice, one could handle the difficulty caused by such requirements on n by first performing a linear principal components analysis and then pursuing nonlinear analysis, using only the first few linear principal components (i.e., the ones with largest variances).

The magnitude of the eigenvector computation grows rapidly, both with the degree of the polynomial coordinate system considered and with the dimension of the response. With $p=5$ a completely general cubic

Exhibit 10b. Eigenanalysis for quadratic principal components analysis of data in Exhibit 10a

EIGENVALUES

96.3697	62.7167	61.6238	49.5563	34.4642	2.5909	0.9510	0.0779	0.0000

EIGENVECTORS

0.0496	0.2371	0.0000	0.0000	-0.3170	-0.0000	0.8900	-0.2209	-0.0001
0.0467	0.1698	-0.0000	0.0000	0.0494	-0.0000	0.2082	0.9608	-0.0006
-0.0000	-0.0000	0.2737	0.3037	-0.0000	0.9126	0.0000	0.0000	-0.0000
0.2259	0.2665	0.0000	0.0000	-0.8573	0.0000	-0.3723	0.0667	0.0001
0.0000	-0.0000	0.4116	-0.8946	-0.0000	0.1742	0.0000	0.0000	-0.0000
-0.0000	-0.0000	0.8693	0.3280	0.0000	-0.3699	-0.0000	-0.0000	0.0000
0.5745	-0.5639	0.0000	0.0000	-0.0665	-0.0000	0.1073	0.0515	-0.5774
0.1867	0.7103	0.0000	-0.0000	0.3121	-0.0000	-0.1196	-0.1251	-0.5773
-0.7612	-0.1464	0.0000	0.0000	-0.2457	-0.0000	0.0119	0.0726	-0.5774

Equation obtained by QPCA

$$0.577x^2 + 0.577y^2 + 0.577z^2 = 14.4$$

principal components analysis would involve a 55-dimensional eigenanalysis. For the numerical computations in these eigenanalyses, it would therefore be desirable to use the singular value decomposition method (see Businger & Golub, 1969) mentioned earlier in connection with linear principal components.

Certain advantages of developing a function-fitting approach to generalized principal components analysis have been mentioned already. A further, possibly intangible, advantage of viewing generalized principal components analysis in the framework of function fitting is that the latter area has received considerable attention in statistics, both conceptually and methodologically, under categories such as linear and nonlinear regression. The available methodology of these familiar areas may, after appropriate modifications, be relevant for further extending the usefulness of the generalized principal components approach. For instance, with a projected class of coordinates that involves arbitrary functions of the response variables, with some unspecified coefficients that may occur *nonlinearly*, the mathematical problem is still just that of finding members of the class which give closest fit, and nonlinear-fitting ideas and procedures may prove useful. Concepts and methods for linearizing the nonlinear problem and for iterative solutions may carry over. The eigenvector algorithms used earlier cannot be applied simply to yield the solution, although one may be able to use them iteratively to develop an approximate solution.

REFERENCES

Section 2.2.1 Blackith (1960), Blackith & Roberts (1958), Businger (1965), Businger & Golub (1969), Golub (1968), Hotelling (1933), Pearson (1901), Rao (1965), Seal (1964).

Section 2.2.2 Anderson & Rubin (1956), Bartlett (1951), Carroll (1969), Devlin et al. (1975), Fletcher & Powell (1963), Harman (1967), Hoerl & Kennard (1970), Howe (1955), Imbrie (1963), Imbrie & Kipp (1971), Imbrie & Van Andel (1964), Jöreskog (1967), Jöreskog & Lawley (1968), Lawley (1940, 1967), Lawley & Maxwell (1963), McDonald (1962, 1967), Seal (1964), Theil (1963), Thomson (1934), Thurstone (1931).

Section 2.3 Abelson & Tukey (1959), Bartholomew (1959), Barton & Mallows (1961), Bennett (1965), Boynton & Gordon (1965), Carroll & Chang (1970), Coombs (1964), Ekman (1954), Kruskal (1964a, b), Miles (1959), Rothkopf (1957), Shepard (1962a, b, 1963), Shepard & Carroll (1966), Van Eeden (1957a, b).

Section 2.4 Businger & Golub (1969), Gnanadesikan & Wilk (1966, 1969), Van de Geer (1968), Wilk, Gnanadesikan, & Huyett (1962a).

CHAPTER 3

Development and Study of Multivariate Dependencies

3.1. GENERAL

The general concern here is with the study of dependencies, both association and relationship, among several responses. It is possible to delineate two broad categories of multivariate dependencies: (i) those that involve only one set of multivariate responses, and (ii) dependencies of one set of responses on other sets of responses, or on extraneous design or regressor variables. The first category of dependencies may be called *internal*; the second, *external.*

3.2. INTERNAL DEPENDENCIES

For a case with n observations on a p-dimensional response vector, the familiar techniques of computing and studying simple and various partial correlation coefficients are examples of methods for studying the relative degrees of association among the p responses. A pictorial technique for displaying association, which can be useful at times, has been discussed by Anderson (1954, 1957, 1960) under the name *glyphs*, including generalized glyphs and metroglyphs. An illustrative example follows.

Example 11. Exhibit 11, taken from Anderson (1960), pertains to an example involving observed measures of five $(=p)$ qualities as possessed by each of four $(=n)$ individuals. The table of data is shown at top left. For each individual, a graphical representation called a glyph may be obtained in which each quality is pictured as a ray emanating from a circle corresponding to an individual (see top right of Exhibit 11). The position of a ray in such a glyph corresponds to one of the qualities, while the length of the ray reflects the level of the quality—a long ray indicating

Exhibit 11. Metroglyph (Anderson, 1960)

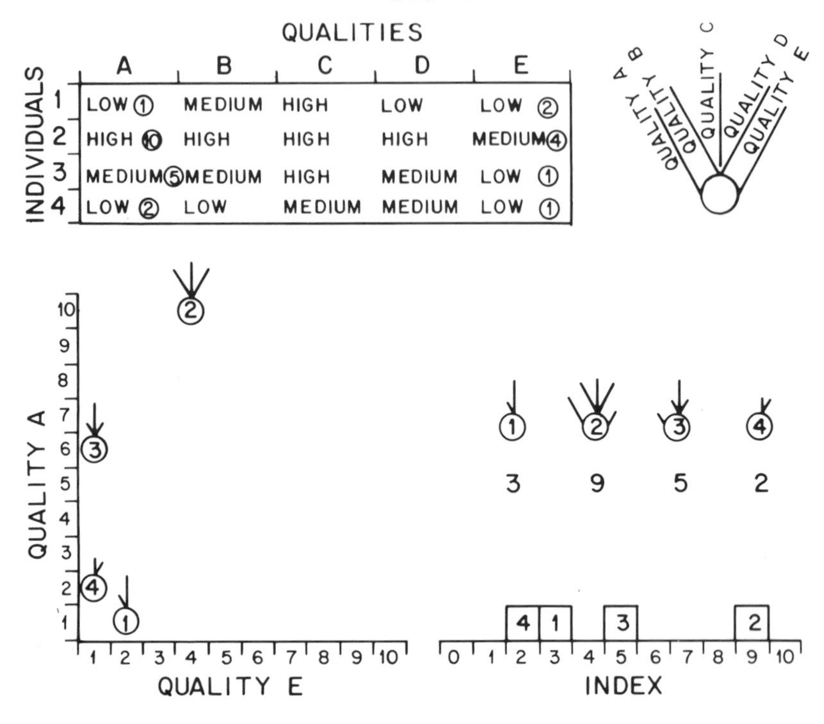

high level, a short one representing medium level, and no ray at all corresponding to low level. Thus, in the metroglyphs for all four individuals shown in the middle on the right side of Exhibit 11, at a glance one can see that individual 2 is high on most qualities whereas individual 1 is low on many.

An alternative representation, which contains fewer rays and may be particularly useful for depicting associations among the qualities, is the scatter diagram shown at bottom left of Exhibit 11. Here the rays corresponding to two of the qualities (A and E) have been dropped from the glyphs, and coordinate axes with a finer degree of quantization of levels (10 instead of just 3) of A and E are introduced. The values of A and E actually observed on scales ranging from 1 to 10 are shown encircled for each of the four individuals in the data table at the top left of Exhibit 11. These values provide the coordinates for the placement of the glyphs in the bottom left picture. As stated by Anderson (1957, 1960), this representation of the four pentavariate observations in this example shows that qualities B and C, as also B and D, are associated. Moreover, one can see a fairly strong association between B, C, D (both individually and concurrently) and quality A, while only a weak association is manifest between the former set of qualities and quality E.

As an overall numerical summary, Anderson (1957, 1960) obtains an index for each glyph by scoring 2 for each long ray (i.e., high level of a quality), 1 for each short ray (i.e., medium level), and 0 for absence of a ray. Values of this index are given below the metroglyphs, and a histogram is shown at the bottom right of Exhibit 11.

Two other pictorial techniques, of somewhat the same spirit as metroglyphs, for representing multidimensional data are worth mentioning and illustrating. The first of these, called a *weathervane plot* by Bruntz et al. (1974), was developed for analyzing air pollution data, and the next example illustrates the technique.

Example 12. In analyzing air pollution data it is often appropriate to consider not only the chemical reactions involved in producing the pollutants but also the prevailing meteorological conditions. Specifically, solar radiation, wind speed and direction, and temperature are some of the variables of interest.

Exhibit 12, taken from Bruntz et al. (1974), shows a weathervane plot in which the abscissa is the total solar radiation from 8 A.M. to noon, while the ordinate is the average ozone level observed from 1 P.M. to 3 P.M. The plot pertains to a specific site, and the points (centers of the circles) correspond to different days. As a two-dimensional scatter plot, the centers of the circles in Exhibit 12 provide information on the relationship between ozone levels and solar radiation. In addition, the diameter of the circle plotted has been coded to be proportional to the observed daily maximum temperature, while the line emanating from the circle is coded with information regarding the wind. The length of the line segment is inversely proportional to an average wind speed. If the lines are considered as arrows whose heads are at the centers of the circles, the orientations of the lines correspond to average wind directions. For instance, for the point at the top of Exhibit 12 (i.e., the day with highest ozone), the average wind direction is from the northwest.

One indication of the plot is that ozone levels do not become high when there is low solar radiation. However, the presence of points in the lower right-hand part of Exhibit 12 suggests that high solar radiation alone does not guarantee high ozone levels, and for the days in this part of the picture the temperature is generally low and the wind speed is generally high. Also, at a given level of solar radiation, ozone seems to increase as temperature increases and wind speed decreases. Wind direction does not seem to be a dominant factor in influencing patterns. Thus the five-dimensional display in Exhibit 12 enables one to get a "feeling"

Exhibit 12. Weathervane plot of air pollution data (Bruntz et al., 1974)

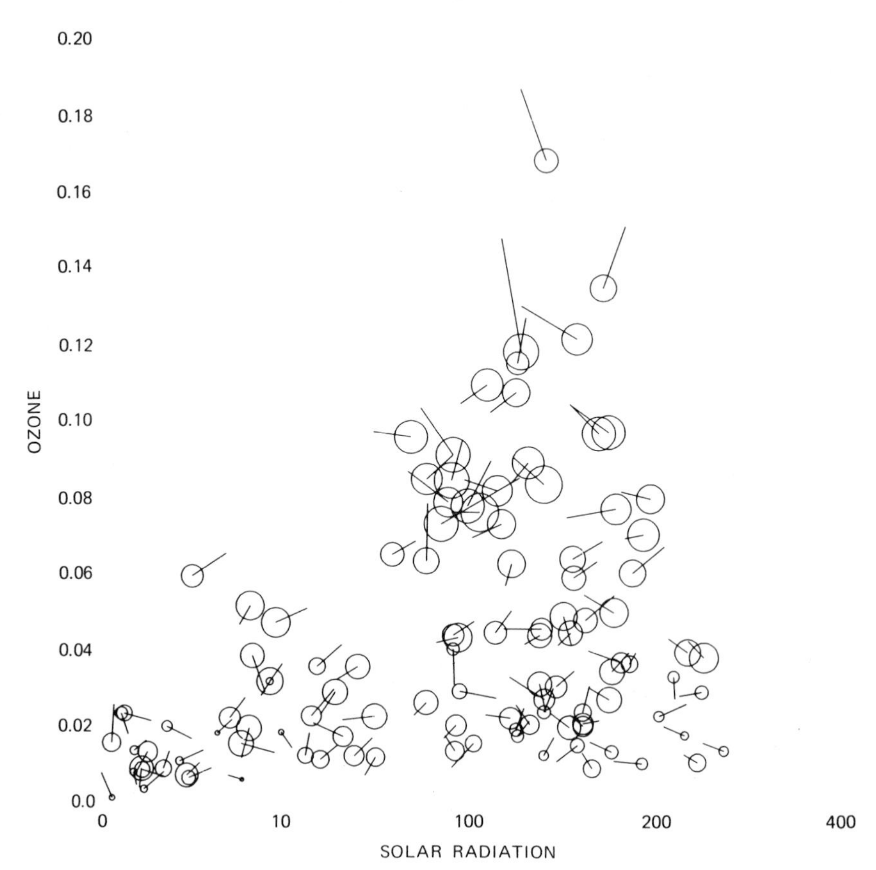

for some of the interrelationships among the five variables involved in this example.

The second pictorial representation that is analogous to glyphs in spirit is a novel scheme proposed by Chernoff (1973). The idea is to code the values of the variables by associating them with different characteristics of a human face. An important issue in connection with Chernoff's scheme that needs further study is how to go about associating the variables with different aspects of a face in any specific application. Developing guidelines for doing this would be valuable. The next example illustrates the procedure.

Example 13. This example, taken from Chernoff (1973), pertains to 12-dimensional data on mineral contents assayed on 53 equally spaced

Exhibit 13a. Data for faces (Chernoff, 1973)

DATA ON 12 VARIABLES REPRESENTING MINERAL CONTENTS FROM A 4500-FOOT CORE DRILLED FROM A COLORADO MOUNTAINSIDE

ID	Z_1	Z_2	Z_3	Z_4	Z_5	Z_6	Z_7	Z_8	Z_9	Z_{10}	Z_{11}	Z_{12}
200	320	105	057	050	001	001	001	060	020	250	210	370
201	280	150	040	050	001	001	001	060	040	210	130	420
202	260	165	033	050	001	001	001	060	010	250	090	440
203	305	110	044	040	001	001	001	050	050	260	140	250
204	290	160	035	035	001	001	001	050	020	210	060	510
205	275	130	047	035	001	001	001	050	020	230	090	570
206	280	155	035	035	001	001	001	080	020	270	170	400
207	300	115	050	060	001	001	001	120	010	280	190	300
208	250	130	041	030	005	001	001	070	030	250	110	330
209	285	120	047	040	001	001	001	070	010	240	170	280
210	280	105	047	070	001	001	001	060	020	370	070	300
211	300	135	050	040	001	001	001	120	060	250	160	200
212	280	110	056	050	001	001	001	150	010	280	270	280
213	305	080	065	080	005	001	001	130	010	300	260	260
214	230	175	029	035	001	001	001	270	030	250	140	240
215	325	060	052	090	001	001	001	160	010	280	260	170
216	270	170	025	040	001	001	001	160	010	290	070	330
217	250	185	031	025	001	001	001	120	001	260	080	330
218	260	185	030	015	001	001	001	270	080	480	010	330
219	270	185	032	010	005	001	001	180	040	450	020	220
220	325	045	053	005	020	001	001	600	080	660	020	250
221	315	090	047	005	020	001	001	410	200	600	060	260
222	335	100	047	010	040	001	001	360	080	590	110	170
223	310	010	049	005	080	018	001	640	240	630	060	190
224	410	001	049	001	075	032	001	760	440	800	001	001
225	360	001	048	001	080	055	001	770	260	770	010	010
226	310	015	051	001	105	036	001	660	380	640	001	010
227	420	005	049	001	095	056	001	620	520	680	001	001
228	415	020	049	005	025	036	001	370	220	340	001	001
229	420	005	041	001	070	060	001	630	510	580	001	001
230	450	005	040	001	090	070	001	690	570	630	001	001
231	395	001	025	015	100	071	001	580	530	560	001	010
232	380	010	027	025	035	039	001	350	320	400	001	270
233	430	010	025	030	030	025	001	340	340	360	001	200
234	410	075	022	010	005	015	001	170	170	170	001	060
235	520	055	024	040	005	001	001	210	190	190	001	180
236	385	135	018	010	005	008	001	140	200	260	001	020
237	535	065	010	020	001	001	001	110	230	270	001	070
238	550	095	001	010	001	001	001	050	230	270	001	030
239	510	100	001	001	001	001	001	190	150	230	001	110
240	510	095	001	040	001	001	001	140	100	150	001	040
241	385	180	010	001	001	001	001	050	050	300	001	050
242	505	125	001	001	001	001	001	001	200	130	001	030
243	470	090	001	020	001	001	001	160	300	380	001	060
244	465	110	001	035	001	001	001	260	440	500	001	060
245	400	140	001	015	001	023	001	330	400	390	001	040
246	415	105	015	025	040	032	001	220	190	270	001	010
247	435	075	010	015	001	069	001	370	360	500	001	010
248	370	145	010	010	005	012	040	130	080	330	001	030
249	380	210	001	001	001	001	020	070	001	050	001	030
250	430	065	001	005	020	001	075	130	070	300	001	020
251	420	080	030	001	005	026	001	050	100	350	001	050
252	425	060	035	005	001	001	030	100	010	340	001	010
min	250	001	001	001	001	001	001	001	001	050	001	001
max	520	210	065	090	105	071	075	770	570	800	270	570

specimens taken from a 4500-foot core drilled into a Colorado mountainside. Exhibit 13*a* shows the numerical data, while Exhibit 13*b* shows the 53 faces as obtained by Chernoff. Some major changes in the values of the variables are noticed even on inspection of the numbers in Exhibit 13*a* (e.g., values of Z_5 for specimens 220–233). The breakpoints are clearly visible in the sequence of faces too. For instance, there is an abrupt change in the overall shape of the head and the location and shape of the eyes after the face for specimen 219. Also striking are the smile and the small, high eyes of the faces for specimens 224–231. In the absence of more specific information regarding the coding employed in obtaining the

Exhibit 13b. The faces obtained by Chernoff (1973)

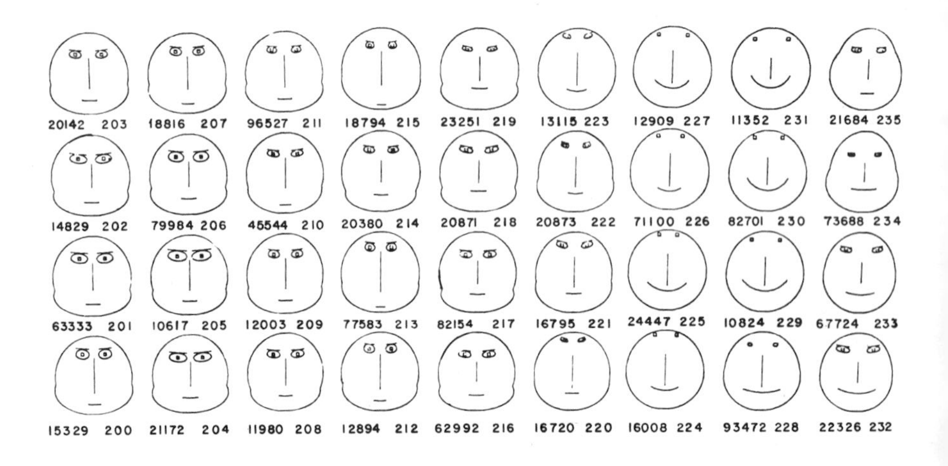

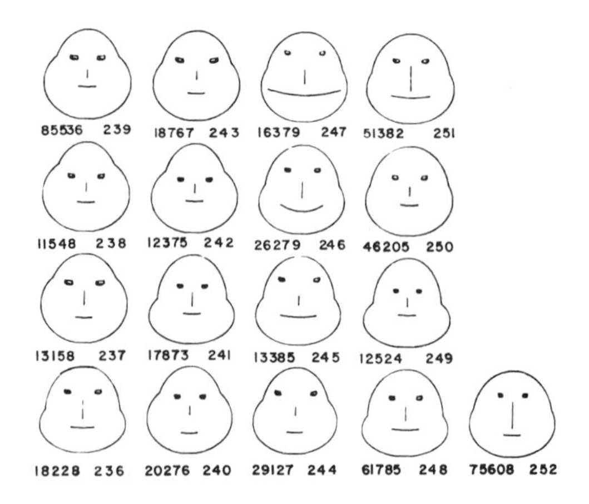

faces in this example, however, it is difficult to provide any further detailed interpretations of the data here.

The detection and description of relationships (as against associations) among a set of response variables overlap, in obvious ways, the concerns and methods of reduction of dimensionality discussed in Chapter 2. The detection of linear and nonlinear singularities and the characterization of them are useful not only for uncovering redundancies among the response variables but also for studying the nature of the interrelationships among the variables. Thus most of the techniques discussed in Chapter 2 are also relevant to the objective of studying internal relationships.

3.3. EXTERNAL DEPENDENCIES

The multiple correlation coefficient, often discussed in the context of regressing a single variable on a set of extraneous variables, is one example of a measure of association between two sets of variables wherein one of the sets contains just a single variable. Canonical correlation analysis, developed by Hotelling (1936), is another classical technique for studying associations between two sets of variables. Given a set of variables, **x**, and another set, **y**, the basic idea is to find the two linear combinations, one of the x-variables and one of the y-variables, that have maximal correlation; then, from among the two sets of linear combinations orthogonal to those already determined, to select the two with maximal correlation, and so on. In general, if p and q are, respectively, the numbers of x- and y-variables, and if $p \le q$, one can extract p pairs of linear combinations by this process. The derived linear functions will be called *canonical variates* (see also the remarks in Section 4.2).

More recently (Steel, 1951; Horst, 1965; Kettenring, 1969, 1971), the concepts and techniques of canonical correlation analysis have been extended to the case of more than two sets of variables. Kettenring (1969, 1971) provides a unifying discussion of the various approaches, and the following material relies heavily on his treatment.

Given m sets of variables, $\mathbf{y}_j(p_j \times 1)$ for $j = 1, 2, \ldots, m$, suppose that $p_1 \le p_2 \le \cdots \le p_m$ and $p = \sum_{j=1}^{m} p_j$. It is assumed, without any loss of generality for present purposes, that $\mathscr{E}(\mathbf{y}_j) = \mathbf{0}$ for all j and that the $p_j \times p_j$ covariance matrix of $\mathbf{y}_j$ is nonsingular and denoted as $\mathbf{\Sigma}_{jj}$. The Cholesky decomposition, $\mathbf{\Sigma}_{jj} = \boldsymbol{\tau}_j \boldsymbol{\tau}_j'$, may then be utilized to obtain a linear transformation of $\mathbf{y}_j$ to $\mathbf{x}_j = \boldsymbol{\tau}_j^{-1} \mathbf{y}_j$ such that the covariance matrix of $\mathbf{x}_j$ is the identity matrix of order p_j. Now, if

$$\mathbf{y}' = (\mathbf{y}_1' \vdots \mathbf{y}_2' \vdots \cdots \vdots \mathbf{y}_m') \qquad \text{and} \qquad \mathbf{x}' = (\mathbf{x}_1' \vdots \mathbf{x}_2' \vdots \cdots \vdots \mathbf{x}_m'),$$

the covariance matrices of $\mathbf{y}$ and $\mathbf{x}$, denoted as $\mathbf{\Sigma}$ and $\mathbf{\Gamma}$, respectively, are

$$\mathbf{\Sigma}=\begin{pmatrix}\mathbf{\Sigma}_{11} & \mathbf{\Sigma}_{12} & \cdots & \mathbf{\Sigma}_{1m}\\ \mathbf{\Sigma}'_{12} & \mathbf{\Sigma}_{22} & \cdots & \mathbf{\Sigma}_{2m}\\ \cdots & \cdots & \cdots & \cdots\\ \mathbf{\Sigma}'_{1m} & \mathbf{\Sigma}'_{2m} & \cdots & \mathbf{\Sigma}_{mm}\end{pmatrix} \tag{42}$$

and

$$\mathbf{\Gamma}=\begin{pmatrix}\mathbf{I} & \mathbf{\Gamma}_{12} & \cdots & \mathbf{\Gamma}_{1m}\\ \mathbf{\Gamma}'_{12} & \mathbf{I} & \cdots & \mathbf{\Gamma}_{2m}\\ \cdots & \cdots & \cdots & \cdots\\ \mathbf{\Gamma}'_{1m} & \mathbf{\Gamma}'_{2m} & \cdots & \mathbf{I}\end{pmatrix}, \tag{43}$$

where $\mathbf{\Gamma}_{ij}=\boldsymbol{\tau}_i^{-1}\mathbf{\Sigma}_{ij}(\boldsymbol{\tau}_j^{-1})'$.

Loosely speaking, one wishes to find linear functions of the variables (i.e., canonical variates) in each of the m sets so as to satisfy criteria that are specified in terms of the intercorrelations among the linear functions. Let ${}_jz_1={}_j\boldsymbol{\alpha}'_1\mathbf{x}_j={}_j\boldsymbol{\beta}'_1\mathbf{y}_j$ (where ${}_j\boldsymbol{\beta}'_1={}_j\boldsymbol{\alpha}'_1\boldsymbol{\tau}_j^{-1}$). for $j=1, 2, \ldots, m$, denote the m linear functions, one from each of the m sets, at the first stage. Let the coefficients of the linear combinations be required to satisfy the normalizing constraints ${}_j\boldsymbol{\beta}'_1\cdot\mathbf{\Sigma}_{jj}\cdot{}_j\boldsymbol{\beta}_1={}_j\boldsymbol{\alpha}'_1\cdot{}_j\boldsymbol{\alpha}_1=1$, so that the variance of ${}_jz_1$ is 1 for all j. Suppose that $\mathbf{z}'_1=({}_1z_1, {}_2z_1, \ldots, {}_mz_1)$ denotes the set of m first-stage canonical variates whose correlation (and covariance) matrix is

$$\mathbf{\Phi}(1)=\begin{pmatrix}1 & \phi_{12}(1) & \cdots & \phi_{1m}(1)\\ & \cdot & & \cdot\\ & \cdot & & \cdot\\ & \cdot & & \cdot\\ \phi_{1m}(1) & \phi_{2m}(1) & \cdots & 1\end{pmatrix}=\mathbf{D}_{\beta_1}\mathbf{\Sigma}\mathbf{D}'_{\beta_1}=\mathbf{D}_{\alpha_1}\mathbf{\Gamma}\mathbf{D}'_{\alpha_1}, \tag{44}$$

where the $m\times p$ matrices, $\mathbf{D}_{\alpha_1}$ and $\mathbf{D}_{\beta_1}$, are block-diagonal and are defined by

$$\mathbf{D}_{\alpha_1}=\begin{pmatrix}{}_1\boldsymbol{\alpha}'_1 & \mathbf{0}' & \cdots & \mathbf{0}'\\ \mathbf{0}' & {}_2\boldsymbol{\alpha}'_1 & \cdots & \mathbf{0}'\\ \cdots & \cdots & \cdots & \cdots\\ \mathbf{0}' & \cdots & & {}_m\boldsymbol{\alpha}'_1\end{pmatrix} \tag{45}$$

and

$$\mathbf{D}_{\beta_1}=\begin{pmatrix}{}_1\boldsymbol{\beta}'_1 & \mathbf{0}' & \cdots & \mathbf{0}'\\ \mathbf{0}' & {}_2\boldsymbol{\beta}'_1 & \cdots & \mathbf{0}'\\ \cdots & \cdots & \cdots & \cdots\\ \mathbf{0}' & \cdots & & {}_m\boldsymbol{\beta}'_1\end{pmatrix}. \tag{46}$$

The criteria for choosing the first-stage canonical variates $\mathbf{z}_1'$ (i.e., for choosing ${}_j\boldsymbol{\alpha}_1'$ or, equivalently, ${}_j\boldsymbol{\beta}_1'$) are all specifiable in terms of the matrix $\boldsymbol{\Phi}(1)$.

For instance, Horst (1965) proposed the criterion of maximizing the sum,

$$\sum_{i<j=1}^{m}\sum \phi_{ij}(1),$$

of the intercorrelations among the elements of $\mathbf{z}_1$, which is equivalent to maximizing the quadratic form,

$$\mathbf{1}'\boldsymbol{\Phi}(1)\mathbf{1}\left(=m+2\sum_{i<j=1}^{m}\sum \phi_{ij}(1)\right).$$

The method based on this criterion, which takes into account both the magnitudes and the signs of ϕ_{ij}'s, will be called the *SUMCOR* method. A second criterion, also due to Horst (1965), is to maximize the variance of the first principal component of $\mathbf{z}_1$. If $m > {}_1\lambda_1 \geq {}_2\lambda_1 \geq \cdots \geq {}_m\lambda_1 > 0$ denote the ordered eigenvalues of $\boldsymbol{\Phi}(1)$, the second criterion amounts to maximizing ${}_1\lambda_1$, and the method associated with this criterion will be called the *MAXVAR* procedure. Kettenring (1971) proposed two additional criteria leading to methods termed *SSQCOR* and *MINVAR*, respectively. The first of these attempts to maximize the sum of squares,

$$\sum_{i<j=1}^{m}\sum \phi_{ij}^2(1),$$

of the off-diagonal elements of $\boldsymbol{\Phi}(1)$, which is equivalent to maximizing the trace of

$$\boldsymbol{\Phi}^2(1)\left(=m+2\sum_{i<j}\sum \phi_{ij}^2(1)\right),$$

or the sum of squares of the eigenvalues, $\sum_{j=1}^{m} {}_j\lambda_1^2$. Unlike SUMCOR, SSQCOR takes account only of the magnitudes of the intercorrelations among the canonical variates. The SSQCOR criterion is also interpretable as maximizing the "distance" between $\boldsymbol{\Phi}(1)$ and the identity matrix of order m. MINVAR uses the criterion of minimizing the variance of the "smallest" principal component of $\mathbf{z}_1$, i.e., minimizing ${}_m\lambda_1$. The first attempt at generalizing Hotelling's two-set canonical correlation analysis is due to Steel (1951), who used the criterion of minimizing the so-called generalized variance of $\mathbf{z}_1$, namely, $|\boldsymbol{\Phi}(1)|$ or, equivalently, the product, $\prod_{j=1}^{m} {}_j\lambda_1$, of the eigenvalues of $\boldsymbol{\Phi}(1)$. The method associated with this criterion will be called the *GENVAR* technique.

Each of the five methods mentioned in the preceding paragraph employs a particular unidimensional summary of the matrix of intercorrelations among the canonical variates and determines the set of canonical variates optimally with respect to that summary. In each approach the higher-stage canonical variates, $\mathbf{z}_2, \mathbf{z}_3, \cdots, \mathbf{z}_{p_1}$, are chosen by using the same criterion at each stage and imposing additional constraints (e.g., mutual orthogonality of the different linear combinations within any given set of the original m sets of variables) to ensure that new canonical variates are being found at each stage. When $m=2$, all five of the methods reduce to Hotelling's (1936) treatment of the problem.

Kettenring (1971) uses "factor-analytic" types of models (see Section 2.2.2) to motivate each of the five criteria and to discuss the similarities and differences that one might anticipate among the results of using the five methods for analyzing a given body of data. Thus, for instance, the SUMCOR and MAXVAR methods may be motivated by a single-common-factor model for the canonical variates:

$$\mathbf{z}_1 = \boldsymbol{\gamma}_1 f_1 + \mathbf{e}_1,$$

where f_1 is the single standardized common factor, and $\mathbf{e}_1$ is an m-dimensional vector of unique factors, which can also be considered in more familiar terms as a vector of residual errors, with mean $\mathbf{0}$ and covariance matrix $\boldsymbol{\Psi}$. Then, assuming that $\boldsymbol{\gamma}_1$ is known and is proportional to the vector $\mathbf{1}$, it can be shown that choosing f_1 so as to minimize $\text{tr}(\boldsymbol{\Psi})$ is equivalent to the SUMCOR procedure. In other words, the SUMCOR method generates a $\mathbf{z}_1$ having the best fitting (in the sense of minimizing the sum of the variances of the residual errors) single common factor, assuming that the factor contributes equally to each of the first-stage canonical variates. On the other hand, with the same single-factor model, choosing both $\boldsymbol{\gamma}_1$ and f_1 so as to minimize $\text{tr}(\boldsymbol{\Psi})$ turns out to be equivalent to the MAXVAR method.

The SSQCOR and GENVAR methods may be motivated by using an m-factor model:

$$\mathbf{z}_1 = \sum_{j=1}^{m} \boldsymbol{\gamma}_{1j} f_j + \mathbf{e}_1.$$

Since the number of common factors is equal to the dimensionality of $\mathbf{z}_1$, the approach of fitting factors so as to minimize a criterion such as $\text{tr}(\boldsymbol{\Psi})$ is no longer adequate for distinguishing between different sets of $\mathbf{z}_1$. However, suppose that one were to choose the f_j's to be the principal components transformations of $\mathbf{z}_1$ (see Section 2.2.1) so that they account for decreasing amounts of variance. In fact, if ${}_j\boldsymbol{\varepsilon}_1$ is the eigenvector corresponding to the (ordered) eigenvalue ${}_j\lambda_1$ of $\mathbf{\Phi}(1)$, then suppose that

$$\boldsymbol{\gamma}_{1j} = \sqrt{{}_j\lambda_1}\,{}_j\boldsymbol{\varepsilon}_1 \qquad \text{and} \qquad f_j = \frac{1}{\sqrt{{}_j\lambda_1}} \cdot {}_j\boldsymbol{\varepsilon}_1' \cdot \mathbf{z}_1 \qquad \text{for } j = 1, \ldots, m,$$

so that the f_j's are the standardized and mutually uncorrelated principal components derived from $\mathbf{\Phi}(1)$. The factors derived from the largest and the smallest eigenvalues will be of particular interest, and seeking a $\mathbf{z}_1$ that corresponds to large separations among the eigenvalues will therefore be useful. One way of obtaining such a $\mathbf{z}_1$ is to choose them so as to maximize the measure of spread, $\sum_{j=1}^m {}_j\lambda_1^2$, subject to the constraint that the sum of the ${}_j\lambda_1$'s has to equal m. This, of course, is what the SSQCOR method attempts to do. Hence the SSQCOR method will tend to produce a $\mathbf{z}_1$ such that its first few principal components account for most of the variability. On the other hand, since the GENVAR method attempts to minimize $\prod_{j=1}^m {}_j\boldsymbol{\lambda}_1$, it will be expected to focus on the smallest eigenvalues and to minimize the contribution of the last few f_j's.

The inequalities, $m \leq \sum_{j=1}^m {}_j\lambda_1^2 \leq m^2$, can be established, and, furthermore, the lower bound can be shown to be attained when ${}_j\lambda_1 = 1$ for all j (i.e., $\mathbf{\Phi}_{(1)} = \mathbf{I}$), while the upper bound is attained when ${}_1\lambda_1 = m$ and ${}_2\lambda_1 = \cdots = {}_m\lambda_1 = 0$. This result suggests that MAXVAR and SSQCOR will yield similar $\mathbf{z}_1$'s whenever most of the variability in $\mathbf{z}_1$ can be accounted for by a single factor.

The MINVAR method may be studied by using a $(m-1)$-factor model:

$$\mathbf{z}_1 = \sum_{j=1}^{m-1} \boldsymbol{\gamma}_{1j} f_j + \mathbf{e}_1.$$

For a given $\mathbf{z}_1$, choosing $\boldsymbol{\gamma}_{1j}$'s and f_j's so as to minimize the trace of the covariance matrix of the residual error variables, $\mathbf{e}_1$, leads in this case to

$$\boldsymbol{\gamma}_{1j} = \sqrt{{}_j\lambda_1}\,{}_j\boldsymbol{\varepsilon}_1 \qquad \text{and} \qquad f_j = \frac{1}{\sqrt{{}_j\lambda_1}} \cdot {}_j\boldsymbol{\varepsilon}_1' \cdot \mathbf{z}_1, \qquad \text{for } j = 1, \ldots, (m-1),$$

so that the f_j's are the first $(m-1)$ standardized principal components of $\mathbf{\Phi}(1)$. The residual variance after fitting all $(m-1)$ factors is $m - \sum_{j=1}^{m-1} {}_j\lambda_1$ $(= {}_m\lambda_1)$. Now choosing $\mathbf{z}_1$ to optimize the fit by such a set of $(m-1)$ factors amounts to choosing $\mathbf{z}_1$ so as to minimize ${}_m\lambda_1$, i.e., the MINVAR method. This suggests that MINVAR and GENVAR may be expected to yield similar $\mathbf{z}_1$'s whenever the smallest eigenvalue, ${}_m\lambda_1$, is very small, i.e., whenever almost all of the variability in $\mathbf{z}_1$ is confined to an $(m-1)$-dimensional linear subspace (see the discussion in Example 14).

The above discussion has been presented in terms of "population" entities. With a sample of size $n\ (>p=\sum_{j=1}^{m} p_j)$ on the m sets of variables, one would have the following correspondences between the population entities and statistics computed from the observations:

$$\mathbf{\Sigma} \leftrightarrow \mathbf{S} = ((\mathbf{S}_{ij})); \qquad \mathbf{\Gamma} \leftrightarrow \mathbf{R} = ((\mathbf{R}_{ij})),$$

where $\mathbf{R}_{ij} = \mathbf{T}_i^{-1}\mathbf{S}_{ij}(\mathbf{T}_j^{-1})'$ and $\mathbf{S}_{jj} = \mathbf{T}_j\mathbf{T}_j'$;

$$\{{}_1\boldsymbol{\alpha}_1, \ldots, {}_m\boldsymbol{\alpha}_1\} \leftrightarrow \{{}_1\mathbf{a}_1, \ldots, {}_m\mathbf{a}_1\};$$

$$\{{}_1\boldsymbol{\beta}_1, \ldots, {}_m\boldsymbol{\beta}_1\} \leftrightarrow \{{}_1\mathbf{b}_1, \ldots, {}_m\mathbf{b}_1\};$$

$$\mathbf{\Phi}(1) \leftrightarrow \hat{\mathbf{\Phi}}(1); \qquad {}_j\lambda_1 \leftrightarrow {}_j\hat{\lambda}_1; \qquad {}_j\boldsymbol{\varepsilon}_1 \leftrightarrow {}_j\hat{\boldsymbol{\varepsilon}}_1.$$

Kettenring (1969) describes algorithms associated with each of the five methods that may be used with these sample statistics. SUMCOR, SSQCOR, and GENVAR involve iterative techniques, whereas MAXVAR and MINVAR depend only on a single eigenanalysis of the $p \times p$ matrix $\mathbf{R}$. Also, when $m=2$, which is the case considered by Hotelling (1936), no iterative methods are involved, and all five methods reduce to utilization of the results from an eigenanalysis of $\mathbf{R}$.

In fact, for the MAXVAR method, the first-stage canonical variates, for instance, are obtained from the eigenvector corresponding to the largest eigenvalue of $\mathbf{R}$. If $c_1 \geq c_2 \geq \cdots \geq c_p > 0$ denote the ordered eigenvalues of $\mathbf{R}$ with corresponding p-dimensional eigenvectors $\mathbf{v}_1, \mathbf{v}_2, \ldots, \mathbf{v}_p$, where $\mathbf{v}_k'$ in partitioned form is $\{{}_1\mathbf{v}_k' \vdots {}_2\mathbf{v}_k' \vdots \cdots \vdots {}_m\mathbf{v}_k'\}$ for $k=1, \ldots, p$, then the required solution for the coefficients of the first-stage MAXVAR canonical variates is

$${}_j\mathbf{a}_1 = \pm \frac{{}_j\mathbf{v}_1}{\|{}_j\mathbf{v}_1\|}, \qquad j=1, \ldots, m, \tag{47}$$

where $\|\mathbf{x}\|$ denotes the Eulidean norm (i.e., square root of the sum of squares of the elements) of $\mathbf{x}$. Similarly, the first-stage MINVAR canonical variates are derived from

$${}_j\mathbf{a}_1 = \pm \frac{{}_j\mathbf{v}_p}{\|{}_j\mathbf{v}_p\|}, \qquad j=1, \ldots, m. \tag{48}$$

Canonical variates may also be obtained at additional stages and will depend on the nature of the constraints imposed on them to ensure that they are different from ones determined at earlier stages. (See Kettenring, 1969, for a more detailed discussion.)

Example 14. This example, taken from Thurstone & Thurstone (1941) and also used by Horst (1965) and Kettenring (1971), deals with three ($=m$) sets of scores by several people on three batteries of three tests each, i.e., $p_1 = p_2 = p_3 = 3$. The three tests in each battery were intended to measure, respectively, the verbal, numerical, and spatial abilities of the persons tested. Exhibit 14*a* shows the 9×9 covariance matrix of the standardized scores or, equivalently, the correlation matrix $\mathbf{R}_0$ of the original scores. Also shown in Exhibit 14*a* is the 9×9 correlation matrix, $\mathbf{R}$, of the internally "sphericized" standardized variables (the $\mathbf{x}_j$-variables in terms of the earlier descriptions) derived from the standardized scores. The diagonal elements of the matrices in the off-diagonal blocks of $\mathbf{R}_0$ are all relatively large. Thus the scores on tests intended to measure the same ability tend to be highly correlated, whereas the correlations between scores on tests (even within the same battery) measuring different abilities, although positive, are not as high. After the internal transformations of the three sets, the off-diagonal terms of the matrices in the off-diagonal blocks tend to be even smaller (compare $\mathbf{R}$ with $\mathbf{R}_0$).

When the five methods of multiset canonical correlation analysis were applied in this case, with the exception of the MINVAR method the results were similar; i.e., differences in the numerical answers occurred only in the third or higher decimal places. The results for the four methods other than MINVAR which lead to similar answers are shown in

Exhibit 14a. Correlation matrices of the original and the internally sphericized variables (Horst, 1965; Kettenring, 1971)

$$\mathbf{S} = \mathbf{R}_0 = \left[\begin{array}{ccc|ccc|ccc}
1 & 0.249 & 0.271 & 0.636 & 0.183 & 0.185 & 0.626 & 0.369 & 0.279 \\
 & 1 & 0.399 & 0.138 & 0.654 & 0.262 & 0.190 & 0.527 & 0.356 \\
 & & 1 & 0.180 & 0.407 & 0.613 & 0.225 & 0.471 & 0.610 \\
\hline
 & & & 1 & 0.091 & 0.147 & 0.709 & 0.254 & 0.191 \\
 & & & & 1 & 0.296 & 0.103 & 0.541 & 0.394 \\
 & & & & & 1 & 0.179 & 0.437 & 0.496 \\
\hline
 & & & & & & 1 & 0.291 & 0.245 \\
 & & & & & & & 1 & 0.429 \\
 & & & & & & & & 1
\end{array}\right]$$

$$\mathbf{R} = \left[\begin{array}{c|ccc|ccc}
 & 0.636 & 0.126 & 0.059 & 0.626 & 0.195 & 0.059 \\
\mathbf{I} & -0.021 & 0.633 & 0.049 & 0.035 & 0.459 & 0.129 \\
 & 0.016 & 0.157 & 0.521 & 0.048 & 0.238 & 0.426 \\
\hline
 & & & & 0.709 & 0.050 & -0.002 \\
 & & \mathbf{I} & & 0.039 & 0.532 & 0.190 \\
 & & & & 0.067 & 0.258 & 0.299 \\
\hline
 & & & & & \mathbf{I} &
\end{array}\right]$$

[*Note*: Values in the blocks below the diagonal blocks are obtained by symmetry.]

Exhibit 14b. Results of five methods of multiset canonical correlation analysis (Kettenring, 1971)

$$_1z_1 = (0.73, 0.51, 0.45)\mathbf{x}_1$$
$$_2z_1 = (0.66, 0.62, 0.42)\mathbf{x}_2$$
$$_3z_1 = (0.68, 0.64, 0.36)\mathbf{x}_3$$
$$\hat{\mathbf{\Phi}}(1) = \begin{pmatrix} 1 & 0.735 & 0.756 \\ & 1 & 0.743 \\ & & 1 \end{pmatrix}$$
$$_1\hat{\lambda}_1 = 2.49, \quad _1\hat{\boldsymbol{\varepsilon}}_1' = (0.578, 0.574, 0.580)$$
$$_2\hat{\lambda}_1 = 0.27, \quad _2\hat{\boldsymbol{\varepsilon}}_1' = (-0.535, 0.803, -0.262)$$
$$_3\hat{\lambda}_1 = 0.24, \quad _3\hat{\boldsymbol{\varepsilon}}_1' = (-0.616, -0.159, 0.771)$$

$$_1z_1 = (0.68, 0.57, 0.45)\mathbf{x}_1$$
$$_2z_1 = (0.96, -0.22, 0.16)\mathbf{x}_2$$
$$_3z_1 = (-0.78, -0.53, -0.33)\mathbf{x}_3$$
$$\hat{\mathbf{\Phi}}(1) = \begin{pmatrix} 1 & 0.345 & -0.736 \\ & 1 & -0.517 \\ & & 1 \end{pmatrix}$$
$$_1\hat{\lambda}_1 = 2.082, \quad _1\hat{\boldsymbol{\varepsilon}}_1' = (0.591, \; 0.493, -0.638)$$
$$_2\hat{\lambda}_1 = 0.683, \quad _2\hat{\boldsymbol{\varepsilon}}_1' = (0.513, -0.839, -0.517)$$
$$_3\hat{\lambda}_1 = 0.235, \quad _3\hat{\boldsymbol{\varepsilon}}_1' = (0.621, \; 0.228, \; 0.751)$$

the upper portion of Exhibit 14*b*, while those for MINVAR are given in the lower portion. Each set of results in the exhibit pertains only to the first stage of analysis and contains the three first-stage canonical variates, their correlation matrix $\hat{\mathbf{\Phi}}(1)$, and the eigenanalysis of $\hat{\mathbf{\Phi}}(1)$.

The following features emerge from an inspection of the results in the upper portion of Exhibit 14*b*:

1. The largest eigenvalue $_1\hat{\lambda}_1$ is about 83% of $\text{tr}\{\hat{\mathbf{\Phi}}(1)\}$, and the corresponding eigenvector $_1\hat{\boldsymbol{\varepsilon}}_1$ is approximately proportional to the vector $\mathbf{1}$; the latter may be considered an indication that the three sets of variables (viz., the batteries of tests) are so much alike that the three canonical variates which are derived, one from each of them, are contributing equally to the first principal component transformation of $\mathbf{z}_1$.
2. Eigenvalues $_2\hat{\lambda}_1$ and $_3\hat{\lambda}_1$ are approximately equal, each accounting for 8–9% of $\text{tr}\{\hat{\mathbf{\Phi}}(1)\}$; i.e., $\hat{\mathbf{\Phi}}(1)$ has one large eigenvalue and the other two eigenvalues are essentially equal, a result that may also be surmised

from the equicorrelational nature of $\mathbf{\Phi}(1)$, as indicated by the near constancy of its off-diagonal elements (see the discussion of Example 1).

Combining the indications from features (1) and (2) and recalling the earlier discussion pertaining to the similarities and differences among the methods, one can understand the reasons why the SUMCOR, MAXVAR, and SSQCOR methods yield similar results. The reason why GENVAR is not as similar to MINVAR but is more similar to SSQCOR in this example lies, perhaps, in the fact that the smallest eigenvalue is not "small enough." The product function, $\prod_{j=1}^{m} {}_j\lambda_1$, is especially sensitive to the smallest eigenvalue only for extremely small values of it, and in the present example this is not the case. The results for the MINVAR method in Exhibit 14*b* show that ${}_3\hat{\lambda}_1$ accounts for about 8% (not negligible) of $\mathrm{tr}\{\hat{\mathbf{\Phi}}(1)\}$, while ${}_1\hat{\lambda}_1$ and ${}_2\hat{\lambda}_1$ contribute approximately 70% and 22%, respectively.

One use of eigenvectors, such as ${}_1\hat{\boldsymbol{\varepsilon}}_1$ for the MAXVAR method and ${}_3\hat{\boldsymbol{\varepsilon}}_1$ for the MINVAR method, is to study them for selecting "important" subsets of the sets of variables for further analysis. This is generally done by looking at the relative magnitudes of the elements of the eigenvector involved. Thus, in this example, an examination of ${}_3\hat{\boldsymbol{\varepsilon}}_1$ associated with the MINVAR method (see Exhibit 14*b*, lower portion) reveals that the first and third elements are much larger than the second. If one decides, on the basis of this indication, to choose the first and third sets of variables for doing a pairwise canonical correlation analysis, then in this example it does indeed turn out that one would have selected the two sets with the highest two-set canonical correlation. [*Note*: One could also have utilized $\hat{\mathbf{\Phi}}(1)$ for this, since the element in its top right corner indicates that the first and third canonical variates at the first stage have a large (in magnitude) correlation.]

An interesting alternative analysis in this example (left as an exercise to the reader) would be to regroup the nine variables into three sets corresponding to the three abilities measured rather than the three batteries of tests. A quite different approach with somewhat different objectives would be to use an analysis-of-variance approach (see Chapters 5 and 6) for studying the relative importance of various "effects" (e.g., differences of batteries, or a time or trend effect if the tests were administered across time). This, however, would require the original scores on the tests.

From the viewpoint of data analysis, analyzing subsets of responses is very important and should not be replaced by a single overall multiresponse analysis. In the context of canonical correlation analysis for m sets of multiple responses, analyses of subsets of the m sets, as well as the study of subsets (pairs, triplets, etc.) of the canonical variates from the m-set analysis, are important. Specifically, plots of the original observations transformed according to the canonical variate transformations taken two, and three, at a time may be valuable. In the case of two-set canonical correlation analysis, such plots are actual "displays" of the computed canonical correlations and may lead to uncovering possibly aberrant observations or peculiar relationships.

Example 15. The use of pairwise canonical variate plots is illustrated with data from a questionnaire study, which was concerned with assessing employees' readership of, and attitudes toward, a company magazine published by their employer for communicating general information.

Exhibit 15a shows a plot for the two canonical variates corresponding to the largest canonical correlation, derived from the answers of 645 employees to two subsets consisting of four questions each. The questions in one subset pertained to the expectations of the respondent regarding the publication, while the other subset was concerned with the respondent's evaluation of its actual performance. Each of the eight questions was answered on a six-point scale, and, as indicated in Exhibit 15a, the observed largest canonical correlation was 0.4023. The striking features about the configuration are the "bunching" of points on the right-hand boundary of the plot and the vertical striations evident in it. A subsequent inspection of the data, stimulated by these indications of

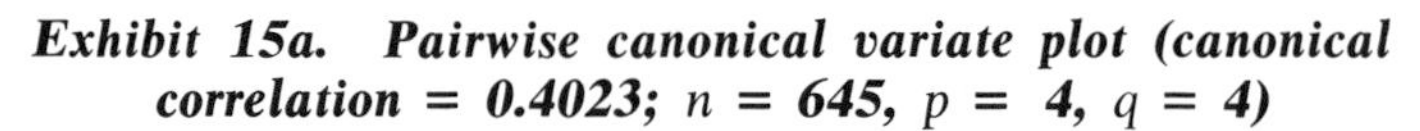

Exhibit 15a. Pairwise canonical variate plot (canonical correlation = 0.4023; n ***= 645,*** p ***= 4,*** q ***= 4)***

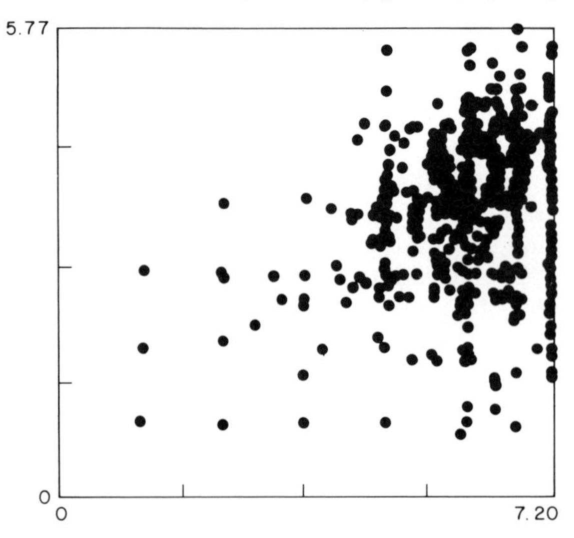

Exhibit 15b. Pairwise canonical variate plot (canonical correlation = 0.4833; $n = 580$, $p = 14$, $q = 14$)

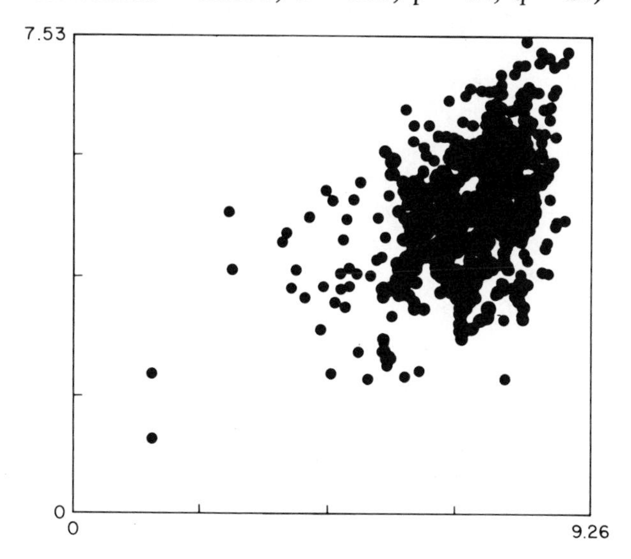

peculiarities, revealed that a large proportion of the respondents tended to use only the higher values of the six-point scale when dealing with their expectations and only the middle values of the scale in evaluating the performance of the publication. Such tendencies would lead to the peculiarities indicated in the plot of canonical variates, although one could detect their existence by other methods (e.g., histograms of original observations) of displaying the data as well.

Exhibit 15*b* shows a plot derived from a canonical correlation analysis of two other subsets of questions in the same study. Each subset consisted of 14 questions, and answers from 580 respondents were used in the analysis. The observed value of the largest canonical correlation was 0.4833. The scatter of the points appears to be bounded above by a straight line parallel to a "diagonal line" drawn through the configuration, thus suggesting possible asymmetry in, and departure from normality of, the joint distribution of the canonical variates. There is also a mild suggestion of two outliers in the lower left-hand corner of the plot.

With multiple sets of multiresponse observations, one can also define and determine canonical correlational analogues of partial correlations between scalar variables. Thus, for instance, if $\mathbf{y}_j$ denotes a set of p_j responses, for $j = 1, 2, 3$, one can use, as measures of the first-order partial canonical correlations between any pair of sets $\mathbf{y}_j$ and $\mathbf{y}_k$, given the third set $\mathbf{y}_l$, just the two-set canonical correlations between the "residuals," $\mathbf{r}_j$ and $\mathbf{r}_k$, from the multivariate multiple regressions of $\mathbf{y}_j$ and $\mathbf{y}_k$, respectively, on $\mathbf{y}_l$ ($j \neq k \neq l = 1, 2, 3$). Similarly, with more than three sets, one can define higher-order partial canonical correlations as well. At each stage only a two-set canonical correlational analysis is involved between sets of "residuals" derived from multivariate multiple regressions of pairs of the original sets of responses on the remaining sets.

The concepts and methods involved in multivariate multiple regression mentioned in the preceding paragraph are utilized widely for studying relationships between a set of response variables, $\mathbf{y}$, and a set of so-called independent variables or regressor variables, $\mathbf{x}$. The multivariate multiple regression model, or the so-called multivariate general linear model (see Roy et al., 1971), is usually specified as follows:

$$\mathbf{Y}' = \mathbf{X} \cdot \boldsymbol{\Theta} + \boldsymbol{\varepsilon}, \tag{49}$$

where the n rows of $\mathbf{Y}'$ are the n observations on the p-dimensional response variable; the rows of the $n \times k$ matrix, $\mathbf{X}$, are the corresponding observations on k regressor variables; the elements of the $k \times p$ matrix, $\boldsymbol{\Theta}$, are the unknown regression coefficients; and the n rows of $\boldsymbol{\varepsilon}$ are p-dimensional error variables which are generally assumed to have a mean vector, $\mathbf{0}$, and a common $p \times p$ unknown covariance matrix, $\boldsymbol{\Sigma}$. The rows of $\boldsymbol{\varepsilon}$ are also generally assumed to be mutually uncorrelated and, for some purposes of formal statistical inference, p-variate normally distributed as well. Thus the n p-dimensional observations are considered to be mutually uncorrelated with means specified by the regression relationships, $\mathscr{E}(\mathbf{Y}' \mid \mathbf{X}) = \mathbf{X}\boldsymbol{\Theta}$, and a common unknown covariance matrix, $\boldsymbol{\Sigma}$.

The multivariate multiple regression model of Eq. 49 may be rewritten in its equivalent form,

$$\mathbf{Y}' = [\mathbf{Y}_1\mathbf{Y}_2 \cdots \mathbf{Y}_p] = \mathbf{X}[\boldsymbol{\theta}_1\boldsymbol{\theta}_2 \cdots \boldsymbol{\theta}_p] + [\boldsymbol{\varepsilon}_1\boldsymbol{\varepsilon}_2 \cdots \boldsymbol{\varepsilon}_p], \tag{50}$$

where $\mathbf{Y}_j$, the jth column of $\mathbf{Y}'$, consists of the n observations on the jth response, $\boldsymbol{\theta}_j$ consists of the regression coefficients in the univariate multiple linear regression of the jth response variable on the k regressor variables, and $\boldsymbol{\varepsilon}_j$ is an n-dimensional vector of mutually uncorrelated errors pertaining to the jth response variable ($j = 1, 2, \ldots, p$). In this form it is clear that the multivariate model is merely a simultaneous statement of p univariate multiple regression models. In particular, in this treatment the matrix $\mathbf{X}$ is assumed to be the same for all p response variables. When the regressor variables are dummy variables corresponding to factors or treatments in a designed experiment, this means that all p responses are observed under the same design.

In the usual treatment of multivariate multiple regression, the estimate of $\mathbf{\Theta}$ is taken to be $\hat{\mathbf{\Theta}} = [\hat{\boldsymbol{\theta}}_1 \hat{\boldsymbol{\theta}}_2 \cdots \hat{\boldsymbol{\theta}}_p]$, where $\hat{\boldsymbol{\theta}}_j = (\mathbf{X}'\mathbf{X})^{-1}\mathbf{X}'\mathbf{Y}_j$, for $j = 1, \cdots, p$, are the least squares estimates of the regression coefficients for the jth response analyzed individually. A more detailed discussion of the formal issues, such as the statistical estimation, involved in this approach is presented later in Chapter 5 and may also be found in Roy et al. (1971). For present purposes, however, it is probably worth reiterating that adapting a multivariate view in multiresponse multiple regression situations may be important because the estimated regression coefficients may be statistically dependent because of the intercorrelations of the responses. In other words, although the $\hat{\boldsymbol{\theta}}_j$'s are obtained from separate analyses of the responses, the corresponding elements of the $\mathbf{Y}_j$'s (viz., all their first elements, all second elements, etc.) are assumed to be simultaneously observed on an experimental unit and may therefore be expected to be statistically correlated in many situations. Recognition of this may play an important role in the subsequent analysis and interpretation of the results. (See Example 36 in Chapter 6.)

A slightly more general form of the above multivariate general linear model is provided by $\mathscr{E}(\mathbf{Y}' \mid \mathbf{X}, \mathbf{G}) = \mathbf{X\Xi G}$, where $\mathbf{\Xi}$ is a $k \times q$ matrix of unknown parameters and $\mathbf{G}$ is a $q \times p$ matrix, with known elements, of rank $q \leq p$. This generalization enables one to include polynomial growth-curve models in the class of general linear models (see, e.g., Section 6 of Chapter IV in Roy et al., 1971). The application of nonlinear (in the parameters) models for studying multivariate relationships has been considered recently, but, perhaps because of the inherent difficulties of nonlinear modeling even in uniresponse problems, the use of these models in practice is not widespread.

REFERENCES

Section 3.2 Anderson (1954, 1957, 1960), Bruntz et al. (1974), Chernoff (1973).

Section 3.3 Horst (1965), Hotelling (1936), Kettenring (1969, 1971), Roy, Gnanadesikan, & Srivastava (1971), Steel (1951), Thurstone & Thurstone (1941).

CHAPTER 4

Multidimensional Classification And Clustering

4.1. GENERAL

A wide variety of objectives, concepts, and techniques is encompassed under the heading "multidimensional classification and clustering." Loosely speaking, the concern is with respect to categorization of objects or experimental units, and a dichotomy into two broad types of approaches is possible. First, there are situations in which the categorization is based on prespecified groups. The term used here for this case will be *classification;* other terms used in the literature for describing it include "discriminant analysis," "classificatory analysis," "pattern recognition," and "allocation." Secondly, there are situations in which the categorization is done in terms of groups that are themselves determined from the data. The term used for describing the concern in such situations, wherein one is seeking meaningful data-determined groupings of objects, is *clustering.* There are, of course, many problems that tend to fall somewhere between classification and clustering, rather than entirely into either one of these two cases (see Example 17 in this chapter).

Typically, problems both of classification and of clustering tend, in their primitive form, to be multidimensional in nature. The discussion in this chapter will first be concerned with classification problems and procedures and will then consider cluster analysis.

4.2. CLASSIFICATION

Even when the concern is with classifying an object in terms of prespecified groups, distinctions will usually exist in regard to the kind and amount of background information in individual problems. For example, given a set of fingerprints of some unknown person, it is one problem to check on whether they do or do not correspond to a specific individual. It is quite another problem to attempt to determine to which one, if any, of a large population of alternative possibilities the unknown might correspond. Clearly, the strategy of the procedures, including the characteristics used, may differ between the verification and the identification problems.

From the viewpoint of data analysis, apart from correct formulation of the problem and initial choice of variables, there appear to be two other basic aspects of multidimensional classification: (i) the choice of an effective space, or representation, for discrimination, and (ii) the choice of a distance measure or metric for use in such a space.

Perhaps the simplest guise of the classification problem, although not usually considered as such, is the one-group problem wherein one wishes to decide whether or not an item belongs to a particular group. A test of significance and methods for assessing whether an observation is an outlier are simple examples of this case. A multivariate quality control procedure suggested by Hotelling (1947) is essentially a test of significance viewed as a one-group classification problem. Jackson (1956) has suggested a bivariate graphical implementation of Hotelling's procedure involving the plotting of points in an elliptical frame defined by Hotelling's T^2 statistic.

The classical form of the classification problem is the two-group case considered by Fisher (1936, 1938), leading to the so-called discriminant function. Suppose that, given two groups, G_1 and G_2, one has a reference set of observations, $\mathbf{Y}_1$ and $\mathbf{Y}_2$, respectively, from them, that is, the n_1 columns of $\mathbf{Y}_1$ are p-dimensional observations on n_1 units known to come from G_1, and, similarly, the n_2 columns of $\mathbf{Y}_2$ are observations on n_2 units from G_2. Utilizing the observations in the reference set, one can obtain the sample mean vectors, $\bar{\mathbf{y}}_1$ and $\bar{\mathbf{y}}_2$, as well as the sample covariance matrices, $\mathbf{S}_1$ and $\mathbf{S}_2$. Fisher's discriminant function is that linear combination of the p original responses which exhibits the largest ratio of variance between the two groups relative to that within the groups. More explicitly, if the linear combination of the original variables is denoted as $z = a_1y_1 + a_2y_2 + \cdots + a_py_p = \mathbf{a}'\mathbf{y}$, a two-sample t statistic for the variable z may be written as

$$t_{\mathbf{a}} = \frac{\mathbf{a}'(\bar{\mathbf{y}}_1 - \bar{\mathbf{y}}_2)}{\{\mathbf{a}'\mathbf{S}\mathbf{a}(1/n_1 + 1/n_2)\}^{1/2}},$$

where $(n_1 + n_2 - 2)\mathbf{S} = (n_1 - 1)\mathbf{S}_1 + (n_2 - 1)\mathbf{S}_2$. Fisher's discriminant function is obtained by choosing $\mathbf{a}$ so as to maximize $|t_{\mathbf{a}}|$ or, equivalently,

$$t_{\mathbf{a}}^2 = \left(\frac{n_1n_2}{n_1 + n_2}\right)\left\{\frac{\mathbf{a}'(\bar{\mathbf{y}}_1 - \bar{\mathbf{y}}_2)(\bar{\mathbf{y}}_1 - \bar{\mathbf{y}}_2)'\mathbf{a}}{\mathbf{a}'\mathbf{S}\mathbf{a}}\right\}.$$

The required solution for $\mathbf{a}$ can be shown to be proportional (i.e., equal except for a multiplicative constant) to $\mathbf{S}^{-1}(\bar{\mathbf{y}}_1-\bar{\mathbf{y}}_2)$. In the p-dimensional space of the responses $y_1, y_2, \ldots, y_p$, such a vector $\mathbf{a}$ defines the direction of maximal group separation in the sense that the means of the projections of the observations from the two groups are maximally apart relative to the variance of the projections around their respective means. Choosing $\mathbf{a} \propto \mathbf{S}^{-1}(\bar{\mathbf{y}}_1-\bar{\mathbf{y}}_2)$ leads to the maximum value, $(n_1n_2/n_1+n_2)\times(\bar{\mathbf{y}}_1-\bar{\mathbf{y}}_2)'\mathbf{S}^{-1}(\bar{\mathbf{y}}_1-\bar{\mathbf{y}}_2)$, for $t_{\mathbf{a}}^2$, and this maximum is thus seen to be the value of the two-sample Hotelling's T^2 statistic. Also, the quadratic form, $(\bar{\mathbf{y}}_1-\bar{\mathbf{y}}_2)'\mathbf{S}^{-1}(\bar{\mathbf{y}}_1-\bar{\mathbf{y}}_2)$, is just the so-called Mahalanobis' D^2 statistic.

For the two-group problem, one can consider the unidimensional space of the derived variable $z=\mathbf{a}'\mathbf{y}$ (with $\mathbf{a}$ chosen as above) as an effective space for discriminating between the two groups. Since the dimensionality of the space is 1, the issue of selecting a distance measure is relatively simple in this case. Specifically, given an "unknown" object which is known only to belong to either G_1 or G_2 and for which the values of the p variables are observed to be $\mathbf{u}'=(u_1, \ldots, u_p)$, one can project the points $\bar{\mathbf{y}}_1$, $\bar{\mathbf{y}}_2$, and $\mathbf{u}$ onto the unidimensional space corresponding to z (viz., the vector $\mathbf{a}$) and assign the unknown to G_1 or G_2 according as the projection of $\mathbf{u}$ is closer to the projection of $\bar{\mathbf{y}}_1$ or of $\bar{\mathbf{y}}_2$. Algebraically, this amounts to calculating the value of the discriminant function for the unknown, viz., $\mathbf{a}'\mathbf{u}=(\bar{\mathbf{y}}_1-\bar{\mathbf{y}}_2)'\mathbf{S}^{-1}\mathbf{u}$, and classifying the unknown in G_1 or G_2 according as $\mathbf{a}'\mathbf{u} \gtrless \mathbf{a}'(\bar{\mathbf{y}}_1+\bar{\mathbf{y}}_2)/2$.

A generalization of the two-group procedure to several groups is described, for example, by Rao (1952, Section 9c). Suppose that one has g groups, $G_1, \ldots, G_g$, with the reference set of observations consisting of n_i p-dimensional observations (constituting the columns of a $p\times n_i$ matrix $\mathbf{Y}_i$) from G_i $(i=1, \ldots, g)$. Using the observations from the ith group, one can compute the sample mean vector, $\bar{\mathbf{y}}_i$, and the sample covariance matrix, $\mathbf{S}_i$, for $i=1, \ldots, g$. For the total set of $n=\sum_{i=1}^{g} n_i$ observations, one can calculate an overall mean vector, $\bar{\mathbf{y}}=\sum_{i=1}^{g} n_i\bar{\mathbf{y}}_i/n$, and a $p\times p$ pooled *within-groups* covariance matrix,

$$\mathbf{W}=\frac{1}{n-g}\sum_{i=1}^{g}(n_i-1)\mathbf{S}_i. \tag{51}$$

Furthermore, one can define a $p\times p$ *between-groups* covariance matrix,

$$\mathbf{B}=\frac{1}{g-1}\sum_{i=1}^{g} n_i(\bar{\mathbf{y}}_i-\bar{\mathbf{y}})(\bar{\mathbf{y}}_i-\bar{\mathbf{y}})', \tag{52}$$

which provides a summary of the dispersion among the group means, $\bar{\mathbf{y}}_i$'s, in p-space. In some situations, when the n_i's are extremely disparate, one may wish not to weight the deviations of the group centroids from the overall centroid by the n_i's as in $\mathbf{B}$ but rather to use the $p \times p$ matrix,

$$\mathbf{B}^* = \frac{1}{g-1} \sum_{i=1}^{g} (\bar{\mathbf{y}}_i - \bar{\mathbf{y}})(\bar{\mathbf{y}}_i - \bar{\mathbf{y}})', \tag{53}$$

in place of $\mathbf{B}$ for the subsequent analysis.

Next, exactly as in the two-group problem, if $z = \mathbf{a}'\mathbf{y}$ denotes a linear combination of the original variables, a one-way analysis of variance for the derived variable z will lead to the following F-ratio of the between-groups mean square to the within-groups mean square:

$$F_{\mathbf{a}} = \frac{\mathbf{a}'\mathbf{B}\mathbf{a}}{\mathbf{a}'\mathbf{W}\mathbf{a}}. \tag{54}$$

If now one were to choose $\mathbf{a}$ so as to maximize this F-ratio, the required $\mathbf{a}$ would be the eigenvector, $\mathbf{a}_1$, corresponding to the largest eigenvalue, c_1, of $\mathbf{W}^{-1}\mathbf{B}$. The maximum value of the F-ratio would be $F_{\mathbf{a}_1} = \mathbf{a}_1'\mathbf{B}\mathbf{a}_1/\mathbf{a}_1'\mathbf{W}\mathbf{a}_1 = c_1$. Having determined $\mathbf{a}_1$, one can seek a second linear combination of the original variables which has the next largest F-ratio. The required solution for the coefficients in the second linear combination turns out to be the eigenvector, $\mathbf{a}_2$, corresponding to the second largest eigenvalue, c_2, of $\mathbf{W}^{-1}\mathbf{B}$. The process may be repeated for determining additional linear combinations. To ensure that new linear combinations are being found at each stage, some constraints (e.g., linear independence) have to be imposed on the sets of coefficients. All that is involved computationally is an eigenanalysis of $\mathbf{W}^{-1}\mathbf{B}$, leading to the ordered eigenvalues $c_1 \geq c_2 \geq \cdots \geq c_r > 0$ and the corresponding eigenvectors, $\mathbf{a}_1$, $\mathbf{a}_2, \ldots, \mathbf{a}_r$, which will satisfy the constraints $\mathbf{a}_j'\mathbf{W}\mathbf{a}_k = \delta_{jk}$, the Kronecker delta, for $j, k = 1, \ldots, r$. The eigenanalysis may be performed by using a singular-value decomposition algorithm which is appropriate for this case involving the two matrices $\mathbf{B}$ and $\mathbf{W}$ (see Chambers, 1974, Section 2.6c). The computations involved may also be viewed in terms of an initial transformation to sphericize the within-groups dispersion, followed by an eigenanalysis of the between-groups dispersion in this transformed space. More explicitly, one can first linearly transform the initial variables, $\mathbf{y}$, to p new variables, $\mathbf{x} = \mathbf{T}^{-1}\mathbf{y}$, where $\mathbf{W} = \mathbf{T}\mathbf{T}'$ is the so-called Cholesky decomposition of $\mathbf{W}$. Then the within-groups covariance matrix for the x-variables will be the identity matrix, and the between-groups covariance matrix will be $\mathbf{B}_x = \mathbf{T}^{-1}\mathbf{B}(\mathbf{T}^{-1})'$, where $\mathbf{B}$ is defined in Eq. 52. Next an eigenanalysis on the $p \times p$ symmetric matrix, $\mathbf{B}_x$, may be performed. The eigenvalues of $\mathbf{B}_x$ are, in fact, also the eigenvalues of $\mathbf{W}^{-1}\mathbf{B}$, and the eigenvectors, $\{\mathbf{a}_j\}$, of $\mathbf{W}^{-1}\mathbf{B}$ are related to the eigenvectors, $\{\mathbf{l}_j\}$, of $\mathbf{B}_x$ by the equations, $\mathbf{a}_j = (\mathbf{T}')^{-1}\mathbf{l}_j$, for $j = 1, 2, \ldots, r$.

In general, if there are g groups and the problem is p-dimensional, the number, r, of positive eigenvalues of $\mathbf{W}^{-1}\mathbf{B}$ will be equal to the smaller of $(g-1)$ and p. This is a consequence of the fact that, if g is less than p, the g group means are contained in a $(g-1)$-dimensional hyperplane. In particular, when $g=2$, the analysis is exactly equivalent to the two-group discriminant analysis, considered earlier, leading to a single discriminant function. More generally, with $g>2$, one can determine up to r linear combinations, $z_i=\mathbf{a}_i'\mathbf{y}$, for $i=1, 2, \ldots, r$, and the z_i's will be called *discriminant coordinates* or *CRIMCOORDS*. [*Note*: other authors have referred to these as "canonical variates" (e.g., Rao, 1952; Seal, 1964), but the present author's preference is to use the term "canonical variates" only in the context of canonical correlational analysis, discussed in Chapter 3.] The space defined by the *CRIMCOORDS*, or by a subset of the first $t\,(\leq r)$ of them, will be called the *discriminant space*. Since the CRIMCOORDS are determined so that they account for group separation in decreasing order, there is an issue of how many of them (viz., choice of a value for t) one ought to use. The nature of the diminishing returns from using the later CRIMCOORDS has to be studied in any given problem, and often t has to be chosen by trying several alternative values.

In the multigroup case, the discriminant space, which is a specifically chosen linear transformation of the original space, may be considered as an effective space for use in classifying "unknown" objects. The original reference set of observations, as well as the observations corresponding to the (unknown) objects which are to be classified into one of the g groups, may be represented in the discriminant space of dimension $t\,(\leq r)$. The representation of the original data consists in making the transformation

$$\mathbf{Z}=\mathbf{A}_t'\mathbf{Y}, \tag{55}$$

where $\mathbf{A}_t'$ is a $t\times p$ matrix whose rows are the eigenvectors $\mathbf{a}_1', \ldots, \mathbf{a}_t'$ $(t\leq r)$, and $\mathbf{Y}=[\mathbf{Y}_1 \vdots \mathbf{Y}_2 \vdots \cdots \vdots \mathbf{Y}_g]$ is the $p\times n$ set of all the reference observations. The columns of $\mathbf{Z}$ may, of course, be partitioned according to the partitioning of $\mathbf{Y}$ so as to provide the original group identities for the representations in the discriminant space. If $\mathbf{u}'=(u_1, \ldots, u_p)$ denotes the p-dimensional observation on an object which is to be classified as belonging to one of the g groups, a representation of $\mathbf{u}$ in the t-dimensional discriminant space is given by $\mathbf{A}_t'\cdot\mathbf{u}$.

For data-analytic purposes, two- and three-dimensional graphical representations of the columns, respectively, of $\mathbf{Z}(2\times n)$ and $\mathbf{Z}(3\times n)$ may be obtained. Such plots are often useful for studying the degree and nature of group separations, for suggesting possible metrics for use in the discriminant space, and for indicating stray or outlying observations. The approach is illustrated by the next two examples.

Example 16. This example, which pertains to the talker-identification problem (for details see Becker et al., 1965; Bricker et al., 1971), involves data from 10 talkers, each of whom repeated a given word six times. The initial representation of each utterance in this particular example was a 16-dimensional summary derived from raw data whose dimensionality was much higher. Exhibit 16 shows a representation of the 60 resultant utterances in the two-dimensional discriminant space of the first two CRIMCOORDS, i.e., coordinates that are obtained from the eigenvectors corresponding to the two largest eigenvalues of a $\mathbf{W}^{-1}\mathbf{B}$ matrix calculated from the initial 16-dimensional data. The utterances are labeled by the 10 digits 0 through 9 to correspond to the talkers with whom they are known to be associated.

Exhibit 16. Representation of utterances in the space of first two CRIMCOORDS

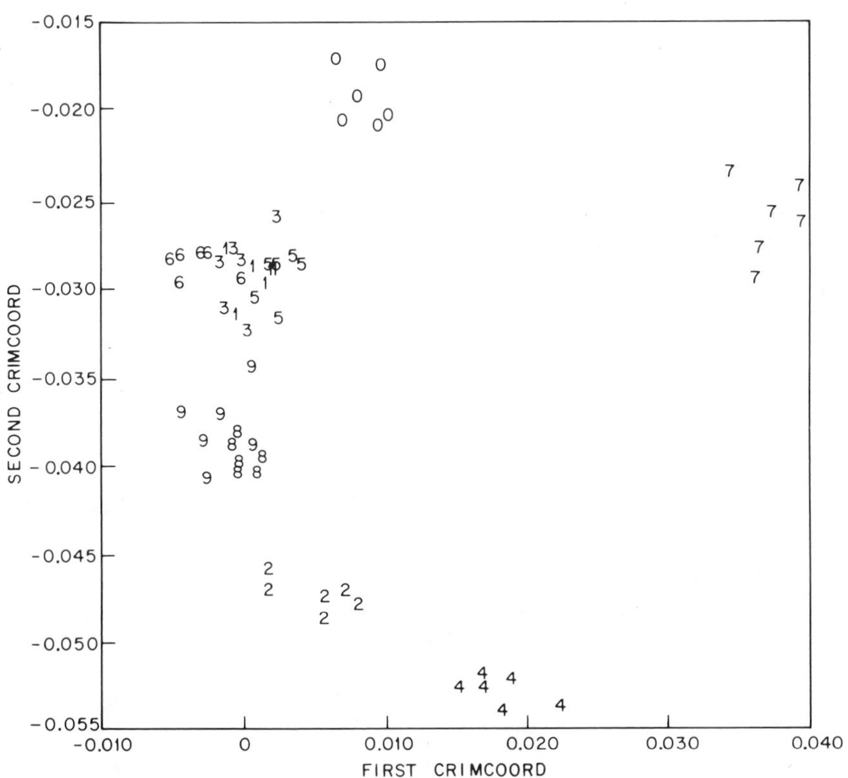

The clustering of the points in Exhibit 16 corresponds generally to the known categorization of the utterances and indicates the clear separations of and among talkers 0, 2, 4, and 7, as well as the considerable overlapping of talkers 1, 3, 5, and 6 and of talkers 8 and 9. There are no indications of outlying observations. Also, despite some indications of possible differences in the dispersions of the utterances when they are represented in the space of the first two CRIMCOORDS, one may feel that it is not unreasonable to use a simple Euclidean metric in the two-dimensional discriminant space (see the discussion in Section 4.2.1 on distance measures).

Example 17. A second example of the value of graphical representations in discriminant space is taken from a study of Chen et al. (1970, 1974) concerned with developing empirical bases for grouping industrial corporations into categories such as chemicals, drugs, oils, etc. One part of the study, using observations on 14 economic and financial variables, was concerned with the validity and appropriateness of such prespecified categories. [*Note*: Since one is interested both in utilizing useful prior groups where these are appropriate and in evolving data-determined groups when these are meaningful, this problem really does not fall totally under either classification or cluster analysis.] A four-group analysis of the chemical, drug, oil, and steel groups of companies in this investigation led to three CRIMCOORDS, and Exhibit 17 shows a representation of the companies in the discriminant space defined by the first two CRIMCOORDS. [*Note*: In this problem an initial analysis of each of the groups internally led to the identification of a few outliers (see also Example 39 in Section 6.4.1), and the determination of the CRIMCOORDS was then based only on the companies retained in the core groups.]

Apart from its usefulness in studying group separations, the configuration in Exhibit 17 indicates a relatively tight grouping of the oil companies and a very widely dispersed chemical group, thus hinting at possibly large disparities among the covariance matrices of the different groups in the space of the original 14-dimensional observations. The pooling involved in obtaining **W** might then be questionable, and other approaches (cf. Section 4.2.1) might prove more appropriate.

As a further aid in using plots such as Exhibits 16 and 17, one can draw circular "confidence regions," defined by

$$n_i(\bar{\mathbf{z}}_i - \boldsymbol{\mu}_i)'(\bar{\mathbf{z}}_i - \boldsymbol{\mu}_i) \le \chi_2^2(\alpha) \qquad \text{for } i = 1, \ldots, g, \tag{56}$$

where $\bar{\mathbf{z}}_i = \mathbf{A}_2'\bar{\mathbf{y}}_i$ is the centroid (i.e., mean) of the representations of the n_i observations in group G_i in terms of the first two CRIMCOORDS, $\boldsymbol{\mu}_i$ is

Exhibit 17. Representation of core group companies in the space of first two CRIMCOORDS

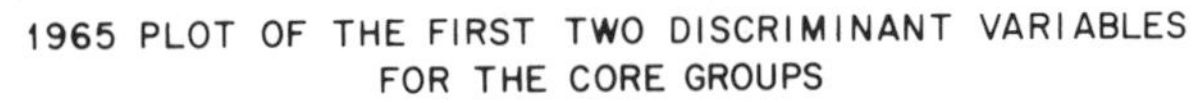

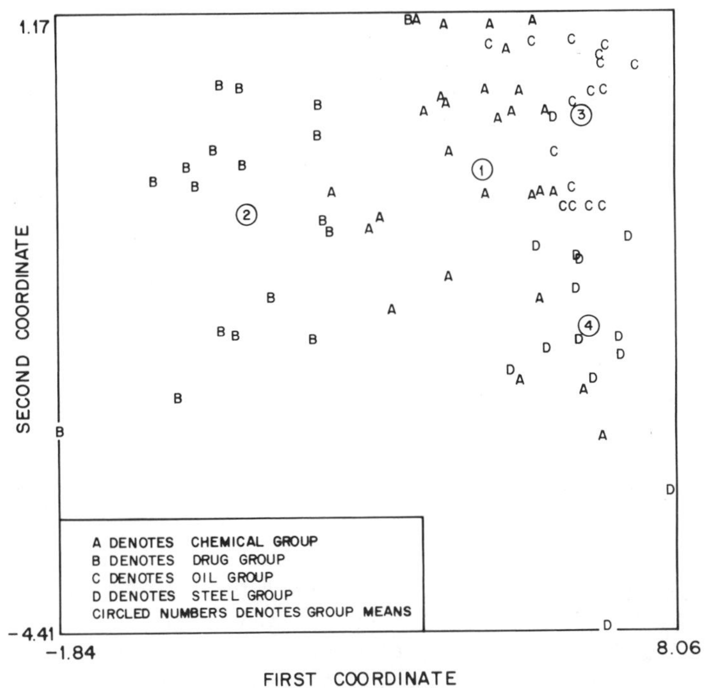

the unknown expected value of $\bar{\mathbf{z}}_i$, and $\chi_2^2(\alpha)$ denotes the upper $100\alpha\%$ point of the chi-squared distribution with 2 degrees of freedom. For three-dimensional representations in the space of the first three CRIM-COORDS, one can define spheres centered again at centroids by analogy with the two-dimensional case. The required percentage point would be from a chi-squared distribution with 3 degrees of freedom in this case. These circular and spherical regions may help in assessing the degree of group separation.

The pooling of the individual group dispersions to obtain the within-groups covariance matrix **W** merits a few comments. First, there is the question of inappropriateness of averaging across dissimilar covariance structures and the statistical effects of such averaging on the details of the classification or discriminant procedures. Suppose, for instance, that **W** includes one covariance matrix from a very widely dispersed group (see Example15). Then the determination of the CRIMCOORDS, and hence the associated discriminant space, may be distorted considerably by the inclusion in **W** of the "large" covariance matrix. A second question raised by the presence of widely varying covariance structures among the groups is the general issue of the meaningfulness of looking for location types of differences in the presence of such dispersion disparities—the so-called Behrens-Fisher problem of statistical inference. Third, particularly important from the point of view of data analysis is the following question: if, in fact, there are large discrepancies in the dispersion characteristics of the groups, and one is interested in discriminating among the groups, should not one attempt to use the dispersion information for the discrimination? As a partial answer to this question, one way of incorporating dispersion differences in the analysis is described in Section 4.2.1.

4.2.1. Distance Measures

Given a space (either the one for the original variables or a derived discriminant space) for representing the objects, the fundamental problem in classification is reduced to choosing a metric or a distance measure. For, if such a metric is available, an object which needs to be assigned to one of the groups may be identified with the group to which it is closest as judged by the metric.

Theoretical formulations have, by and large, been confined to the derivation of specific distance measures to satisfy narrowly defined optimality criteria under a body of assumptions, which themselves are often beyond empirical check by the data on hand. For instance, the optimal Bayes discriminant function minimizes expected loss, using prior probabilities, as well as other distributional assumptions.

From the point of view of data analysis, the prescription of a distance function will generally be a trial and error task in which the use of some general techniques needs to be aided by insight, intuition, and perhaps good luck!

One useful general class of squared distance functions is provided by a class of positive semidefinite quadratic forms. Specifically, if $\mathbf{u}' = (u_1, u_2, \ldots, u_p)$ denotes the p-dimensional observation on an object that is to be assigned to one of the g prespecified groups, then, for measuring the squared distance between $\mathbf{u}$ and the centroid of the ith group, one may consider the function

$$D^2(i) = (\mathbf{u} - \bar{\mathbf{y}}_i)'\mathbf{M}(\mathbf{u} - \bar{\mathbf{y}}_i), \tag{57}$$

where $\mathbf{M}$ is a positive semidefinite matrix to ensure that $D^2(i) \geq 0$. The object will be assigned to the group for which $D^2(i)$ is smallest as i takes on the values from 1 through g. Different choices of the matrix $\mathbf{M}$ lead to different metrics, and the class of squared distance functions represented by Eq. 57 is not unduly narrow.

Thus, when $\mathbf{M}=\mathbf{I}$, one obtains the familiar Euclidean squared distance between the "unknown" and the centroid of the ith group in the p-dimensional space of the responses. Geometrically, as shown in Figure 3*a* for the case when $p=2$, the use of such a measure of squared distance amounts to measuring distances by circles (or spheres when $p>2$)—points A_1 and A_2 lying on the same circle are considered to be the same distance away from the center C, while points B_1 and B_2 lying on the outer circle are considered to be farther away from C than are A_1 and A_2. For statistical uses, when the different responses are noncommensurable and likely to have very different variances, the use of this unweighted Euclidean metric may be inappropriate. For instance, if $p=2$ and y_1 has a larger variance than y_2, one may wish to weight a deviation in the y_1-direction less than an equal deviation in the y_2-direction. A way of accomplishing this would be to use "elliptical" (or ellipsoidal) distance measures as shown in Figure 3*b*—again A_1 and A_2 are considered to be equidistant from C, while B_1 and B_2 are considered to be farther from C than A_1 and A_2. Algebraically, this measure of squared distance corresponds to specifying $\mathbf{M}$ in Eq. 57 to be a diagonal matrix with diagonal elements equal to the reciprocals of the variances of the different variables. Still another extension of the distance measure may be made to accommodate intercorrelations among the responses as well as possible differences among their variances. When $p=2$ and the statistical correlation between y_1 and y_2 is positive, Figure 3*c* shows how one may use "elliptical" distance measures by tilting the ellipses so that their major axis is oriented in a direction reflecting the positive correlation—once again, points on the same ellipse are considered equidistant from C, while points, such as A_1 and B_1, on the different ellipses are

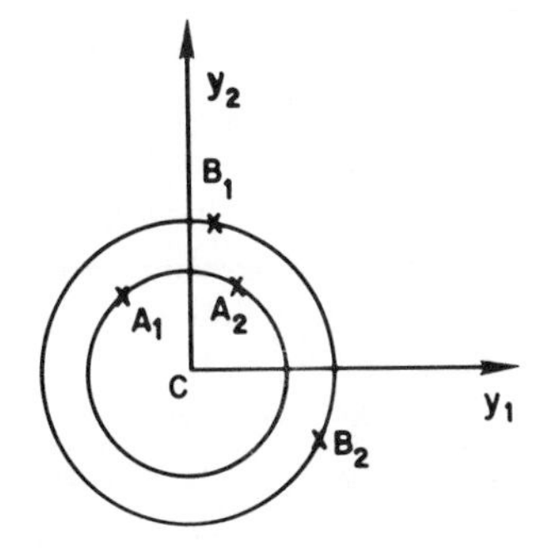

Fig. 3*a*. Euclidean measure of squared distance.

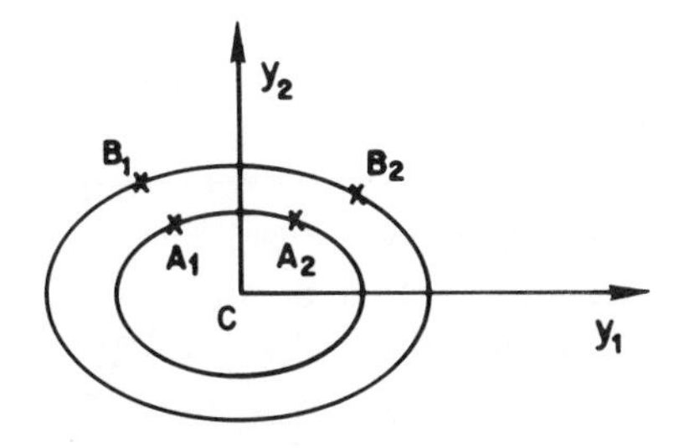

Fig. 3*b*. Measure of squared distance with different weights for the variables.

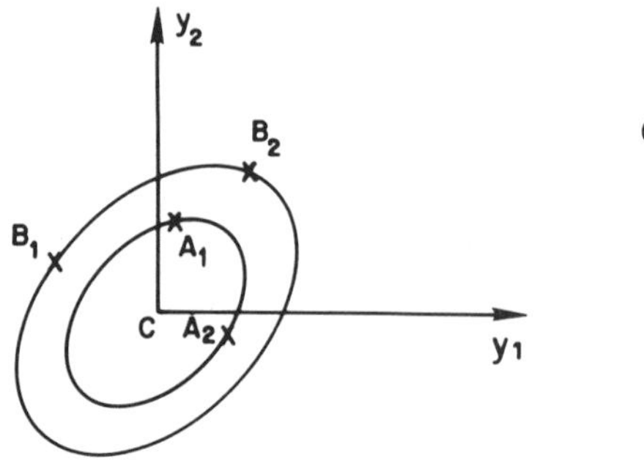

Fig. 3*c*. Generalized squared distance measure.

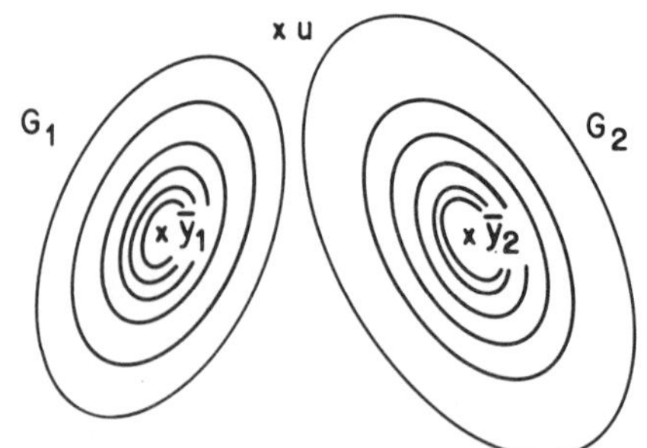

Fig. 4. Classification when within-group dispersions are different.

considered to be at increasing distances away from C. A way of reflecting this choice formally in Eq. 57 is to use for $\mathbf{M}$ the inverse of the covariance matrix of the variables.

Three specific choices for $\mathbf{M}$ in Eq. 57 are worth considering in more detail. The first is $\mathbf{M}=\mathbf{S}_i^{-1}$, yielding

$$D_1^2(i)=(\mathbf{u}-\bar{\mathbf{y}}_i)'\mathbf{S}_i^{-1}(\mathbf{u}-\bar{\mathbf{y}}_i), \tag{58}$$

where $\mathbf{S}_i$ is the covariance matrix derived from the reference set of observations, $\mathbf{Y}_i$, in the ith group, $i=1, \ldots, g$. An important practical constraint which needs to be met to ensure the nonsingularity of $\mathbf{S}_i$ is that $n_i>p$, so that to be able to use the metric $D_1(i)$ for classifying the "unknown" object one would, in general, require the number of reference observations in *every* group to exceed the dimensionality p. Also, since $\mathbf{M}$ changes from group to group, the use of $D_1(i)$ implies a considerable increase in the computational effort involved in classifying several "unknowns." Despite these limitations, however, one appealing feature of $D_1(i)$ is that it uses a dispersion standard that is internal to the group being considered as a possibility for assignment of an "unknown," and hence it may be able to exploit differences in the dispersion characteristics of the different groups. If $p=2$ and one has two groups, G_1 and G_2, Figure 4 illustrates the geometry involved in using the metric D_1 in the presence of dispersion differences. In this example, although the "unknown" (shown as an $\times$) is closer, in Euclidean distance, to the centroid of G_1 than to that of G_2, in terms of D_1 it is likely to be assigned to G_2 rather than G_1. An important feature in using D_1 is that, if one looks for boundaries dividing the p-dimensional space of the responses into regions, one for each of the g groups, such boundaries are nonlinear. The use of a likelihood-ratio approach (see, Anderson, 1958; Rao, 1952) to classification in the presence of heterogeneity of covariance matrices of the groups would lead to a similar but not identically the same procedure. For instance, with two groups, the likelihood-ratio approach based on assuming multivariate normality for the distributions of the observations would lead to classifying $\mathbf{u}$ in G_1 or G_2 according as $D_1^2(1) - D_1^2(2) \lesseqgtr \ln[\,|\mathbf{S}_2|\,/\,|\mathbf{S}_1|\,]$. On the other hand, the procedure described above would assign $\mathbf{u}$ to G_1 or G_2 according as $D_1^2(1) - D_1^2(2) \lesseqgtr 0$. More generally, with g groups, the likelihood-ratio approach would assign $\mathbf{u}$ to the ath group if

$$D_1^2(a) + \ln|\mathbf{S}_a| = \min_{i=1,\ldots,g} \{D_1^2(i) + \ln|\mathbf{S}_i|\},$$

whereas the procedure based on the metric D_1 would do so merely if

$$D_1^2(a) = \min_{i=1,\ldots,g} D_1^2(i).$$

A second choice for $\mathbf{M}$ in Eq. 57 is associated with the derivation of the discriminant space. If $\mathbf{M} = \mathbf{A}_t\mathbf{A}_t'$, where $\mathbf{A}_t'$ is defined following Eq. 55, then

$$D_2^2(i) = (\mathbf{u} - \bar{\mathbf{y}}_i)'\mathbf{A}_t\mathbf{A}_t'(\mathbf{u} - \bar{\mathbf{y}}_i) \tag{59}$$

is the measure of the squared distance of the "unknown" from the ith group. Using the metric D_2 in the p-dimensional space of original responses can be seen to be exactly equivalent to using the simple unweighted Euclidean metric in the t-dimensional discriminant space. The constraint on the eigenvectors of $\mathbf{W}^{-1}\mathbf{B}$ used in obtaining the CRIMCOORDS is that $\mathbf{A}_t'\mathbf{W}\mathbf{A}_t = \mathbf{I}$. Hence, under the assumptions used in deriving the discriminant space (including homogeneity of the group covariance structures), the CRIMCOORDS would be mutually uncorrelated and have unit variance each. This is a reason for using the simple Euclidean metric in the discriminant space, though not in the original space. The choice of $\mathbf{M}$ that leads to the metric D_2 does not vary from group to group. However, it does depend on the number, t, of eigenvectors to be employed from the eigenanalysis of $\mathbf{W}^{-1}\mathbf{B}$.

A third choice of $\mathbf{M}$ leads to the so-called generalized distance of the "unknown" from the centroid of the ith group in the p-dimensional space of the original variables. Specifically, choosing $\mathbf{M} = \mathbf{W}^{-1}$ leads to

$$D_3^2(i) = (\mathbf{u} - \bar{\mathbf{y}}_i)'\mathbf{W}^{-1}(\mathbf{u} - \bar{\mathbf{y}}_i). \tag{60}$$

This choice of $\mathbf{M}$ also does not change from group to group. To ensure nonsingularity of the within-groups covariance matrix, $\mathbf{W}$, the constraint $p \le (n - g)$ must be met, where $n = \sum_{i=1}^{g} n_i$ is the total number of observations from all g groups in the reference set. This constraint on the relationship between the dimensionality of response and the number of observations is less restrictive than the one underlying the choice of **M** that led to D_1. If pooling the dispersions of the different groups is reasonable and justified, one can thus have significant gains in the dimensionality to be used for the initial representation. A method of assessing the homogeneity of the dispersions of the groups is described in Section 6.3.2.

In the sense that both D_1^2 and D_3^2 use inverses of covariance matrices of the responses, one can think of D_3^2 as a generalization of D_1^2 when all the groups have similar dispersion characteristics. However, in the sense that D_1^2 is applicable when the groups have dissimilar dispersion characteristics, it is a generalization of D_3^2. In a somewhat less obvious sense, D_3^2 is interpretable in terms of a discriminant analysis approach which leads to D_2^2. In fact, performing a two-group discriminant analysis (with **W** in place of **S** in the earlier description of Fisher's two-group procedure) for every possible pair of groups [i.e., $g(g-1)/2$ analyses in all] is equivalent to using D_3^2. Also, using the maximum number, r, of eigenvectors corresponding to the nonzero eigenvalues of $\mathbf{W}^{-1}\mathbf{B}$ in D_2^2 would be entirely equivalent to using D_3^2.

These equivalences and relationships between D_2^2 and D_3^2 are easier to see in terms of the sphericized coordinates $\mathbf{x}=\mathbf{T}^{-1}\mathbf{y}$, where $\mathbf{W}=\mathbf{TT}'$ (see the discussion on p. 85). If $\bar{\mathbf{x}}_i$ and $\bar{\mathbf{x}}_j$ denote the centroids of the ith and jth groups, respectively, in this space, Figure 5 provides a geometrical demonstration of the equivalence between D_3^2 and the $g(g-1)/2$ pairs of two-group discriminant analyses. For the two-group analysis involving the ith and jth groups, the unknown **u** is projected onto the line joining $\bar{\mathbf{x}}_i$ and $\bar{\mathbf{x}}_j$ to obtain $\mathbf{u}^*$ and is assigned to the ith or jth group according as $[D_2^*(i)]\lesseqgtr[D_2^*(j)]$. From Figure 5, however, it is clear that for the metric D_3 the relationship $D_3(i)\lesseqgtr D_3(j)$ holds according as $D_2^*(i)\lesseqgtr D_2^*(j)$, so that $g(g-1)/2$ comparisons in terms of D_2^* are equivalent to a comparison of g values of D_3.

Also, in the x-space, $D_3^2(i)$ will be just the Euclidean squared distance of $\mathbf{u}_0$ $(=\mathbf{T}^{-1}\mathbf{u})$ from $\bar{\mathbf{x}}_i$ $(=\mathbf{T}^{-1}\bar{\mathbf{y}}_i)$. Hence, if $r=p$, D_3^2 is not only equivalent to D_2^2 but also identical with it, since $D_2^2(i)=(\mathbf{u}_0-\bar{\mathbf{x}}_i)'\mathbf{LL}'(\mathbf{u}_0-\bar{\mathbf{x}}_i)$, where **L** is now a $p\times p$ orthogonal matrix (i.e., $\mathbf{LL}'=\mathbf{I}$) whose columns are the eigenvectors of $\mathbf{B}_x$, the between-groups covariance matrix in the x-space.

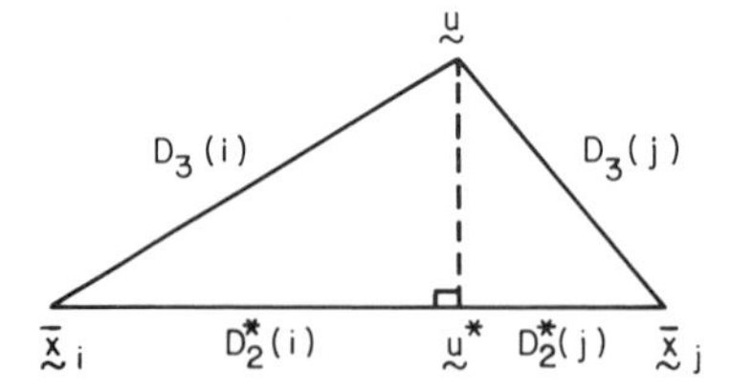

Fig. 5. Relationship between the uses of D_2 and D_3.

If, however, $r=(g-1)<p$, then, to establish the equivalence between D_2^2 and D_3^2, it has to be shown that $D_3^2(i) \leq D_3^2(j)$ if and only if $D_2^2(i) \leq D_2^2(j)$, where D_2^2 is based on all r eigenvectors corresponding to the nonzero eigenvalues of $\mathbf{W}^{-1}\mathbf{B}$. This follows from Pythagoras' theorem in p-space since

$$D_3^2(i) = D_2^2(i) + \begin{array}{l}\{\text{squared length of the perpendicular} \\ \text{from } \mathbf{u}_0 \text{ to the } (g-1)\text{-dimensional} \\ \text{hyperplane containing } \bar{\mathbf{x}}_1, \ldots, \bar{\mathbf{x}}_g\},\end{array}$$

and the second term on the right-hand side of this equation is seen not to depend on i.

Using D_3^2 has the merit of conceptual simplicity, avoidance of the eigenvector computations involved in D_2^2, and a performance in accurately classifying "unknowns" that may be as good as the result obtained by the use of any subset of the eigenvectors in D_2^2. On the other hand, the computation of the eigenvectors for use in D_2^2 may lead to reduction in dimensionality of the problem and perhaps some insight. Also, sometimes when the last few CRIMCOORDS are merely reflecting "noise," using a subset consisting of the first few eigenvectors for calculating D_2^2 may improve its performance over that of D_3^2.

One can also consider the use of squared distance measures that are approximations, in varying degrees of appropriateness, to D_1^2, D_2^2, and D_3^2 respectively. For instance, when the number of observations in the reference set is not sufficiently large for obtaining nonsingular estimates of the covariance matrices involved, one may decide merely to incorporate in the distance measures the differences in the variances of the coordinates and to neglect intercorrelations. Thus, in such a case, one may obtain an "approximation" to D_2^2, for example, by using for $\mathbf{M}$ in Eq. 57 a diagonal matrix whose diagonal elements are the ratios of between-groups to within-groups sums of squares for each of the p variables. (See Becker et al., 1965, for further discussion of these "approximations" in the context of a specific application.)

Example 18. The relative performances of the three metrics D_1, D_2, and D_3 may be illustrated in the context of the corporation-grouping study (see Chen et al., 1970, 1974) used also in Example 17. Exhibit 18 shows, for a particular year, the proportion of companies from each of the four core groups that are classified into their "proper" (i.e., according to the prespecified identification of a company as chemical, drug, oil, or steel) groups when D_1^2, D_2^2, and D_3^2 are used for the assignment. There is an element of bias in the classification procedure in this example since each of the companies being classified has influenced the estimates of the group centroids and covariance matrices, as well as the matrices $\mathbf{B}$ and $\mathbf{W}$

***Exhibit* 18. *Proportion of core-group companies classified into their initial groups for* 1965**

Metric	Initial Group				Overall Proportion
	Chemical	Drug	Oil	Steel	
D_1	26/27	18/18	16/16	8/14	68/75
D_2 with $t=2$	16/27	17/18	15/16	12/14	60/75
D_3	21/27	17/18	15/16	12/14	65/75

used in the eigenanalysis for deriving the discriminant space. Thus there is no clear separation of "unknowns" from the reference set of observations in this example. Nevertheless, since the core groups were determined after an initial elimination of extreme outliers, the proportions in Exhibit 18 may be viewed as indicators of "percent correctly identified" by the three measures of distance.

The metric D_1 has a better overall performance and is seen to be particularly good in handling the chemical group, which happens also to be the most dispersed. Using the first two CRIMCOORDS for the metric D_2 is, of course, equivalent to assigning companies to groups on the basis of Euclidean distance in Exhibit 17. The use of an additional CRIMCOORD, which would amount to employing D_3, is seen not to make any difference for three of the four groups, although for the chemical group the use of D_3 results in a noticeable improvement over the performance of D_2. In fact, it turns out that the third CRIMCOORD mainly pulls the chemical and oil groups apart so that this improvement is explainable.

4.2.2. Classification Strategies for Large Numbers of Groups

When the number of groups, g, is large, classifying an "unknown" by comparing its distances from all of the group centroids can become prohibitively expensive even with present-day high-speed computers. Some means of initially limiting the number of contenders to which an "unknown" may be assigned have to be developed. The rest of this subsection describes an ad hoc procedure based on using the first few CRIMCOORDS for this purpose. The essential ideas are developed in the context of the talker-identification problem, but their general applicability whenever g is large will also, it is hoped, emerge from their description.

Since the first few CRIMCOORDS provide a linear transformation of the original variables so as to maximally separate the groups, one natural approach would be to use a representation of the observations, together with an "unknown," in the space of the first few CRIMCOORDS as the basis for delineating "most likely" contenders for the "unknown." As seen in Exhibit 16 pertaining to the talker-identification example, with only 10 talkers one can see both separations and clusterings among the talkers even in the two-dimensional representation with respect to the first two CRIMCOORDS. When the number of talkers increases, however, such indications may not be as clear. Thus in Figure 6*a*, which shows a representation of only the centroids of the utterances of a given word by 172 talkers in the space of the first two CRIMCOORDS, there are no obvious clusters.

One approach here is to divide the two-dimensional discriminant space arbitrarily into boxes as a first step. The boundaries of the boxes may be determined by using specified quantiles of the distributions of the group centroids along the two CRIMCOORDS, and it would be appropriate to employ a larger number of quantiles for the distribution along the first CRIMCOORD than for the one along the second. For the talker-identification example, Figure 6*b* shows a division of the space in Figure 6*a* into 40 boxes, using nine deciles (i.e., values that divide the distribution into 10 equal parts) of the distribution of the 172 centroids along the first CRIMCOORD and three quartiles (i.e., values that divide the distribution into four quarters) of the distribution with respect to the second CRIMCOORD. Using such an arbitrarily partitioned two-dimensional discriminant space, one can determine the box into which an unknown under consideration for assignment falls (cf. Figure 6*c*), and then can initially limit the comparison of the "unknown" to only the groups whose centroids fall in the same box or a few nearby ones. Figure 6*d* shows a case in which the initial comparison is limited to nine boxes, with the one containing the "unknown" in the center. In the particular example used for Figures 6*a–d*, while the 0 denotes the "unknown," the × corresponds to the centroid of the talker from whom the "unknown" arose. Although the × is not in the same box as the 0 in this example, it is seen to be in a neighboring box, which is included for comparison. This may not always happen, however, and sometimes additional boxes may have to be included for picking up the "true" contender. A statistical strategy for expanding the base of comparisons by considering additional boxes is described below.

The actual comparison of the "unknown" with the groups whose centroids are in nearby boxes is made by calculating distances *not* just in the space of the first two CRIMCOORDS but in terms of all the t CRIMCOORDS that one has decided to include. In other words, the metric D_2 defined by Eq. 59 is used with the chosen value of t, but the

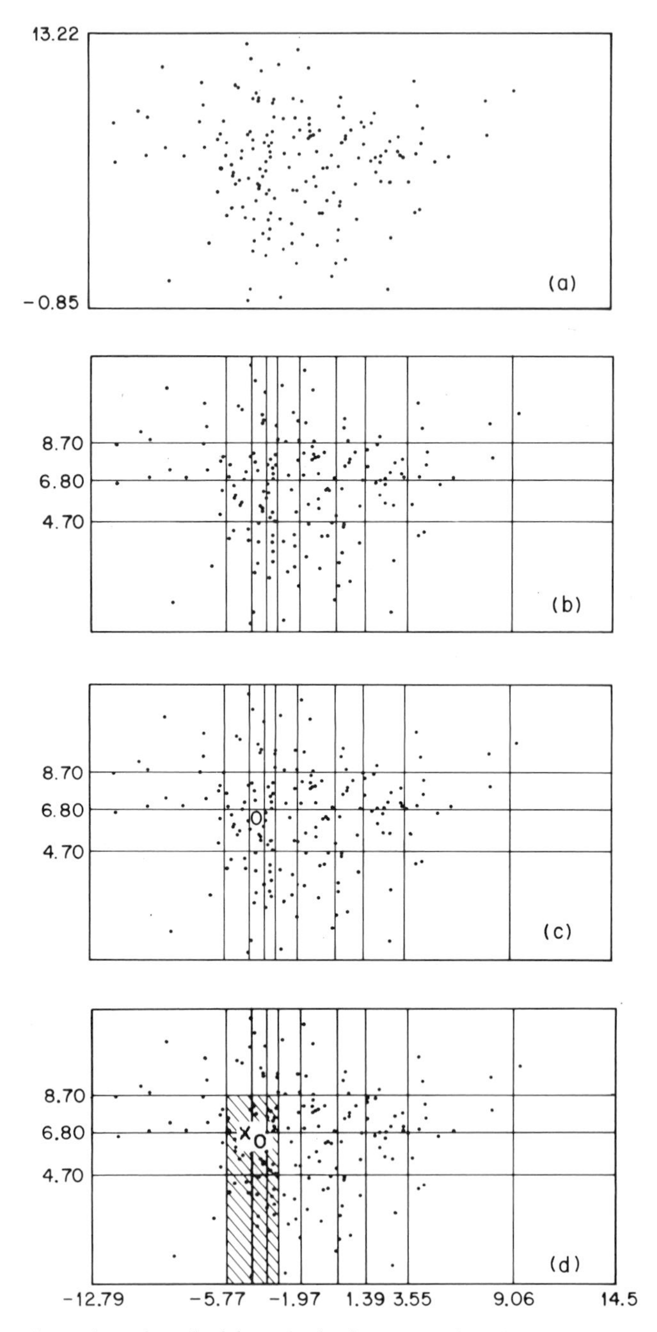

Fig. 6. Illustration of method for subselecting groups for classifying an unknown.

centroids, $\bar{\mathbf{y}}_i$, are initially limited to those of groups that are nearby in the space of the first two CRIMCOORDS.

The decision to include additional boxes may be based on two considerations: (i) the number of groups considered for the assignment of an "unknown" is inadequately small, a situation that may, for instance, occur when the "unknown" falls in a box toward the outer edges in Figure 6*b*; and (ii) the evidence for assigning the "unknown" to a group included in the initial set of boxes is not sufficiently strong.

To evaluate the strength of the evidence for associating an "unknown" with a group, two statistics that depend on the observed values of D_2^2 may be used. For a given set of contending groups, the ratio of the second smallest value of D_2^2 to the smallest value, as well as the latter value by itself, is a useful indicator. Thus, while the smallest value of D_2^2 determines the group to which the "unknown" is assigned, its numerical magnitude is an indicator of actual closeness between the "unknown" and the group. The ratio of the second smallest value to the smallest value is a measure of the closeness of the "unknown" to the group it is assigned to, as compared to its closeness to the next nearest group. Thus a large value of the ratio and/or a small value of the minimum observed D_2^2 lend strength to an assignment. Statistical benchmarks are needed for comparing the observed values of statistics such as the ratio and the smallest distance. If one is dealing with a situation in which there are sufficient data under "null" conditions (i.e., correct classification), one can obtain adequate estimates of the "null" statistical distributions of the statistics; that is, using only the reference set of observations, one can "simulate" the classification procedures, obtain the values of the statistics when the procedures lead to a correct assignment, and study the empirical distribution of these values. Such empirical distributions and their percentage points may then be used for comparing observed values of the statistics in assigning an "unknown" to decide whether they are large (or small) enough to confirm a "safe" assignment.

The essential features in the above type of strategy are, first, initial limitation of contenders by including for the primary comparisons only groups that are near the "unknown" in the space of the first two CRIMCOORDS; and second, enlargement of the population of contenders only when the assignment based on the primary comparisons is suspected of not being statistically sufficiently unequivocal. The hope is that for most of the "unknowns" one will not need to include groups from very many boxes to arrive at a satisfactorily clear assignment and that for only a few of the "unknowns" will one need to consider a large number of groups (possibly even all of them). Obviously, the properties of the strategy depend on various facets, including the number of CRIMCOORDS used initially (two is simplest and is generally recommended), the number and size of the boxes, the cut-off values for comparing statistics, such as the smallest distance and the ratio of the second smallest to the smallest distance, etc. In any example where g is very large, the specific values for these quantities may have to be chosen on a trial and error basis.

In the talker-identification example, which was used to motivate the strategy for large g, for the case of 172 talkers with one utterance from each serving as an "unknown," the use of the above type of strategy led to 81% (140/172) correct identification. An exhaustive comparison of each "unknown" against every talker, at a computing cost almost twice that for this strategy, led only to an improvement of 3%, viz., 84% (144/172) correct identification. A more detailed discussion of the talker-identification problem, including additional means that were employed to increase the percentage of correct identifications, is provided by Bricker et al. (1971).

4.2.3. Classification in the Presence of Possible Systematic Changes Among Replications

The classification process described in the preceding subsections of this chapter may be summarily described as follows: given g group centroids and an "unknown," $\mathbf{u}$, assign $\mathbf{u}$ to the group to whose centroid it is closest in terms of some metric. However, in some situations there may be an arbitrary or systematic change, for artifactual or other reasons, from observation to observation even within a specified group. For example, in repeated utterances of a word by a given talker, the general level of the jointly observed energies may shift because of varying proximity to the microphone. A second example would be a situation in which the groups are different species and the observations within a group are made on members at different stages of growth. In such circumstances a modified view of the classification problem is in order.

Thus suppose that with repeated utterances of a specific word by the same person one observation leads to the vector $\mathbf{x}$ and the next to $\mathbf{x}+\mathbf{c}$, where all the components of the vector $\mathbf{c}$ are equal but unknown. Of course, this is perhaps an oversimplified model for the true effect of proximity to the microphone. However, the essential point is that, when such possibilities exist, it is no longer wise or proper to classify the unknown with the group to whose center it is closest. Now each group is represented, in concept, not by a point, but by the line joining the group center, $\bar{\mathbf{y}}$, and the point, $\bar{\mathbf{y}}+\mathbf{c}$, for any $\mathbf{c}$. The proper classification is then based on the shortest generalized distance to such lines.

Similarly, one may need to allow for possible joint scale change, or even higher-order change, affecting all the coordinates of the multiresponse vector identically. For instance, if both scale and origin are artifactual, so that one observation in a group is $\mathbf{x}$ and another is $b\mathbf{x}+\mathbf{c}$, each group is defined as a plane and classification is based on shortest generalized distances to these group planes.

A simple version of these problems and one approach to them have been considered by Burnaby (1966) and by Rao (1966). A different approach is suggested here by casting the problem in a more familiar and suggestive form.

Instead of considering the ith group centroid, $\bar{\mathbf{y}}_i = (\bar{y}_{i1}, \ldots, \bar{y}_{ip})'$, and the unknown, $\mathbf{u} = (u_1, \ldots, u_p)'$ as two points in p-space, consider them as p points in two-dimensional space with coordinates $(\bar{y}_{ij}, u_j)$ for $j = 1, 2, \ldots, p$. One can then make a scatter plot of these p points.

Clearly, in the absence of any systematic changes between replications within a group, perfect correspondence between the unknown and the ith group will lead to a linear configuration having unit slope and passing through the origin. If the unknown is a member of the group, one expects a good linear configuration, and, indeed, the generalized distance in p-dimensional space between the unknown, $\mathbf{u}$, and the ith group centroid [i.e., $D_3^2(i)$ of Eq. 60] is just an appropriately defined quadratic form in the residuals of the observations from the line of unit slope through the origin in this two-dimensional representation. A joint additive shift and common scale change, if present, will show as a nonzero intercept and a slope not equal to unity.

The above type of scatter plot can be made for each unknown against the centroid of every group, and for classification purposes a linear regression line may be determined corresponding to each plot and the unknown may be assigned to a group by comparing the magnitudes of the g residual sums of squares in the g regressions. In general, since the p points are associated with p responses that may have widely differing variances in addition to being intercorrelated, the fitting may have to be performed by generalized (i.e., weighted) least squares rather than by simple least squares. The classification will then be based on a comparison of g values of a quadratic form in the residuals of the observations from the generalized linear least squares fits in each of the scatter plots. For an initial exploratory analysis in many problems, the simpler approach through ordinary least squares may be adequate.

Example 19. The approach is illustrated by application to data from the talker-identification problem. One summary employed in this problem consisted of a 57-dimensional vector of energies for characterizing each utterance of a given word by each talker.

Exhibit 19 shows a scatter plot of the values of the 57 components for two "unknown" utterances of a word against the corresponding values in the average of four "known" utterances of the same word by a specific talker. One of the unknowns was chosen from the same talker, and the points for this are shown as □'s; the other was from another talker, and the corresponding points are shown in Exhibit 19 as ○'s. Also shown in Exhibit 19 are the simple least squares linear fits to the two sets of points.

The existence in these data of artifacts of the type discussed above is evident in this plot. The configuration of the □'s, although quite linear, has a nonzero (small positive) intercept. Also the slope of the fitted line is very slightly smaller than unity. Thus there is some evidence of a shift (and perhaps no scale) artifact.

The configuration of the ○'s exhibits poor linearity, with considerably more scatter about the linear fit. A comparison of the two configurations suggests the possible utility of a classification procedure based on a quadratic form in the residuals from a least squares linear fit. In the present example the ordinary sums of squares of the residuals, for instance, turn out to be about 8260 for the configuration of the □'s and over 180,000 for that of the ○'s. Also, in this example the use of simple least squares fits and a comparison of the associated residual sums of

Exhibit 19. ***Linear regression of unknown versus centroid: residual sum of squares for □ is 8258 and for ○ is 180,747***

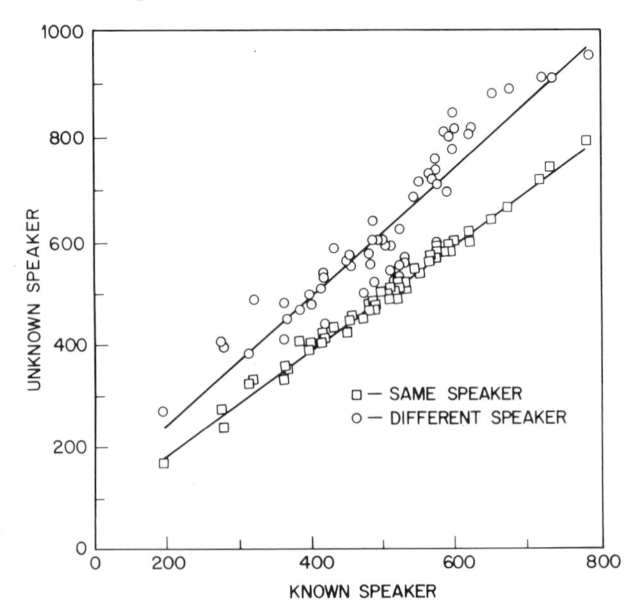

squares led to almost 70% correct identifications, and the utilization of weighted least squares employing estimates of variances (and neglecting the correlational aspects) improved the percentage to about 75%. Of course, in other examples, wherein the variances may be more disparate and the intercorrelations perhaps high, the performance of the approach based on simple least squares may not be as good.

The approach just illustrated has a number of attractions. First, it involves familiar regression ideas. Second, it permits a graphical representation. Third, largely as a consequence of the second point, the procedure enables the use of a flexible internal comparisons process, in that the data themselves may help to suggest the nature of the possible corrections which may be desirable, such as the detection of coordinate outliers or the form of the regression (e.g., quadratic or other nonlinear regressions) which may be more appropriate to use.

In this approach there can be additional methodological problems when *both* the covariance matrix (needed for the generalized least squares fitting) and the regression function have to be estimated from the data. In that case some iterative process is possible, if necessary. There are also problems of strategy and implementation of any iterative technique.

4.3. CLUSTERING

In recent years, under the stimulus of computer and computing technology, several algorithms for cluster seeking have been suggested (see Cormack, 1971,† for a survey of developments). The essential concern of these techniques is to find groupings of n units (objects, experimental units, etc.) such that the units within groups are more"similar" (in some sense to be indicated by the measurements on the units) than the units across groups. In virtually all of these procedures, the groups or clusters are determined by the (iterative) seeking of "neighborhoods" that are defined in terms of some metric; that is, similar units are conceptualized as those that are close together in terms of some metric. Although the work of Neely (1967) is initially motivated by thinking of clusters in terms of the density of points in a high-dimensional space, the actual measure of density used is average reciprocal interpoint distance, and hence metric concepts are ultimately introduced. The work of Hartigan (1967) does not depend explicitly on a metric, but it appears that one can recover an implicit metric.

The available methods for cluster seeking may be categorized broadly

†Other, more recent and relevant references are books by Everitt (1974) and Hartigan (1975).

as being *hierarchical* (e.g., Hartigan, 1967; Johnson, 1967; Sokal & Sneath, 1963) or *nonhierarchical* (e.g., Ball & Hall, 1965; Friedman & Rubin, 1967). The former class is one in which every cluster obtained at any stage is a merger of clusters at previous stages. In this case, therefore, it is possible to visualize not only the two extremes of clustering, viz., n clusters with one unit per cluster (*weak clustering*) and a single cluster with all n units (*strong clustering*), but also a monotonically increasing strength of clustering as one goes from one level to another. In the nonhierarchical procedures, on the other hand, new clusters are obtained by both lumping and splitting of old clusters, and, although the two extremes of clustering are still the same, the intermediary stages of clustering do not have this natural monotone character of strength of clustering.

A crucial question, especially in nonhierarchical clustering, is the computational feasibility of any specific algorithm. Looking at all possible partitions of the data for determining a clustering or grouping that is optimal with respect to some criterion is prohibitively expensive and may even be impossible despite the speed of today's computers. Gower (1967), for instance, points out that the computations involved in looking at the $(2^{n-1}-1)$ possible partitions of n units into two sets for choosing the partition with minimum within-sets sum of squares would take approximately $(n-1)^2 2^{n-11}$ seconds on a 5-microsecond-access-time machine, so that with $n=21$ units the time involved would be approximately 114 hours and with $n=41$ it would be approximately 54,000 years! (See also Scott & Symons, 1971.) The problems here are reminiscent of (although perhaps more extreme than) those in stepwise and steered multiple regression analysis, and it is not surprising that much of the difference between currently available clustering schemes lies in their relative computational efficiencies.

The format of the input data for clustering procedures may be metric or nonmetric, i.e., as a representation of n points in p-space or only as rank order information regarding the similarities of pairs of the n units. The descriptions of most nonhierarchical schemes seem to assume a metric input with an implied choice of p as well. This, however, is not a necessary limitation since the observed ordering of the similarities may be utilized as input to multidimensional scaling (see Section 2.3) for obtaining a representation of the n units in a Euclidean space whose dimensionality is data determined. Also, even with metric data inputs, if redundancy among the p coordinates is suspected, the original data may first be transformed to a reduced dimensional space by using linear or generalized principal components analyses (see Sections 2.2.1 and 2.4.2), and then the clustering may be performed in the lower-dimensional linear or nonlinear subspace of the original p-dimensional space.

In any specific application, whether one uses hierarchical or nonhierarchical methods is largely dependent on the meaningfulness, in the particular situation, of the tree structure imposed by hierarchical clustering procedures. For instance, in biological applications concerned with groupings of species, clusters of species, subclusters of subspecies, and so on may be of interest, and hierarchical clustering may then be a sensible approach to adopt. Even the area of numerical taxonomy, however, is not without controversy as to the biological meaningfulness of clusters (hierarchical or otherwise) determined by the use of statistical data-analytic techniques.

Section 4.3.1. will discuss hierarchical clustering methods, and Section 4.3.2 will be concerned with nonhierarchical clustering.

4.3.1. Hierarchical Clustering Procedures

The discussion of methods for hierarchical clustering in this subsection follows closely the development due to Johnson (1967). The essential idea of a hierarchical clustering scheme is that n units are grouped into clusters in a nested sequence of, say, $(m+1)$ clusterings, $C_0, C_1, \ldots, C_m$, where C_0 is the weak clustering, C_m is the strong clustering, and every cluster in C_i is the union or merger of some clusters in C_{i-1} for $i=1, \ldots, m$. Also, corresponding to C_i we have its "strength," α_i, where $\alpha_0=0$ and $\alpha_i<\alpha_{i+1}$ for $i=0, 1, \ldots, (m-1)$. The α's, therefore, are an increasing sequence of nonnegative numbers.

Johnson (1967) demonstrates that, for any such hierarchical clustering scheme, a metric for measuring the distance between every pair of the n units is implied, and, conversely, that, given such a metric, one can recover the hierarchical clustering scheme from it. Given two units, x and y, and the above hierarchical clustering scheme, let j be the smallest integer in the set $[0, 1, \ldots, m]$ such that in clustering C_j the units x and y belong to the same cluster; then define the distance between x and y, $d(x, y)$, to be the strength, α_j, of the clustering C_j. In other words, the distance between any pair of units is defined as the strength of the clustering at which the units first appear together in the same cluster. This definition leads to a distance measure with properties generally associated with metrics. Thus $d(x, y)=0$ if and only if x and y first appear together in C_0, which means that x and y are not distinct units or that $x=y$. Also, from the definition it follows that $d(x, y)=d(y, x)$. Finally, if x, y, and z are three units, the triangle inequality $d(x, z)\leq d(x, y)+d(y, z)$ may be shown to hold. In fact, a stronger inequality, viz., $d(x, z)\leq \max\{d(x, y), d(y, z)\}$, which implies the triangle inequality, may be shown to be satisfied by d. This stronger inequality, which states that the distance between x and z cannot exceed the larger of the two distances, $d(x, y)$ and $d(y, z)$, has been called the *ultrametric inequality* by Johnson (1967). That the above definition of distance between pairs of units in a hierarchical clustering scheme satisfies the ultrametric inequality may be established as follows. Let $d(x, y)=\alpha_i$ and $d(y, z)=\alpha_j$, so that the units x and y appear together in the same cluster for the first time in clustering C_i, and the units y and z do so in clustering C_j. Then, because of the hierarchical nature, one of these clusters includes the other, viz., the one which appears in the clustering whose index corresponds to the larger of i and j includes the other. Hence, if $k=\max(i, j)$, in clustering C_k the units x, y, and z are all in the same cluster and, clearly, $d(x, z) \leq \alpha_k = \max(\alpha_i, \alpha_j)$. Thus, given a hierarchical clustering scheme such as the one in the preceding paragraph, one may derive a metric satisfying the ultrametric inequality (and hence the triangle inequality).

The converse—viz., given a set of $n(n-1)/2$ interunit values of a metric that satisfies the ultrametric inequality, one may recover a hierarchical clustering of the n units—is also demonstrated by Johnson (1967). The equivalence between hierarchical clustering and a metric that satisfies the ultrametric inequality is perhaps most easily shown by the following simple example, taken from Johnson (1967).

Example 20. Exhibit 20*a* shows a hierarchical clustering of six $(=n)$ units involving five $(=m+1)$ stages of clustering, $C_0, \ldots, C_4$, with respective associated strengths $\alpha_0, \ldots, \alpha_4$ ranging from 0 to 0.31. [*Note*: The value of the strength of each clustering is, for the moment, assumed to be specified, and the later discussion in this subsection will deal with how these strengths are actually obtained in various hierarchical clustering algorithms.]

Using the earlier-mentioned definition of a distance between a pair of units, one may derive the matrix of interunit distances shown in Exhibit 20*b*. Thus, since every unit "appears with itself in the same cluster" for the first time in C_0 and $\alpha_0=0$, the diagonal elements are all 0. Also, for

Exhibit 20a. Example of hierarchical clustering tree

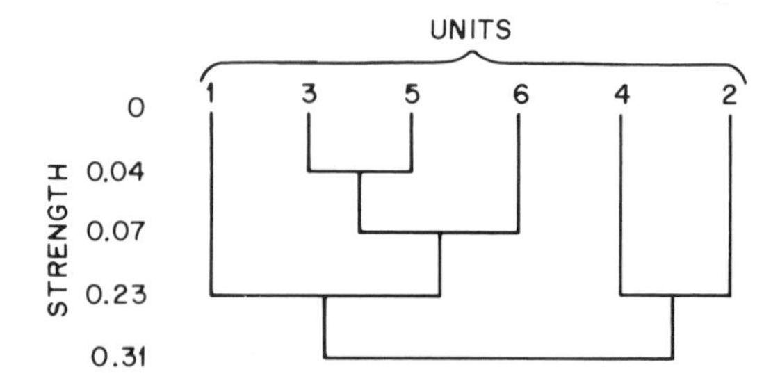

Exhibit 20b. Initial distance matrix for the example in Exhibit 20a

6×6 DISTANCE MATRIX

	1	2	3	4	5	6
1	0	0.31	0.23	0.31	0.23	0.23
2	0.31	0	0.31	0.23	0.31	0.31
3	0.23	0.31	0	0.31	0.04	0.07
4	0.31	0.23	0.31	0	0.31	0.31
5	0.23	0.31	0.04	0.31	0	0.07
6	0.23	0.31	0.07	0.31	0.07	0

instance, since units 3 and 5 are clustered for the first time in C_1 with $\alpha_1=0.04$, the distance between these units is 0.04, while the distance between units 1 and 2 is 0.31, the strength of C_4, the strong clustering, which is the first stage in which units 1 and 2 are clustered together. All of the metric properties claimed for the definition of distance used in obtaining Exhibit 20*b* from Exhibit 20*a* can be verified in this example in terms of the elements of the distance matrix shown in Exhibit 20*b*.

Next, the inverse process of going from Exhibit 20*b* to Exhibit 20*a* may be demonstrated in terms of this simple example. To start the process, at the first level we form the weak clustering C_0 with six clusters containing one unit each. Next, by scanning the elements of the distance matrix in Exhibit 20*b*, we identify the smallest interunit distance as being 0.04, the distance between units 3 and 5. In a natural manner, suppose that we decide to create a cluster (3, 5) and to leave the remaining four units by themselves, thus leading to five clusters at level C_1 with associated strength α_1 equal to the smallest distance, 0.04, in the distance matrix of Exhibit 20*b*. To repeat the process for constructing the higher-level clusterings, we now need a way of defining distances between cluster (3, 5) and the four units 1, 2, 4, and 6. An interesting property (shown below to be a consequence of the distance function here satisfying the ultrametric inequality) of the values in Exhibit 20*b* is that $d(3, x)=d(5, x)$ for $x=1$, 2, 4, and 6. Hence a natural measure of the distance between cluster (3, 5) and unit x would be $d([3, 5], x)=d(3, x)=d(5, x)$ for $x=1$, 2, 4, and 6. Exhibit 20*c* shows a 5×5 distance matrix obtained by using this definition.

Scanning Exhibit 20*c* for the smallest element, we find that it is $0.07=d([3, 5], 6)$, and we can then form a cluster (3, 5, 6) while leaving units 1, 2, and 4 by themselves, thus obtaining four clusters in stage C_2 with associated strength $\alpha_2=0.07$. It is now clear that by repeating this

Exhibit 20c. Distance matrix after forming first cluster in Exhibit 20a

5×5 DISTANCE MATRIX

	1	2	(3,5)	4	6
1	0	0.31	0.23	0.31	0.23
2	0.31	0	0.31	0.23	0.31
(3,5)	0.23	0.31	0	0.31	0.07
4	0.31	0.23	0.31	0	0.31
6	0.23	0.31	0.07	0.31	0

process of constructing distance matrices and scanning them for a minimum value, for deciding on which clusters to merge and for determining the value for the strength of clustering, we can recover the entire hierarchical clustering shown in Exhibit 20*a*. At any stage in this process, there may be a tie, i.e., more than one interentity distance [e.g., the distance between cluster (3, 5, 6) and unit 1 and the distance between units 2 and 4 are both 0.23] may correspond to the smallest value in the distance matrix. This merely implies that parallel clusters are being formed at such a stage.

The foregoing simple example provides a basis for the following summarization of the steps involved in going from a matrix of values of a metric satisfying the ultrametric inequality to a hierarchical clustering of n units in a nested sequence of $(m+1)$ clusterings, $C_0, C_1, \ldots, C_m$:

1. Form C_0 of strength 0 by considering each unit as a cluster.
2. Given a clustering C_j with a corresponding matrix of interentity (where an entity may be a single unit or a cluster of units) distances that satisfy the ultrametric inequality, merge the pair of entities with the smallest nonzero distance, α_{j+1}, to create C_{j+1} with strength α_{j+1}.
3. Create a new distance matrix corresponding to C_{j+1}.
4. Starting with $j=0$, by repeated use of steps 2 and 3, generate C_1, $C_2, \ldots$ and C_m (the strong clustering).

The assumption in step 3 is that the distance matrix corresponding to C_{j+1} can be constructed in an *unambiguous* manner. That this is a consequence of the distances satisfying the ultrametric inequality can be demonstrated. Suppose that x and y are two entities in C_j that become clustered together in C_{j+1}, so that $d(x, y)=\alpha_{j+1}$, and suppose that z is any other entity in C_j. Then the unambiguous construction of the distance matrix corresponding to C_{j+1} is possible because $d(x, z)$ necessarily has to be equal to $d(y, z)$ if d satisfies the ultrametric inequality. If $d(x, z) \neq d(y,$

z), then suppose that $d(x, z) > d(y, z)$. But, as a consequence of the ultrametric inequality, $d(x, z) \leq \max\{d(x, y), d(y, z)\}$, so that the inequality in the preceding sentence would imply that $d(y, z) < d(x, z) \leq d(x, y) = \alpha_{j+1}$. But α_{j+1} is the smallest nonzero distance in clustering C_j, and hence $d(y, z)$ cannot be smaller than α_{j+1}. Thus, assuming that $d(x, z) \neq d(y, z)$ leads to a contradiction, and therefore $d(x, z)$ has to be equal to $d(y, z)$.

In practice, the observed measures of distance between pairs of units may be either subjective measures of dissimilarity (or similarity) or measures of distance computed from metric data representing the n units, perhaps as n points in a p-dimensional space. Such measures of distance may not, and need not, satisfy the ultrametric inequality, with the consequence that the distance between cluster (x, y) and entity z may not be definable unambiguously since $d(x, z)$ need not necessarily equal $d(y, z)$ unless d satisfies the ultrametric inequality. Hence ways of defining $d([x, y], z)$, given $d(x, z)$ and $d(y, z)$, need to be devised for most situations. Considering $d([x, y], z)$ as a function, $f\{d(x, z), d(y, z)\}$, which is required to equal the common value of $d(x, z)$ and $d(y, z)$ whenever these are equal, leads to a fairly wide class of functions, f, including any weighted average, the geometric mean, etc. Motivated partially by the considerations underlying multidimensional scaling (viz., a monotone relationship between distance and dissimilarity), Johnson (1967) proposes two specific choices of f which, while satisfying the above constraint, are also invariant under monotone transformations of the distances. The two choices are $f =$ the min (or the smaller of) function and $f =$ the max (or the larger of) function. The methods based on these two choices have been called, respectively, the *minimum method* and the *maximum method* by Johnson (1967). In the numerical taxonomy literature, Sneath (1957) has proposed a hierarchical method called the *single linkage method*, and Sørensen (1948) a method called the *complete linkage method*. These two methods appear to be very similar, respectively, to the minimum and maximum methods suggested by Johnson (1967). At any rate the two methods of hierarchical clustering described by Johnson (1967) may now be summarized.

The Minimum Method. (See Johnson, 1967; also Sneath, 1957)

1. Form C_0, consisting of n clusters with one unit each, with corresponding strength $\alpha_0 = 0$.
2. Given C_j with associated distance (or dissimilarity) matrix (where the observed values at stage C_0, e.g., may not satisfy the ultrametric inequality), merge the entities whose distance, α_{j+1} (>0), is smallest to obtain C_{j+1} of strength α_{j+1}.

3. Create a matrix of distances for C_{j+1}, using the following rules: (a) if x and y are entities clustered in C_{j+1} but not in C_j [i.e., $d(x, y)=\alpha_{j+1}$], then $d([x, y], z)=\min\{d(x, z), d(y, z)\}$; ($b$) if x and y are separate entities in C_j that remain unclustered in C_{j+1}, then do not change $d(x, y)$.
4. Repeat the process until the strong clustering is obtained.

The Maximum Method. (See Johnson, 1967; also Sørensen, 1948). For this method, steps 1, 2, 3b, and 4 are the same as those in the minimum method. For step 3a, however, the following is substituted: if x and y are entities clustered in C_{j+1} but not in C_j, define $d([x, y], z)=\max\{d(x, z), d(y, z)\}$.

The two methods are directed toward different objectives and may not necessarily yield similar results when applied to the same body of data. The maximum method is concerned essentially with minimizing the maximum intracluster distance at each stage and hence tends to find compact clusters, sometimes forming several small clusters in parallel. The minimum method, on the other hand, tends to maximize the "connectedness" of a pair of units through the "intermediary" units in the same cluster (see Johnson, 1967, for a discussion of these interpretations), with a tendency to create fewer distinct clusters than the maximum method. When the initial data consist of p-dimensional representations of the n units and the interpoint distances in the representation are used as the elements of a distance matrix, both methods may tend to be unduly sensitive to outliers. For this reason, defining $d([x, y], z)$ in step 3a of the preceding descriptions as the average of $d(x, z)$ and $d(y, z)$ may be preferable. This method, which may be called the *averaging method*, does not, however, possess the property of invariance under monotone transformations of the distances. A different but critical issue, in the case when the initial representation of the n units is metric, is the dependence of the clusters obtained on the type of distance function used for measuring interunit distance. The results of the clustering techniques described heretofore seem to be highly dependent on the specification of a metric for measuring interunit distance.

Example 21. This example, taken from Johnson (1967), deals with data from Miller & Nicely (1955) pertaining to the confusability of 16 consonant sounds. The observed data were the values of the frequency, $f(x, y)$, with which the consonant phoneme x was heard as the consonant phoneme y by a group of human listeners. Different levels of both filtering and noise were involved in the experiment, and the frequency of confusions for each pair of consonant sounds was observed separately for each experimental condition.

For purposes of hierarchical clustering of the 16 consonants under each experimental condition, Johnson (1967) defines the symmetric measure of similarity, $s(x, y) = f(x, y)/f(x, x) + f(y, x)/f(y, y)$, and considers it as an inverse measure of distance (i.e., the similarity increases as distance decreases, and, in particular, $d = 0$ is taken to correspond to $s = \infty$, which would be the strength of C_0 in terms of s).

Exhibits 21a and b show the solutions obtained by Johnson (1967) by using the minimum and maximum methods, respectively. The two clustering solutions in this example are generally similar, although there are differences both in the numerical values of the strengths of the clusterings and in the stage of clustering when specific clusters are formed. For instance, two of the so-called unvoiced stop consonants, p and t, join the third one, k, earlier in the maximum than in the minimum method. Also, in respect to the unvoiced fricatives, f, θ, s, ʃ, f joins (θ, s, ʃ) earlier in the minimum than in the maximum method.

The similarity between the two solutions also extends to the order in which the consonant phonemes or groups of them come together, with the exception of the manner in which the last three groups merge. In the maximum method, the voiced consonant phonemes (both the voiced stops, b, d, g, and the voiced fricatives, v, ð, z, ʒ) merge with the nasals,

Exhibit 21a. Hierarchical clustering obtained by minimum (or single linkage) method for data on confusability of 16 consonant sounds (Johnson, 1967)

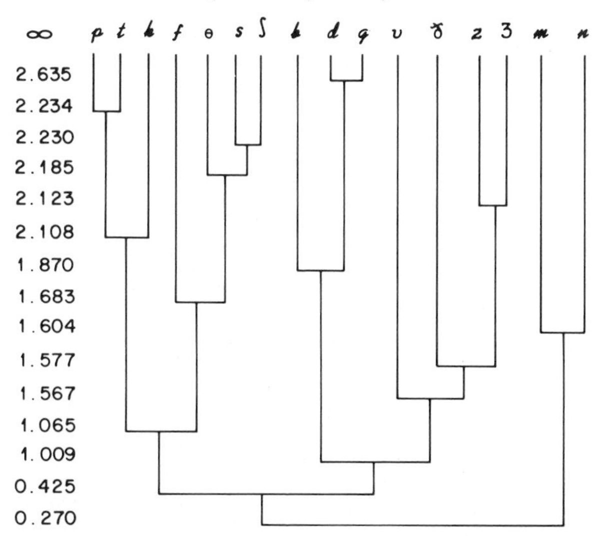

Exhibit 21b. Hierarchical clustering obtained by maximum (or complete linkage) method for data on confusability of 16 consonant sounds (Johnson, 1967)

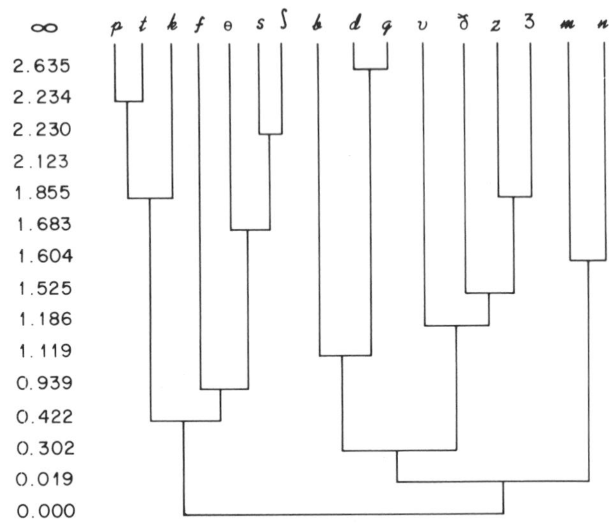

m, n, and their combination then merges with the unvoiced consonant phonemes (the unvoiced stops, p, t, k, and the unvoiced fricatives, f, θ, s, ʃ). With the minimum method, the voiced and the unvoiced consonants merge before their combined group joins the nasals.

In this example the manner of the hierarchical grouping of the consonants makes sense in terms of what is known about the discrimination of consonant phonemes. First, the stops and the fricatives in both the unvoiced and the voiced categories come together in four separate groups, whereas the nasals combine by themselves to form a fifth group. Then the unvoiced and voiced consonants coalesce into two separate groups, while the nasals constitute the third group.

The close similarity between the solutions obtained by the two methods in this example is, although comforting, not necessarily to be expected in all applications, as illustrated by the next example.

Example 22. As a part of the corporation-grouping study (see Examples 17 and 18), clustering procedures were employed to investigate the structure among specific groups of companies as indicated by the observations on the 14 variables involved. Thus, for the year 1967, hierarchical cluster analyses were performed for 18 of the domestic oil companies, and Exhibits 22*a*–*c* show, respectively, the results obtained by the

Exhibit 22a. ***Hierarchical cluster of 18 domestic oil companies obtained by minimum method***

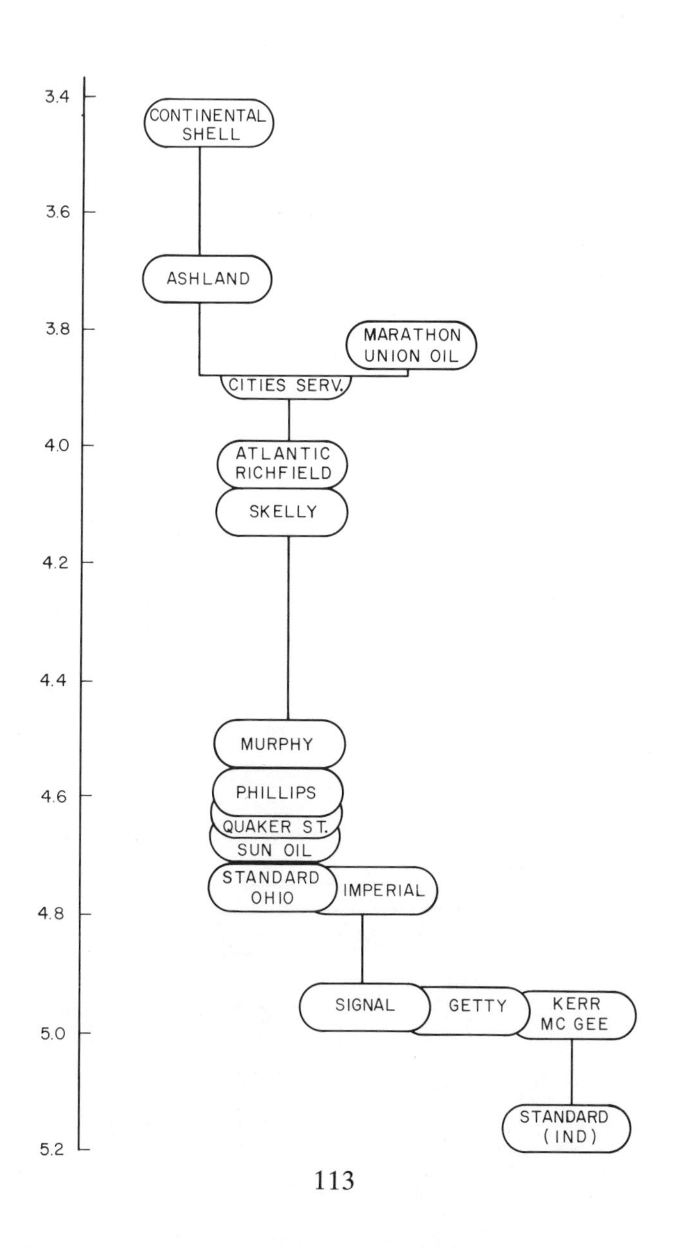

Exhibit 22b. Hierarchical cluster of 18 domestic oil companies obtained by averaging method

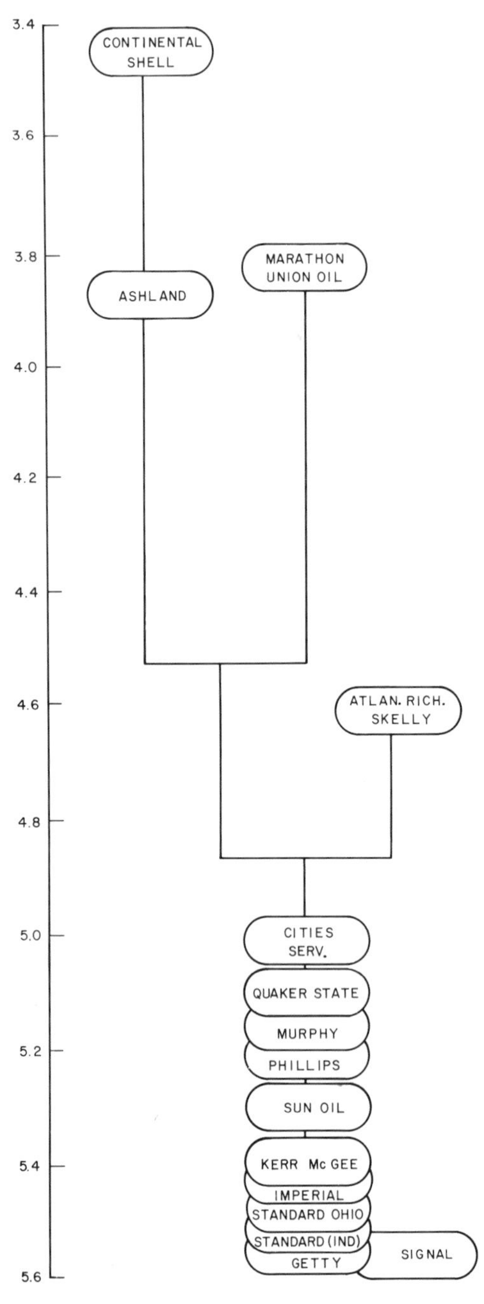

Exhibit 22c. Hierarchical cluster of 18 domestic oil companies obtained by maximum method

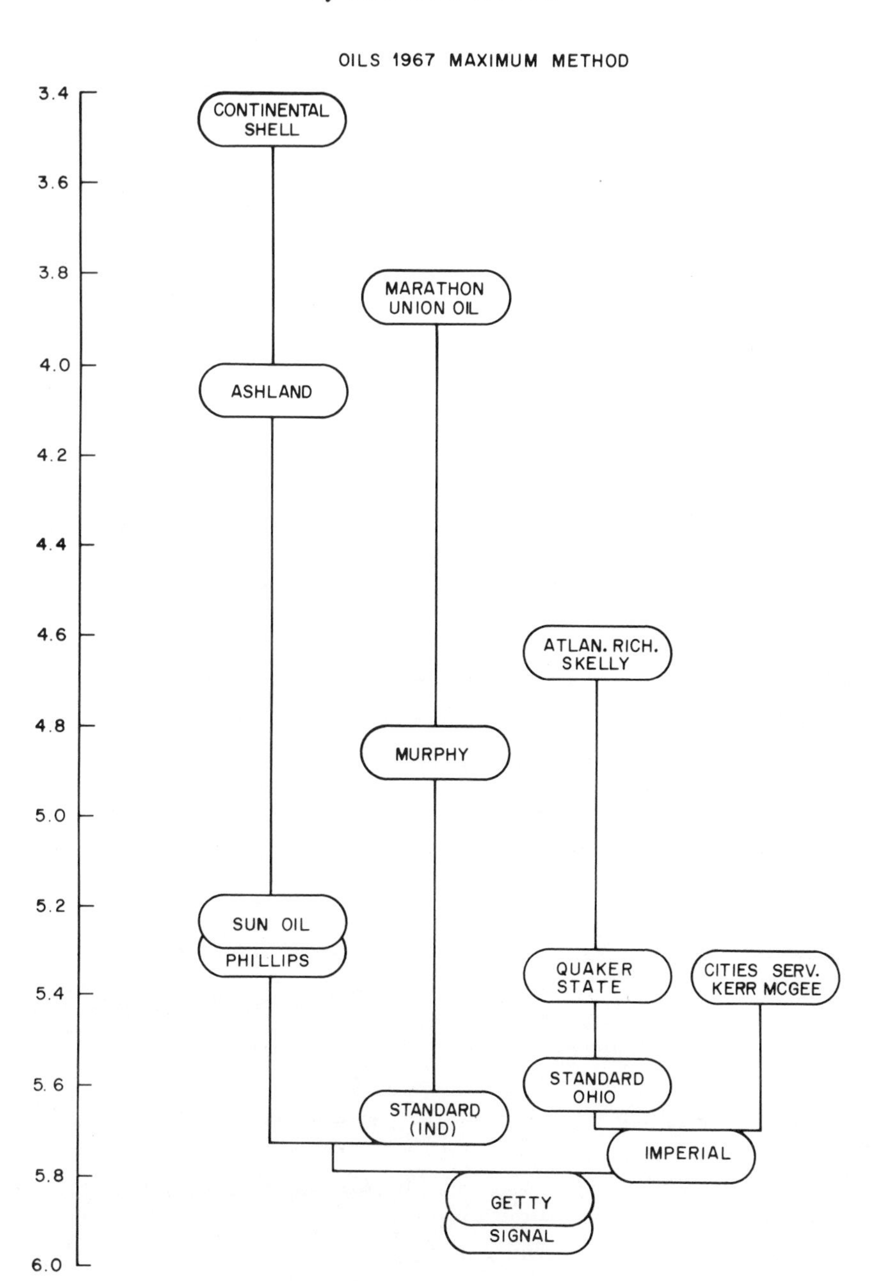

minimum, average, and maximum methods. [*Note*; The type of representation used in these figures is different from the one in Exhibits 21*a* and *b*. Instead of a listing of all the objects at the top, with lines emanating downward from them being joined together by horizontal lines at the different clustering levels, the representation in Exhibits 22*a*–*c* shows the companies by their names as the clusters are formed. Thus, in Exhibit 22*a*, the first cluster to form at a clustering strength of about 3.4 consists of two oil companies, Continental and Shell, which are next joined by Ashland, then merged with a cluster consisting of Marathon and Union Oil, and so on. The scale on the left depicting the values of the strength of clustering is the same for all three figures.]

A striking feature of Exhibits 22*a*–*c* is that, in this example, the minimum method exhibits its characteristic tendency of stringing out the clusters, bringing the companies in one at a time, whereas the maximum method appears to form several compact clusters (see the four branches in Exhibit 22*c*), which come together quite late in the hierarchical clustering, and the average method is intermediary between the minimum and maximum methods. Except for this feature, however, the general indications of the clustering from the three methods are not markedly different in terms of which companies appear to be together and of the strength of clustering at which a particular company is brought into an existent cluster.

4.3.2. Nonhierarchical Clustering Procedures

The nonhierarchical method called ISODATA (Iterative Self-Organizing Data Analysis Techniques A), proposed by Ball & Hall (1965), will be described in this section. The input to the procedure consists of the n p-dimensional observations, $\mathbf{y}_i$, $i = 1, 2, \ldots, n$, on the n units to be clustered. The method also requires an initial specification of the number, k $(<n)$, of clusters desired and a set of so-called cluster points in p-dimensions, $\mathbf{x}_1, \mathbf{x}_2, \ldots, \mathbf{x}_k$.

The first stage of the process is referred to as *sorting* and consists of assigning each of the n units to one of the k clusters, $C_1, C_2, \ldots, C_k$, by using the criterion of closeness (as measured by Euclidean distance) of the observation on the unit to the cluster point. Thus the ith unit is assigned to cluster C_ℓ if

$$(\mathbf{y}_i - \mathbf{x}_\ell)'(\mathbf{y}_i - \mathbf{x}_\ell) = \min_{a=1,\ldots,k} (\mathbf{y}_i - \mathbf{x}_a)'(\mathbf{y}_i - \mathbf{x}_a).$$

If cluster C_r consists of n_r units with corresponding observations denoted, $\mathbf{y}_{rs}(r = 1, \ldots, k;\ s = 1, \ldots, n_r;\ \sum n_r = n)$, then in the sorting stage all clusters with no units assigned to them are discarded, and for the

$\ell\,(\leq k)$ remaining clusters the initially specified cluster points are replaced by the mean vectors, $\bar{\mathbf{y}}_r$ (centroids), of the observations within each of the clusters. Thus the accomplishment of the sorting phase of ISODATA is to form preliminary clusters and to utilize them to define more appropriate cluster points or typical values of the clusters.

At the next stage additional cluster statistics are computed. Specifically, the following statistics are calculated: (i) for each cluster C_r, the average distance of points from the cluster centroid, i.e.,

$$\bar{d}_r = \frac{1}{n_r} \sum_{s=1}^{n_r} d(\mathbf{y}_{rs}, \bar{\mathbf{y}}_r),$$

where $d(\mathbf{x}, \mathbf{z})$ denotes a defined measure of distance between the points $\mathbf{x}$ and $\mathbf{z}$; (ii) for each cluster C_r, the $p \times p$ covariance matrix,

$$\mathbf{S}_r = \frac{1}{n_r - 1} \sum_{s=1}^{n_r} (\mathbf{y}_{rs} - \bar{\mathbf{y}}_r)(\mathbf{y}_{rs} - \bar{\mathbf{y}}_r)';$$

and (iii) the average intracluster distance across all clusters, i.e.,

$$\bar{d} = \frac{1}{n} \sum_{r=1}^{\ell} n_r \bar{d}_r.$$

The statistics computed under (i) and (iii) are of descriptive value in that they provide measures of "tightness" of the clusters. The covariance matrix, $\mathbf{S}_r$, is used for basing decisions pertaining to the formation of new clusters at the next stage.

The third stage of the ISODATA scheme is the formation of new clusters by *splitting* apart, or *lumping* together, existent clusters. Roughly speaking, if one has too few clusters, splitting will be desirable, if there are too many clusters, lumping will be more appropriate. The splitting or lumping is actually accomplished by comparing certain cluster properties against user-specified values (benchmarks) of two parameters, θ_S and θ_L, called, respectively, the splitting and lumping parameters. Specifically, (*a*) if the maximum coordinate variance in a cluster C_ℓ exceeds the specified value of θ_S, i.e., if the largest diagonal element of $\mathbf{S}_\ell > \theta_S$, then C_ℓ is split along that coordinate into two new clusters; or (*b*) if the variance of the first principal component within a cluster C_ℓ exceeds the specified θ_S, i.e., if the largest eigenvalue of $S_\ell > \theta_S$, then C_ℓ is split along the direction of the first principal component into two new clusters. The user has a choice between the two methods of splitting. The decision to lump two clusters, C_r and C_s, is based on comparing the distance between the two cluster points, $\bar{\mathbf{y}}_r$ and $\bar{\mathbf{y}}_s$, with the specified value of the lumping parameter, θ_L. If the distance, $d(\bar{\mathbf{y}}_r, \bar{\mathbf{y}}_s)$, is smaller than the value of θ_L, C_r and C_s are combined into a single cluster whose centroid is then computed and used as the cluster point of the merged cluster.

The remaining step of the ISODATA process is to iterate the three above-mentioned stages. In the early iterations the method tends to alternate between splitting and lumping, but in the later iterations a comparison of the number of clusters found at the end of a given iteration with the initially desired number, k, of clusters also influences the decision to split or lump clusters in the next iteration. The number of iterations is also a specification under the user's control. In fact, the process will terminate at the end of the specified number of iterations, and the number of clusters found may not be exactly k, the initially desired number. Figure 7 shows a summary flowchart of the steps involved in one computer implementation of the ISODATA procedure (see Warner, 1968).

Although ISODATA is nonhierarchical in that it allows splitting (in addition to lumping) of existent clusters to form new ones, because it starts off with all units in a single cluster and then alternates between splitting and lumping there is a tendency in the process to impose a loose tree (hierarchical) structure on the clusters.

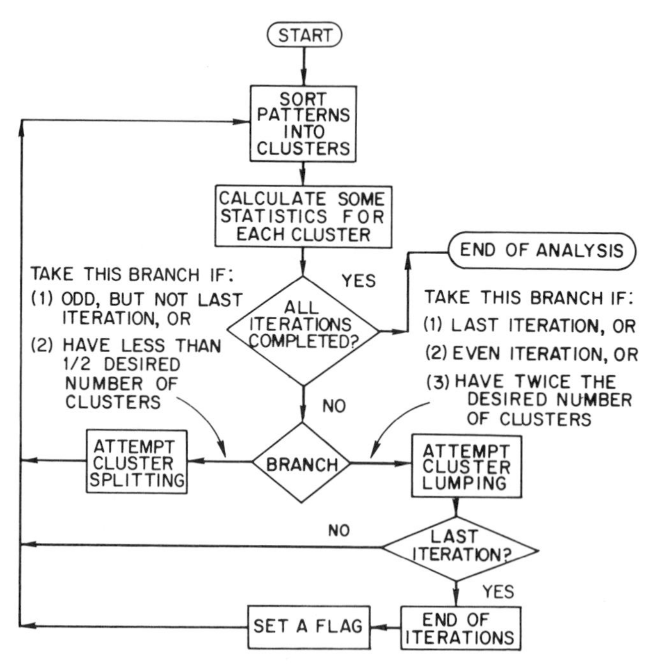

Fig. 7. Flowchart of steps in ISODATA algorithm.

The algorithm requires the user to specify several things (the number of clusters desired, the number of iterations, the initial values of cluster points, and the values of θ_S and θ_L), and, in the present state of the art, such specifications tend to be arrived at by a trial and error process. Little is known regarding the statistical-inferential aspects of cluster analysis techniques, and no general guidance or simple data-dependent method is available for choosing particular values of the quantities that need to be specified. All that one can say is that the number of clusters desired may not be a critical specification in that it is not guaranteed anyway and that ISODATA appears to be reasonably robust to the initial choice of cluster points provided that the number of iterations specified is not inadequate. This robustness implies that, if the user does not wish to specify the initial cluster points, the use of default values will probably not excessively distort the final results.

Since the splitting methods in ISODATA are based on variances (either of original variables or of the first principal component), and since the Euclidean metric is employed as the basis for measuring closeness, the results are scale dependent in that different scalings of the initial variables can in general lead to different clusterings of the objects. The nonhierarchical clustering method proposed by Friedman & Rubin (1967) uses eigenvalues and eigenvectors of a $\mathbf{W}^{-1}\mathbf{B}$ type of matrix ($\mathbf{W}$ would be the pooled within-clusters covariance matrix, and $\mathbf{B}$ the between-clusters covariance matrix for a given set of clusters) for basing the decisions regarding splitting and lumping. Although computationally more involved and expensive, this method does, of course, have the property that a solution obtained by its use will be invariant under all affine transformations of the original set of variables, including simple scaling of each of them separately.

ISODATA seems to be similar to the so-called k-means method investigated by other authors (e.g., MacQueen, 1965), the main difference being in the cluster splitting. Whereas ISODATA bases splitting on intracluster dispersion characteristics, in the k-means method the creation of new clusters depends on the distance of the proposed new cluster centroid from the nearest existent cluster centroid exceeding some prespecified value of a splitting parameter. MacQueen (1965) has established some asymptotic properties of the k-means method, and, in view of the similarities between the methods, it would be interesting to ask whether these properties carry over to ISODATA. As mentioned earlier, however, useful inferential tools (e.g., statistical tests to aid in detecting real clusters) are needed, and this aspect of cluster analysis is still in its infancy (see Ling, 1971).

REFERENCES

Section 4.2 Becker et al. (1965), Bricker et al. (1971), Chambers (1974), Chen et al. (1970, 1974), Fisher (1936, 1938), Hotelling (1947), Jackson (1956), Rao (1952), Seal (1964).

Section 4.2.1 Anderson (1958), Becker et al. (1965), Chen et al. (1970, 1974), Rao (1952).

Section 4.2.2 Bricker et al. (1971).

Section 4.2.3 Burnaby (1966), Rao (1966).

Section 4.3 Ball & Hall (1965), Cormack (1971), Everitt (1974), Friedman & Rubin (1967), Gower (1967), Hartigan (1967, 1975), Johnson (1967), Neely (1967), Scott & Symons (1971), Sokal & Sneath (1963).

Section 4.3.1 Johnson (1967), Miller & Nicely (1955), Sneath (1957), Sørensen (1948).

Section 4.3.2 Ball & Hall (1965), Friedman & Rubin (1967), Ling (1971), Mac Queen (1965), Warner (1968).

CHAPTER 5

Assessment of Specific Aspects of Multivariate Statistical Models

5.1. GENERAL

The major portion of formal multivariate statistical theory has been directed toward the assessment of specific aspects of an assumed (and often unverified) mathematicostatistical model; i.e., one assumes a model and then is concerned with formal statistical inferences about particular aspects of it. Examples include theories and methods of estimating multivariate parameters, and tests of hypotheses concerning location and/or dispersion parameters under either a multivariate normal or a more general model.

The assessment of statistical models is a legitimate and important concern of data analysis. Typically, however, the multivariate procedures for assessment have been developed by mathematical analogy with corresponding univariate methods, and often it is not clear that the complex aspects of the multivariate problem are incorporated in such an analogy or are better appreciated or otherwise benefit from it. For instance, more varied departures from a null hypothesis are possible in a multivariate situation than in the "analogous" univariate problem, and having a test per se of the null hypothesis against a completely general alternative is not of much value for multiresponse data analysis.

The standard results, pertaining to assessments of tightly posed questions in the framework of statistical models, are well organized and easily accessible in the multivariate literature (see, e.g., Anderson, 1958; Puri & Sen, 1971; Rao, 1965; Roy, 1957). Therefore the present chapter provides no more than a cursory review of some of the standard results, whereas it concentrates on some relatively recent work and covers it in greater detail (see Sections 5.2.3, 5.3 and 5.4).

5.2. ESTIMATION AND TESTS OF HYPOTHESES

The classical multivariate theory has been based largely on a multivariate normal distributional assumption. One consequence of this has been the concentration of almost all of the work on just location and dispersion parameters, with relatively little attention paid to questions of shape and other high-order characteristics.

5.2.1. Location and Regression Parameters

The so-called multivariate general linear model, mentioned toward the end of Chapter 3, has been the focus of much of the developments. Specifically, the model, which was defined earlier in Eqs. 49 and 50, is a simultaneous statement of p univariate general linear models:

$$\mathbf{Y}_j = \mathbf{X}\boldsymbol{\theta}_j + \boldsymbol{\varepsilon}_j \qquad \text{for } j = 1, \ldots, p, \tag{61}$$

where $\mathbf{Y}_j$ is the vector of n observations on the jth response, $\mathbf{X}$ is the $n \times k$ matrix of known values of design and/or regressor variables, $\boldsymbol{\theta}_j$ is a $k \times 1$ vector of unknown parameters (treatment effects or regression coefficients) associated with the jth response, and $\boldsymbol{\varepsilon}_j$ is a vector of n random errors associated with the observations on the jth response. The multivariate nature of the formulation is introduced by assuming that the p corresponding elements of the vectors $\boldsymbol{\varepsilon}_1, \boldsymbol{\varepsilon}_2, \ldots, \boldsymbol{\varepsilon}_p$ are not necessarily statistically independent. The usual assumptions, in fact, are that each of the n sets of p elements has a mean (expected) value $\mathbf{0}$ and a common $p \times p$ covariance matrix $\boldsymbol{\Sigma}$, which is generally unknown and has to be estimated.

This general statement of the model subsumes both the multivariate multiple regression case and the more standard multivariate designed experiment situation, although it does not include any of the nonstandard multivariate experimental designs. The term "standard" is used to denote the case in which the design does not vary for the different responses; i.e., one of the familiar univariate designs is used, but on each experimental unit p responses are observed simultaneously. In the multivariate multiple regression case it is generally assumed that the matrix $\mathbf{X}$ is of "full" rank k $(<n)$, whereas in the usual experimental design situations $\mathbf{X}$ will be a design matrix of rank r $(<k<n)$. In the latter case, however, one can reparametrize the problem in terms of a set of r parameters that are linear functions of the θ's and rewrite the general linear model in terms of the derived parameters and a new $n \times r$ design matrix of "full" rank r (see, e.g., Scheffé, 1959, pp. 15–16). Hence, for present purposes, it is assumed that the matrix $\mathbf{X}$ is of full rank k so that $\mathbf{X}'\mathbf{X}$, in particular, will be a nonsingular matrix.

A useful device in the formal treatment of the estimation problem is the so-called rolled-out version of the model stated in Eq. 61 (see Section 4.a of Chapter III in Roy et al., 1971). By stringing out all the elements of the vector $\mathbf{Y}_1$, followed by those of $\mathbf{Y}_2$, and so on, one can obtain an np-dimensional column vector, $\mathbf{y}^*$, for which the following linear model is a consequence of Eq. 61:

$$\mathbf{y}^* = \mathbf{X}^*\boldsymbol{\theta}^* + \boldsymbol{\varepsilon}^*, \tag{62}$$

where (i) $\boldsymbol{\theta}^*$ and $\boldsymbol{\varepsilon}^*$ are, respectively, kp- and np-dimensional column vectors obtained by rolling out the $\boldsymbol{\theta}_j$'s and $\boldsymbol{\varepsilon}_j$'s of Eq. 61 in exactly the same manner as the $\mathbf{Y}_j$'s, (ii) the $np \times kp$ matrix, $\mathbf{X}^*$, can be written compactly as the Kronecker product $\mathbf{I}(p) \otimes \mathbf{X}$, and (iii) the covariance structure for the elements of $\boldsymbol{\varepsilon}^*$ is the $np \times np$ matrix, $\boldsymbol{\Sigma}^*$, which is the Kronecker product $\boldsymbol{\Sigma} \otimes \mathbf{I}(n)$. Then, if ξ is a linear function of the elements of $\boldsymbol{\theta}^*$, i.e., $\xi = \mathbf{c}'\boldsymbol{\theta}^* = \mathbf{c}_1'\boldsymbol{\theta}_1 + \mathbf{c}_2'\boldsymbol{\theta}_2 + \cdots + \mathbf{c}_p'\boldsymbol{\theta}_p$, it can be established (see Section 4.a of Chapter III of Roy et al., 1971) that the minimum variance unbiased linear estimate of ξ is $\hat{\xi} = \mathbf{c}_1'\hat{\boldsymbol{\theta}}_1 + \mathbf{c}_2'\hat{\boldsymbol{\theta}}_2 + \cdots + \mathbf{c}_p'\hat{\boldsymbol{\theta}}_p$, where $\mathbf{c}_j'\hat{\boldsymbol{\theta}}_j$ is the least squares estimate of $\mathbf{c}_j'\boldsymbol{\theta}_j$, obtained by considering only the jth response and ignoring the rest of the variables. Specifically, the unknown covariance matrix $\boldsymbol{\Sigma}$ is not involved in computing $\hat{\xi}$, although the variance of $\hat{\xi}$ would involve $\boldsymbol{\Sigma}$.

This formal result, which is derived within the framework of linear models and the method of least squares, is perhaps one reason for the well-worn and widespread practice of constructing multivariate location estimates simply by assembling together univariate location estimates which have themselves been obtained by separate univariate analyses that ignore the multivariate nature of the data. In particular, as mentioned in the discussion following Eqs. 49 and 50 in Chapter 3, the classical result on linear estimation under the multivariate general linear model is that the estimate of $\boldsymbol{\Theta}$ is

$$\hat{\boldsymbol{\Theta}} = [\hat{\boldsymbol{\theta}}_1, \hat{\boldsymbol{\theta}}_2, \ldots, \hat{\boldsymbol{\theta}}_p]; \qquad \hat{\boldsymbol{\theta}}_j = (\mathbf{X}'\mathbf{X})^{-1}\mathbf{X}'\mathbf{Y}_j \qquad \text{for } j = 1, \ldots, p. \tag{63}$$

For the simple case of an unstructured (or single) sample, the matrix $\mathbf{X}$ would just be a column vector of 1's; $\boldsymbol{\Theta}$ would be the p-dimensional row vector, $\boldsymbol{\mu}'$, the unknown mean vector of the p-variate population from which the sample is presumed to be drawn; and Eq. 63 would yield as an estimate of $\boldsymbol{\Theta}$ the sample mean vector, $\bar{\mathbf{y}}'$, defined in Eq. 1 in Section 2.2.1. An interesting theoretical sidelight, the critical importance of which for analyzing multivariate data is unclear, is the result due to Stein (1956) on the inadmissibility of the sample mean vector as an estimator of $\boldsymbol{\mu}'$ when $p \geq 3$. From the viewpoint of data analysis, a far more significant objection to estimators, such as the sample mean vector, associated with the least squares criterion is their nonrobustness or susceptibility to the influence of a few outliers (see Section 5.2.3 for a discussion of robust estimators).

In regard to the problem of testing linear hypotheses concerning the parameters $\mathbf{\Theta}$, for the classical normal theory treatment of the multiresponse general linear model of Eq. 49 it is assumed that the rows of $\boldsymbol{\varepsilon}$ are p-variate normally distributed (or, equivalently, that in the rolled-out version given in Eq. 62 $\boldsymbol{\varepsilon}^*$ is np-variate normal with mean $\mathbf{0}$ and covariance structure $\mathbf{\Sigma}^*$). Under this additional distributional assumption, the usual null hypothesis considered is

$$H_0\colon \mathbf{C\Theta U}=\mathbf{O}, \tag{64}$$

where the $s\times k$ matrix, $\mathbf{C}$ ($s\le k$), and the $p\times u$ matrix, $\mathbf{U}$ ($u\le p\le n-k$), are matrices of specified constants with ranks s and u, respectively. For testing H_0 of Eq. 64 against the completely general alternative, H_1: $\mathbf{C\Theta U}\ne\mathbf{O}$, several statistics have been proposed in the literature (see, e.g., Anderson, 1958; Roy et al., 1971). The test statistics are different functions of the eigenvalues of the matrix $\mathbf{S}_h\mathbf{S}_e^{-1}$, where

$$\left.\begin{aligned} \mathbf{S}_h &= \mathbf{U'YX(X'X)^{-1}C'[C(X'X)^{-1}C']^{-1}C(X'X)^{-1}X'Y'U}, \\ \text{and}\qquad & \\ \mathbf{S}_e &= \mathbf{U'Y[I}(n)-\mathbf{X(X'X)^{-1}X']Y'U}. \end{aligned}\right\} \tag{65}$$

For instance, if $\infty>c_1\ge c_2\ge\cdots\ge c_t>0$ denote the t [$=\min(u, s)$] ordered positive eigenvalues of $\mathbf{S}_h\mathbf{S}_e^{-1}$, the likelihood ratio test of H_0 is based on the statistic $\Lambda=\prod_{j=1}^{t} 1/(1+c_j)$, whereas the so-called largest-root test proposed by Roy (1953) is based on c_1 and the sum-of-the-roots test on $\sum_{j=1}^{t} c_j$. Only when at least one of the quantities u and s equals 1 are the different tests entirely equivalent, and in situations where this condition does not apply the application of the different tests may indeed lead to different conclusions regarding the tenability of the null hypothesis. A more detailed discussion of tests of hypotheses (including not only the general null hypothesis of Eq. 64 but also more specialized hypotheses) and of formal power properties of such tests will be found in Chapters IV and V of Roy et al. (1971). (See also the work of Pillai and his students, e.g., Pillai and Jayachandran, 1967, on power comparisons of the tests.) For present purposes it suffices to say that these tests of hypotheses tend to be of very limited value in multivariate data analysis, especially of an exploratory nature, when tightly specified models are either unavailable or unreasonable to commit oneself to.

A geometrical description may help to elucidate the concepts and methods associated with the general linear model stated in Eq. 61. One can think of the n observations on each of the p responses, $\mathbf{Y}_1, \ldots, \mathbf{Y}_p$, as the coordinates of p points, $P_1, P_2, \ldots, P_p$, in n-dimensional space (although, as stated in Chapter 1, the usual view is to consider the multiresponse observations as n points in p-space). Let O denote the origin in the n-space. Then the least squares estimate, $\hat{\boldsymbol{\theta}}_j$, from the univariate analysis of the jth response is known (see, e.g., Scheffé, 1959) to be associated with the projection of P_j onto the k-dimensional subspace of n-space spanned by the k columns of $\mathbf{X}$, and the matrix $\mathbf{X}(\mathbf{X}'\mathbf{X})^{-1}\mathbf{X}'$, which relates $\mathbf{Y}_j$ to its projection, is called the projection matrix involved in the operation of obtaining the least squares estimate. In the multivariate linear model which assumes a common design or regression structure, $\mathbf{X}$, for all the responses, $\hat{\boldsymbol{\Theta}}$ of Eq. 63 is then associated with the set of p *projections of* $P_1, \ldots, P_p$ onto the same space, and the same projection operator is applied to each of the p points.

In more specialized situations the matrix $\mathbf{X}$ may derive from an orthogonal design (e.g., a balanced multifactorial experiment), and then its columns will specify a decomposition of the n-space into mutually orthogonal linear subspaces, each associated with meaningful sources of variation (e.g., blocks, main effects, error) incorporated in the experimental design. In such situations a univariate analysis of variance of the observations on a single response, $\mathbf{Y}_j$, for instance, will yield, for the particular variable, a decomposition of the total sum of squares into sums of squares associated with each of the meaningful orthogonal sources of variation, and this process can be viewed geometrically as decomposing (á la Pythagoras' theorem) the squared length of the vector OP_j in terms of the squared lengths of the vectors joining O to the projections of P_j onto the different mutually orthogonal linear subspaces. Similarly, in orthogonal multivariate analyses of variance, the decomposition of the total sum-of-products matrices, associated with orthogonal sources of variation underlying the experiment, may be visualized as a decomposition of the squared lengths of and angles between the p vectors, $OP_1, OP_2, \ldots, OP_p$, in terms of the squared lengths of and the angles between the vectors joining O to the projections of $P_1, \ldots, P_p$ onto the different mutually orthogonal linear subspaces. Sum-of-products matrices associated with tests of hypotheses, such as $\mathbf{S}_h$ in Eq. 65, may be given similar interpretations in these geometrical terms.

In addition to point estimation and tests of hypotheses, under the normal theory various confidence regions and interval estimation schemes have also been proposed (see, e.g., Chapter VI of Roy et al., 1971). For instance, a set of simultaneous confidence intervals can be obtained for bilinear functions of $\boldsymbol{\Theta}$, and these are multiresponse analogues of the univariate result due to Scheffé (1953) and to Roy and Bose (1953).

With regard to carrying out the computations involved in multiresponse multiple regression or analysis of variance, although the algebraic representations in some of the formulae have involved expressions such as inverses of certain matrices (e.g., Eq. 63), they are not intended to suggest that it would be appropriate to develop computational algorithms directly from them. In fact, the recent literature in statistical computing is rich in its emphasis on avoiding not only pitfalls in inverting matrices but also round-off errors in forming sum-of-products matrices. Approaches involving different types of matrix decompositions (e.g., singular value decompositions, Givens rotations) are the ones recommended currently, and Chambers (1974), Fowlkes and Lee (Appendix C in Roy et al., 1971), Golub and Reinsch (1970), and Wilkinson (1970) are a few of the relevant references.

5.2.2. Dispersion Parameters

A familiar example of estimation of dispersion is the estimation of the unknown covariance matrix $\mathbf{\Sigma}$ that specifies the error structure in the multiresponse general linear model stated in Eq. 49 or 62. An unbiased estimate of $\mathbf{\Sigma}$ is

$$\mathbf{S}_{\text{error}} = \frac{1}{n-k}\mathbf{Y}[\mathbf{I}(n) - \mathbf{X}(\mathbf{X}'\mathbf{X})^{-1}\mathbf{X}']\mathbf{Y}', \tag{66}$$

while the (biased) maximum likelihood estimate is

$$\hat{\mathbf{\Sigma}} = \frac{n-k}{n}\mathbf{S}_{\text{error}}.$$

With $p \le (n-k)$, $\mathbf{S}_{\text{error}}$ will be nonsingular with probability 1.

In the special case of an unstructured sample, the expression in Eq. 66 simplifies to yield the sample covariance matrix $\mathbf{S}$, defined earlier in Eq. 2 (see Section 2.2.1). Some (see Lindley, 1972) have worried about the inadmissibility issue à la Stein (1956, 1965) regarding such estimates of dispersion, but once again the nonrobustness of these estimates is perhaps more worrisome for data-analytic purposes than the theoretically fascinating issue of inadmissibility. Section 5.2.3 discusses some preliminary ideas regarding the robust estimation of dispersion parameters.

The literature pertaining to the standard normal theory treatment of multivariate analysis contains much material on formal statistical inference procedures (including estimation, tests of hypotheses, and distribution theory) associated with covariance matrices. The results on tests of hypotheses range from methods for testing the equality of two or more covariance matrices to procedures for testing hypotheses concerning specific structures for covariance matrices (e.g., sphericity, mutual independence of subsets of variables) and concerning eigenvalues and eigenvectors of covariance matrices. Chapters 9–11 and 13 (especially Chapter

10) of Anderson (1958) and various chapters of Rao (1965) contain many of the available results. Some results on confidence bounds for dispersion parameters are given by Roy & Gnanadesikan (1957, 1962). Methods associated with the study of structured covariance matrices have been discussed by several authors, including Srivastava (1966), Anderson (1969), and Jöreskog (1973).

5.2.3. Robust Estimation of Location and Dispersion

For the uniresponse situation, problems and methods of robust estimation have received considerable attention (see, e.g., Tukey, 1960; Huber, 1964, 1972, 1973; Andrews et al., 1972). The univariate work has been largely concerned with the problem of obtaining and studying estimates of location that are insensitive to outliers, and the distributions considered as alternatives to the normal (for purposes of comparing the relative behaviors of the proposed estimates) have almost always been symmetrical with heavier tails than the normal. More recently, the robust estimation of uniresponse multiple regression parameters has been considered by Andrews (1973, 1974), Huber (1973), and Mallows (1973). Gnanadesikan and Kettenring (1972) have discussed some issues of and techniques for the robust estimation of multiresponse location and dispersion, and this section is based on their discussion and proposals. Although these methods are useful in protecting data summaries against certain kinds of outliers, it should be emphasized that the variety, both in kind and in effect, of outliers in multiresponse data can indeed be large, and the routine use of any robust estimate without exploring the data for the existence of specific peculiarities in them is neither wise nor necessary. (See Chapter 6 for a discussion of additional techniques for facilitating the exposure of peculiarities in data.)

The problem of robust estimation of multivariate location has received some attention in the literature (see Mood, 1941; Bickel, 1964; Gentleman, 1965). The usual estimate of multivariate location is the sample mean vector $\bar{\mathbf{y}}$, whose elements are just the uniresponse means, and except for the estimator proposed by Gentleman (1965) the multivariate robust estimators are just vectors of univariate robust estimators obtained by analyzing the observations on each of the response variables separately. Some possibilities for such robust estimators of multivariate location are the following:

1. $\mathbf{y}_M^*$, the vector of medians of the observations on each response, as suggested by Mood (1941).
2. $\mathbf{y}_{HL}^*$, the vector of Hodges-Lehmann estimators (i.e., the median of

averages of pairs of observations) for each response, as proposed and investigated by Bickel (1964).

3. $\mathbf{y}^*_{T(\alpha)}$, the vector of α-trimmed means (i.e., the mean of the data remaining after omitting a proportion, α, of the smallest and of the largest observations) for each response, as considered by Gnanadesikan and Kettenring (1972), or in a similar vein $\mathbf{y}^*_{W(\alpha)}$, the vector of α-Winsorized means.

4. $\mathbf{y}^*_{\text{SINE}}$, the vector of sine estimates for each response, with the sine estimate for location of the jth response defined, following Andrews et al. (1972), as the solution, T_j, of the equation

$$\sum_{i=1}^{n} \psi\left(\frac{y_{(i)j} - T_j}{s_j}\right) = 0,$$

where $y_{(1)j} \leq y_{(2)j} \leq \cdots \leq y_{(n)j}$ are the ordered observations on the jth response, s_j is a simple estimate of scale of the jth response (e.g., median of the absolute deviations of the observations about the location estimate, T_j), and ψ is defined by

$$\psi(x) = \begin{cases} \sin(x/2.1) & \text{for } |x| < 2.1\pi, \\ 0 & \text{otherwise.} \end{cases}$$

5. More generally, a vector each of whose elements is any univariate robust estimator (see Andrews et al., 1972, for a variety of possibilities) of the location for a single response variable.

Unlike the above estimators, each of which is just a collation into vector form of univariate estimators, the estimator proposed by Gentleman (1965) is multivariate in character in that the analysis involves a combined manipulation of the observations on the different responses. The essential idea is to choose the estimator, $\mathbf{y}^* = (y_1^*, y_2^*, \ldots, y_p^*)'$, of $\boldsymbol{\mu} = (\mu_1, \mu_2, \ldots, \mu_p)'$ so as to minimize the criterion, $\sum_{j=1}^{p} |y_j^* - \mu_j|^k$, for a specified value of k in the range $1 \leq k \leq 2$. For $k = 2$ the estimator is the usual sample mean vector, while for $k < 2$ one obtains an estimator less sensitive to possible outliers. When $k = 1$, the estimator turns out to be $\mathbf{y}^*_M$, the vector of medians defined above. For a general value of k between 1 and 2, a closed-form expression of the estimator is not available, but Gentleman (1965) describes a numerical algorithm which can be used to compute the estimator for any set of multiresponse data. Gentleman also discusses some issues of modifying the criterion of kth power deviations to reflect the existence, if any, of both differences in the scales of the responses and intercorrelations among the responses.

Another approach to specifying a location estimate which also involves considering the responses simultaneously is discussed briefly in Section 6.2 in connection with the uses of a high-dimensional plotting technique.

A theoretical issue, which has been raised by Bickel (1964) in connection with $\mathbf{y}_M^*$ and $\mathbf{y}_{HL}^*$ but in fact applies to all of the estimators mentioned above, is the lack of affine commutativity of the robust estimators of multivariate location, in contrast to the usual mean vector $\bar{\mathbf{y}}$, which does possess this property. (A location estimator is affine commutative if the operations of affine transformation and formation of the estimate can be interchanged without affecting the outcome.) From a practical viewpoint the issue may be viewed in terms of commitment to the coordinate system for the observations. At one extreme the interest may be confined entirely to the observed variables, and, if so, any issue of commutativity will perhaps be remote. At the other extreme one may feel that the location problem is intrinsically affine commutative (e.g., one may wish to require that when one works with metric and nonmetric scales the effect of transforming from one scale to the other and then computing the robust location estimate be the same as directly transforming the robust estimate on the former scale) and insist that all location estimators have this property. As an intermediate position one may seek more limited commutativity (e.g., just linear transformations of each variable separately as in the above metric-nonmetric example, or just orthogonal transformations) than the very general affine commutativity.

In regard to the robust estimation of multivariate dispersion, there are at least two aspects of the problem, viz., the facet that depends on the scales (i.e., variances) of the responses and the one that is concerned with orientation (i.e., intercorrelations among the responses). For some purposes it may be desirable to consider the robust estimation of each of these aspects separately, whereas for other purposes a combined view may be in order.

An approach that separates the two aspects has the advantage of using all the available and relevant information for each estimation task, whereas an approach that combines the two will involve retaining only observations (perhaps fewer in number) which pertain to both aspects simultaneously. On the other hand, in many cases the combined approach may be computationally simpler and more economical.

The problem of robust estimation of the variance of a univariate population has been considered (see Tukey, 1960; Johnson and Leone, 1964, Section 6.9; Hampel, 1968), although not as intensively or extensively as the location case. When one leaves the location case, certain conflicting aims seem to emerge in estimating higher-order characteristics of a distribution. Thus for the variance (and maybe even more so for the shape) there is a possible conflict between the desire to protect the estimate from outliers and the fact that the information for estimating the variance relies more heavily on the tails.

This conflict raises certain questions about the routine use of robust estimation procedures for these higher-order characteristics, especially in relatively small samples. Thus, with a sample of size 10, for instance, the use of a 10% (the minimum possible in this case) trimmed sample to provide a robust estimate may lead to an estimator whose efficiency is unacceptably and unnecessarily low when the data are reasonably well behaved. The main point is that, with relatively small, and yet reasonable, samples sizes, it may be both expedient and wise to study the observations more closely, omitting only clearly indicated outliers (see Chapter 6 for outlier-detection methods) or possibly transforming the observations to make them more nearly normally distributed (see Section 5.3).

The usual unbiased estimator of the variance for the jth response ($j = 1, \ldots, p$) based on n observations may be denoted as s_{jj}, and a corresponding robust estimator, s_{jj}^*, may be developed by any of the following three methods:

1. Trimmed variance from an α-trimmed sample, as suggested by Tukey (1960) and further studied by Hampel (1968).
2. Winsorized variance from an α-Winsorized sample, as suggested by Tukey and McLaughlin (1963).
3. The slope of the lower end of a $\chi^2_{(1)}$ probability plot (see Section 6.2 for a brief discussion of probability plots) of the $n(n-1)/2$ squared differences between pairs of observations.

The first two methods need an estimate of location, and a direct suggestion would be to use a trimmed mean for the trimmed variance and a Winsorized mean for the Winsorized variance. Huber (1970), however, suggests using a trimmed mean for getting the Winsorized variance, and for t-statistic types of considerations associated with the trimmed mean this may be appropriate. But even for applying a trimmed mean for the trimmed variance, or a Winsorized mean for the Winsorized variance, because of the considerations mentioned above it would seem advisable to use a smaller proportion of trimming (or Winsorizing) for the variance estimation than for the location estimation in samples even as large as 20.

To obtain unbiased, or even consistent, estimates from a trimmed or Winsorized variance, multiplicative constants are needed. These constants are based on moments of order statistics of the normal distribution, and an underlying assumption in using these constants is that the "middle" of the sample is sufficiently normal. Johnson and Leone (1964, p. 173) give a table of the required constants for small ($n \leq 15$) samples, and tables provided by McLaughlin and Tukey (1961), together with the tabulation by Teichroew (1956) of the expected values of cross products and squares of normal order statistics, may be used for calculating the required constant for samples of sizes up to 20. Unfortunately, asymptotic results do not appear to be adequate at $n=20$, and further work is needed on developing the required multiplicative constant for larger values of n.

One advantage of the third method mentioned above is that it does not involve an estimate of location. A second is that the type of adjustment provided by the multiplicative constant in the trimmed and Winsorized variances is contained in the probability plot itself—viz., the abscissa (or quantile axis) is used to scale the ordinate for determining the slope (which will be an estimate of twice the variance). A third advantage is that, by looking at $n(n-1)/2$ pieces of information (some of which may be redundant because of statistical correlations), the error configuration on the $\chi^2_{(1)}$ probability plot may often be indicated more stably than on a normal probability plot of the n observations. A fourth, and perhaps the most significant, advantage of the approach is its exposure value in facilitating the detection of unanticipated peculiarities in the data. On the negative side a disadvantage of the technique is that it may not be useful, and may even be misleading, for estimating the variance in circumstances where a large proportion of the observations may be outliers.

The multivariate nature of dispersion is introduced inevitably by considering the estimation of covariance and correlation. A simple idea for estimating the covariance between two variables, Y_1 and Y_2, is based on the identity

$$\text{cov}(Y_1, Y_2) = \tfrac{1}{4}\{\text{var}(Y_1+Y_2) - \text{var}(Y_1-Y_2)\}. \tag{67}$$

One robust estimator, s^*_{12}, of the covariance between Y_1 and Y_2 may, therefore, be obtained from

$$s^*_{12} = \tfrac{1}{4}\{\hat{\sigma}^{*2}_1 - \hat{\sigma}^{*2}_2\}, \tag{68}$$

where $\hat{\sigma}^{*2}_1$ and $\hat{\sigma}^{*2}_2$ are robust estimators of the variances of (Y_1+Y_2) and (Y_1-Y_2), respectively, and may be obtained by any of the methods mentioned above.

When such a robust estimator of the covariance is available, a natural way of defining a corresponding robust estimator of the correlation coefficient between Y_1 and Y_2 is

$$r^*_{12} = \frac{s^*_{12}}{\{s^*_{11}s^*_{22}\}^{1/2}}, \tag{69}$$

where s^*_{jj} is a robust estimator of the variance of the jth response.

Since the robust estimators involved in Eqs. 68 and 69 are determined with no considerations of satisfying the well-known Cauchy-Schwarz inequality relationship between the covariance and the variances, therefore, r^*_{12} as obtained from Eq. 69 may not necessarily lie in the admissible range, [−1, +1], for a correlation coefficient. To ensure an estimate of the correlation coefficient in the valid range, while still retaining the above approach of obtaining the covariance estimate as the difference between two variance estimates, a modification may be suggested. Let $Z_j = Y_j/\sqrt{s_{jj}^*}$ denote the "standardized" form of Y_j, where s_{jj}^* is a robust estimate of the variance of Y_j. Then define

$$\hat{\rho}_{12}^* = \frac{\hat{\sigma}_3^{*2} - \hat{\sigma}_4^{*2}}{\hat{\sigma}_3^{*2} + \hat{\sigma}_4^{*2}}, \tag{70}$$

where now $\hat{\sigma}_3^{*2}$ and $\hat{\sigma}_4{}^{*2}$ are robust estimators of the variances of $(Z_1 + Z_2)$ and $(Z_1 - Z_2)$, respectively. One can use any robust estimate of the variances of the standardized sum and difference, but Devlin et al. (1975) have studied the use of trimmed variances in particular and have denoted by r^*(SSD) the associated $\hat{\rho}_{12}^*$. Corresponding to $\hat{\rho}_{12}^*$, which necessarily lies in the range [−1, +1], a covariance estimator may be defined by

$$\hat{\sigma}_{12}^* = \hat{\rho}_{12}^* \{s_{11}^* s_{22}^*\}^{1/2}. \tag{71}$$

An interesting consequence of estimating the correlation coefficient by Eq. 69 or 70 is that the multiplicative constant, which is required for removing the biases involved in trimmed or Winsorized variances, cancels out by appearing in both the numerator and the denominator of the defining equations 69 and 70. Hence, for any sample size, the trimmed (or Winsorized) variances that provide the bases for obtaining r_{12}^* and $\hat{\rho}_{12}^*$ can be used directly without any multiplicative constant. This does not, however, imply that r_{12}^* and $\hat{\rho}_{12}^*$ are unbiased estimators of the population correlation. In fact, just as the usual product moment correlation coefficient is biased, these robust estimates are biased in small (but not large) samples. Devlin et al. (1975) have studied the biases and efficiencies of the above-mentioned as well as other robust estimators of correlation, including some well-known nonparametric estimators such as Kendall's τ.

A full-fledged consideration of multiresponse dispersion would be necessary if one were interested in the estimation of not just a single covariance or correlation but a collection of these, say a covariance or a correlation matrix. The usual estimates of the covariance and correlation matrices are, respectively, the sample covariance matrix, **S**, and the associated sample correlation matrix, **R** (see defining Eqs. 2 and 3 in Section 2.2.1). When robust estimates of the variances and covariances have been obtained by the methods discussed above, a direct method of obtaining a robust estimate of the covariance matrix is just to "put these together" in a matrix. Thus, corresponding to each of the two methods described above (see Eqs. 69 and 70) for obtaining an estimate of the correlation coefficient, a robust estimate of the covariance matrix would be

$$\mathbf{S}_a^* = \mathbf{D}\mathbf{R}_a^*\mathbf{D} \qquad \text{for } a = 1, 2, \tag{72}$$

where $\mathbf{D}$ is a diagonal matrix with diagonal elements $\sqrt{s_{jj}^*}$ $(j = 1, \ldots, p)$, $\mathbf{R}_1^* = ((r_{jj'}^*))$, and $\mathbf{R}_2^* = ((\hat{\rho}_{jj'}^*))$.

For some purposes of analyzing the multiresponse data, when the underlying distribution is not singular, it may be desirable to have a positive definite estimate of the covariance matrix. For instance, in analyzing the configuration of the sample in terms of the generalized squared distance of the observations from the sample centroid (see Example 7 in Section 2.4), the inverse of the estimate of the covariance matrix is used.

If the dimensionality, p, does not exceed the number of independent observations [$(n-1)$ in the case of an unstructured sample], the usual estimator, $\mathbf{S}$, is positive definite with probability 1. However, neither of the estimators, $\mathbf{S}_1^*$ and $\mathbf{S}_2^*$, defined above is necessarily positive definite. The positive definiteness of these estimators is equivalent to the positive definiteness of the corresponding estimators, $\mathbf{R}_1^*$ and $\mathbf{R}_2^*$, of the correlation matrix, and even though each off-diagonal element of $\mathbf{R}_2^*$ necessarily lies in the range $[-1, +1]$, this does not necessarily imply positive definiteness of $\mathbf{R}_2^*$, except for the bivariate case. [*Note*: Positive definiteness of a correlation matrix may be conceptualized as a high-dimensional analogue of the property that a single correlation coefficient lies between -1 and $+1$, and the constraint of positive definiteness seems to introduce the need to consider all the responses simultaneously with respect to their dispersion or orientational summary, although superficially such a summary might appear to be based only on a pairwise consideration of the responses.] Devlin et al. (1975) suggest a way of modifying $\mathbf{R}_2^*$ when $p > 2$ so as to obtain a positive definite estimate of the correlation matrix, which can then be employed in Eq. 72 to obtain a positive definite estimate of the covariance matrix. The essential idea is to "shrink" the estimates of bivariate correlation sufficiently to ensure positive definiteness. Specific schemes for "shrinking" that will pull in high (in absolute value) correlations only slightly and low correlations more drastically are described by Devlin et al. (1975).

Gnanadesikan and Kettenring (1972) tentatively proposed some other methods for obtaining positive definite robust estimators of covariance (and thence correlation) matrices, and these are described next. The essential idea underlying all of them is to base the estimate on a "sufficiently large" number, ν, of the observations (i.e., $\nu > p$), which are, nevertheless, subselected from the total sample so as to make the estimate robust to outliers. A second feature of these estimators is that they are based on a combined consideration of both scale and orientational aspects, unlike $\mathbf{S}_1^*$ and $\mathbf{S}_2^*$, which were built up from separate considerations of these aspects.

The first method for ensuring a nonsingular robust estimator of the covariance matrix is based on an approach suggested by Wilk et al. (1962), who were concerned with developing appropriate compounding matrices for a squared distance function employed in an internal comparisons technique suggested by Wilk and Gnanadesikan (1964) for analyzing a collection of single-degree-of-freedom contrast vectors (see Section 6.3.1). The first step in the procedure is to rank the multiresponse observations, $\mathbf{y}_i$ $(i=1, \ldots, n)$, in terms of their Euclidean distance, $\|\mathbf{y}_i-\mathbf{y}^*\|$ [or, equivalently, the squared Euclidean distance, $(\mathbf{y}_i-\mathbf{y}^*)'(\mathbf{y}_i-\mathbf{y}^*)$], from some robust estimate of location, $\mathbf{y}^*$. Next, a subset of the observations whose ranks are the smallest $100(1-\alpha)\%$ is chosen and used for computing a sum-of-products matrix,

$$\mathbf{A}_0 = \sum_{\substack{l \in \text{chosen subset} \\ \text{of observations}}} (\mathbf{y}_l-\mathbf{y}^*)(\mathbf{y}_l-\mathbf{y}^*)'. \tag{73}$$

(The fraction α of the observations not included in $\mathbf{A}_0$ is assumed to be small enough to ensure that $\mathbf{A}_0$ is positive definite.) After all n observations have been ranked in terms of the values of the quadratic form, $(\mathbf{y}_i-\mathbf{y}^*)'\mathbf{A}_0^{-1}(\mathbf{y}_i-\mathbf{y}^*)$, a subset of the observations whose ranks are the smallest $100(1-\beta)\%$ may be chosen and employed for defining a robust estimator of the covariance matrix,

$$\mathbf{S}_3^* = \frac{k}{n(1-\beta)} \sum_{\substack{r \in \text{chosen subset} \\ \text{of observations}}} (\mathbf{y}_r-\mathbf{y}^*)(\mathbf{y}_r-\mathbf{y}^*)', \tag{74}$$

where k is a constant that will hopefully make the estimator "sufficiently unbiased," and again β has to be small enough so that $[n(1-\beta)]>p$ and $\mathbf{S}_3^*$ is positive definite with probability 1. It may be convenient, but it is not imperative, to have $\alpha=\beta$. The above steps can be repeated using the sum of products on the right-hand side of Eq. 74 in place of $\mathbf{A}_0$, repeating the ranking of the observations, subselecting a major fraction of them for obtaining a further estimate, and iterating the process until a stable estimate is obtained. The limited experience of the authors with this method seems to suggest that, unless α, β, and n are moderately large (viz., α and $\beta \geq 0.2$ and $n \geq 50$) and unless the underlying correlation structure for the observations is nearly singular, many iterations will not be necessary to improve the estimate defined by Eq. 74. On the other hand, the work of Devlin et al. (1975) indicates that some care in the starting point (viz., not starting with ranking on simple Euclidean distances) may yield significant improvements in the estimator obtained.

The scheme involved in obtaining $\mathbf{S}_3^*$ depends on having an estimate, $\mathbf{y}^*$, of location, and any of the location estimators discussed earlier may be utilized. Exactly as in the estimation of univariate variance, however, the location estimation can be circumvented by working with pairwise differences, $(\mathbf{y}_i - \mathbf{y}_{i'})$, of the observations. Specifically, an estimator, $\mathbf{S}_4^*$, can be obtained by repeating each of the steps involved in getting $\mathbf{S}_3^*$ with $(\mathbf{y}_i - \mathbf{y}^*)$ there replaced by $(\mathbf{y}_i - \mathbf{y}_{i'})$, working with rankings of these $n(n-1)/2$ differences, and obtaining as an estimator analogous to $\mathbf{S}_3^*$ the matrix

$$\mathbf{S}_4^* = \frac{k'}{n(n-1)(1-\beta)} \sum_{\substack{r,s \in \text{chosen subset} \\ \text{of observations}}} (\mathbf{y}_r - \mathbf{y}_s)(\mathbf{y}_r - \mathbf{y}_s)'. \tag{75}$$

Just as in the univariate variance situation mentioned earlier, this estimator may be poor when a large fraction of the observations are outliers.

The multiplicative constants k and k' in Eqs. 74 and 75 are not as simply conceptualized or computed as the constants involved in the trimmed or Winsorized variances and covariances. The hope is that, although these constants may depend on n, p, α, and β, they will not depend on the underlying variances and/or correlations and also, for practical convenience, that a single multiplicative constant will be adequate for "blowing up" the estimator to make it sufficiently unbiased or consistent. This aspect of the problem needs further research.

Associated with each of the robust estimators of the covariance matrix, such as $\mathbf{S}_3^*$ and $\mathbf{S}_4^*$, is a robust estimator of the correlation matrix, which may be obtained by pre- and postmultiplying the covariance matrix estimate by a diagonal matrix whose elements are reciprocals of the square roots of the diagonal elements of the covariance matrix estimate. For instance, one such estimate would be $\mathbf{R}_3^* = \mathbf{D}\mathbf{S}_3^*\mathbf{D}$, where the jth diagonal element of $\mathbf{D}$ would be the reciprocal of the square root of the jth diagonal element of $\mathbf{S}_3^*$. One implication of this is that the robust estimators of the correlation matrix derived in this way can be obtained without knowing the multiplicative constants, such as k and k', as long as these do not depend on the underlying (and unknown) variances and/or correlations. An estimate of bivariate correlation obtained in this manner by trimming whole observations is denoted as r^*(BVT) and included in the comparative study of robust estimators by Devlin et al. (1975). An important feature of this robust estimator is its ability to provide protection against asymmetric outliers.

Robust estimators, such as $\mathbf{R}_3^*$, of correlation matrices can serve as starting points for more complex analyses such as principal components and factor analysis discussed in Chapter 2. In fact, a situation in which robust estimators might be extremely useful is one that involves a very large amount of data which are subjected to a series of reasonably complex statistical analyses, with the output of one analysis constituting the input of another. In such a situation one may not want a few observations to influence excessively the final outcome or conclusions.

The preceding discussion has dealt with robust estimation of location and dispersion for unstructured multiresponse data. More important, however, is the case of structured multiresponse data. As mentioned earlier, even for uniresponse data it is only very recently that robust estimation methods have been proposed for doing multiple regression analysis (see Andrews, 1973, 1974; Huber, 1973; Mallows, 1973). Analogously to the treatment of the simple location problem, one could of course approach the multiresponse multiple regression problem by simply considering as a robust estimator of the multiresponse regression coefficient vectors the vectors whose elements are just the univariate robust regression coefficients. In other words, with the multiresponse multiple regression structure

$$\mathbf{Y}' = (\mathbf{Y}_1, \mathbf{Y}_2, \ldots, \mathbf{Y}_p) = \mathbf{XB} + \boldsymbol{\varepsilon} = \mathbf{X}(\boldsymbol{\beta}_1, \boldsymbol{\beta}_2, \ldots, \boldsymbol{\beta}_p) + (\boldsymbol{\varepsilon}_1, \ldots, \boldsymbol{\varepsilon}_p),$$

given a robust estimator $\hat{\boldsymbol{\beta}}_j^*$ of $\boldsymbol{\beta}_j$, obtained by analyzing the observations on the jth response alone considered according to a uniresponse multiple regression model,

$$\mathbf{Y}_j = \mathbf{X}\boldsymbol{\beta}_j + \boldsymbol{\varepsilon}_j \qquad (j = 1, \ldots, p),$$

a straightforward way of developing a robust estimator $\hat{\mathbf{B}}^*$ of $\mathbf{B}$ is to take

$$\hat{\mathbf{B}}^* = \{\hat{\boldsymbol{\beta}}_1^*, \hat{\boldsymbol{\beta}}_2^*, \ldots, \hat{\boldsymbol{\beta}}_p^*\}.$$

The estimators $\hat{\boldsymbol{\beta}}_j^*$ can be obtained by using the univariate methods proposed by Andrews (1973, 1974), by Huber (1973), or by Mallows (1973). This approach to multivariate robust estimation mimics the usual practice of doing separate univariate analyses of the individual responses, "putting together" the univariate results, and considering the amalgamated result as a solution for the multiresponse problem. Once again, although this approach is simple and appealing in certain ways, it seems to be not fully satisfying in the sense that it does not explicitly exploit the multivariate nature of the data. For treating the multiresponse multiple regression problem robustly, however, the current state of the art seems to offer nothing that is better on this score. Even with this limited approach, an interesting problem would be to investigate the multivariate statistical properties of such estimators. For instance, it would be worthwhile investigating when and if one gains sensitivity and stability by analyzing the multiresponse data in terms of the above multiple regression structure, i.e., whether the elements of rows of $\hat{\mathbf{B}}^*$ exhibit better statistical properties when studied simultaneously than when considered as individual uniresponse regression coefficients. Also, there is the very important question of the statistical behavior of the "robustified residuals," $\mathbf{Y}' - \mathbf{X}\hat{\mathbf{B}}^*$, as opposed to the behavior of the usual least squares residuals (see Section 6.4 and Examples 42 and 43 for further discussion).

5.3. DATA-BASED TRANSFORMATIONS

As stated in Section 5.2, the classical multivariate theory has been based largely on the multivariate normal distribution and the paucity of alternative models for the useful guidance of multiresponse data analysis is a well-recognized limitation. One way of handling this limitation has been to develop nonparametric or distribution-free methods for specifically posed problems such as the formal inferential ones mentioned in Section 5.2. (See, e.g., Puri & Sen, 1971, for an extensive treatment of multivariate nonparametric inference.) Although such methods may serve the specific purpose for which they are designed, the statistics employed by them are not always useful for revealingly summarizing the structure in a body of data. On the other hand, the serendipitous value of many classical methods lies in their utility for summarizing the structure underlying data. Hence it is natural and appropriate to inquire about ways of transforming the data so as to permit the use of more familiar statistical techniques based implicitly or explicitly on normal distributional theory. The choice of a transformation, of course, should depend on the nature of the objectives of the data analysis, and transforming to obtain more nearly normally distributed data is only one of several possible reasonable motivations. Moreover, even if a transformation of variables does not accomplish normality, it may often go a long way toward symmetrizing the data, and this can be a significant improvement of the data as a preliminary to computing standard statistical summaries such as correlation coefficients and covariance matrices.

A transformation may be based on theoretical (or à priori) considerations or be bootstrapped (or estimated) from the data that are being analyzed. Examples of the former type are the logistic transformation of binary data proposed by Cox (1970, 1972) and the well-known variance-stabilizing transformations of the binomial, the Poisson, the correlation coefficient, etc. Techniques for developing data-based transformations of univariate observations have also been proposed by several authors (see, e.g., Moore & Tukey, 1954; Tukey, 1957; Box & Cox, 1964). Andrews et al. (1971) have extended the approach of Box & Cox (1964) to the problem of estimating a power transformation of multiresponse data so as to enhance normality, and the present section is a summary of their proposals and results.

If $\mathbf{y}' = (y_1, y_2, \ldots, y_p)$ denotes the set of p response variables, the general problem may be formulated as follows: determine the vector of transformation parameters, $\boldsymbol{\lambda}$, such that the transformed variables $\{g_1(\mathbf{y}'; \boldsymbol{\lambda}), g_2(\mathbf{y}'; \boldsymbol{\lambda}), \ldots, g_p(\mathbf{y}'; \boldsymbol{\lambda})\}$ are "more nearly" p-variate normal, $N[\boldsymbol{\mu}, \boldsymbol{\Sigma}]$, than the original p variables. The elements of $\boldsymbol{\lambda}$ are unknown, as are those of $\boldsymbol{\mu}$ and $\boldsymbol{\Sigma}$. Provided that one can obtain an appropriate estimate, $\hat{\boldsymbol{\lambda}}$, of $\boldsymbol{\lambda}$ (as well as of $\boldsymbol{\mu}$ and $\boldsymbol{\Sigma}$) from the data, the original observations, $\mathbf{y}'_i$ $(i = 1, \ldots, n)$, can be transformed one at a time to yield new observations, $\{g_1(\mathbf{y}'_i; \hat{\boldsymbol{\lambda}}), \ldots, g_p(\mathbf{y}'_i; \hat{\boldsymbol{\lambda}})\}$, which may then be considered as more nearly conforming to a p-variate normal model than the original observations.

The work of Andrews et al. (1971) is concerned with transformation functions, g_j, which are direct extensions of the power transformation of a single nonnegative response, X, to $X^{(\lambda)}$, considered by Moore & Tukey (1954) and by Box & Cox (1964), where

$$X^{(\lambda)} = \begin{cases} (X^{\lambda} - 1)/\lambda & \text{for } \lambda \neq 0, \\ \ln X & \text{for } \lambda = 0. \end{cases}$$

Furthermore, for simplicity of both the exposition and the computations, some of the details are developed only for the bivariate case, i.e., $p = 2$.

For ease of interpretation it may be desirable to look for transformations that operate on each of the original variables separately. A simple family of such transformations is defined by

$$g_j(\mathbf{y}'; \boldsymbol{\lambda}) = y_j^{(\lambda_j)} = \begin{cases} (y_j^{\lambda_j} - 1)/\lambda_j & \text{for } \lambda_j \neq 0, \\ \ln y_j & \text{for } \lambda_j = 0, \end{cases} \tag{76}$$

where $j = 1, 2$ for the bivariate case and $j = 1, 2, \ldots, p$ for the general p-variate case.

A natural starting point is to choose λ_j so as to improve the marginal normality of $y_j^{(\lambda_j)}$. Although it is recognized that marginal normality does not imply joint normality, the choice of transformations to improve marginal normality may in many cases yield data more amenable to standard analyses. The procedure is merely to apply the method proposed by Box & Cox (1964) to each response separately so that only univariate computations are involved and the theory and techniques for each are identical with those of Box & Cox. Specifically, one of the approaches suggested by Box & Cox leads to estimating λ_j by maximum likelihood, using only the observations on the jth response variable. The logarithm of a likelihood function (which has been initially maximized with respect to the unknown mean and variance for given λ_j), $\mathscr{L}_{\max}(\lambda_j)$, is maximized to provide the estimate $\hat{\lambda}_j$. If $\mathbf{Y}' = [\mathbf{Y}_1, \mathbf{Y}_2, \ldots, \mathbf{Y}_p]$ denotes the $n \times p$ matrix of original observations, and if the transformed observations obtained by using Eq. 76 are denoted as

$$(\mathbf{Y}^{(\boldsymbol{\lambda})})' = [\mathbf{Y}_1^{(\lambda_1)}, \ldots, \mathbf{Y}_j^{(\lambda_j)}, \ldots, \mathbf{Y}_p^{(\lambda_p)}],$$

where $\mathbf{Y}_j^{(\lambda_j)}$ denotes the vector of n observations on the jth variable, each of which has been obtained by transforming according to Eq. 76, then

$$\mathscr{L}_{\max}(\lambda_j) = -\frac{n}{2} \ln \hat{\sigma}_{jj} + (\lambda_j - 1) \sum_{i=1}^{n} \ln y_{ij}, \tag{77}$$

where y_{ij} denotes the ith observation on the untransformed jth response, and $\hat{\sigma}_{jj}$ is the maximum likelihood estimate of the variance of the presumed normal distribution of $\mathbf{Y}_j^{(\lambda_j)}$ [i.e.,

$$\hat{\sigma}_{jj} = \frac{1}{n} (\mathbf{Y}_j^{(\lambda_j)} - \hat{\boldsymbol{\xi}}_j)'(\mathbf{Y}_j^{(\lambda_j)} - \hat{\boldsymbol{\xi}}_j),$$

where $\hat{\boldsymbol{\xi}}_j$ is the maximum likelihood estimate of $\boldsymbol{\xi}_j = \mathscr{E}(\mathbf{Y}_j^{(\lambda_j)})$; specifically, for an unstructured sample, $\hat{\boldsymbol{\xi}}_j$ would be an $n \times 1$ vector all of whose elements are equal to the mean of the transformed observations on the jth variable, while for the more general case of a linear model specification, $\boldsymbol{\xi}_j = \mathbf{X}\boldsymbol{\theta}_j$, the appropriate estimate would be $\hat{\boldsymbol{\xi}}_j = \mathbf{X}\hat{\boldsymbol{\theta}}_j$]. In addition to the second term on the right-hand side of Eq. 77, $\hat{\sigma}_{jj}$ is also a function of λ_j, and the required maximum likelihood estimate, $\hat{\lambda}_j$, is the value of λ_j which maximizes $\mathscr{L}_{\max}(\lambda_j)$ as defined by Eq. 77. Despite the complication caused by $\hat{\sigma}_{jj}$ being a function of λ_j, since the maximization is with respect to a single unknown parameter λ_j the computations involved are quite simple. In fact, one can compute the value of $\mathscr{L}_{\max}(\lambda_j)$ for a sequence of values of λ_j and empirically determine the value, $\hat{\lambda}_j$, for which it is a maximum. Also, for this case of a single parameter, one can graph $\mathscr{L}_{\max}(\lambda_j)$ and study its behavior near $\hat{\lambda}_j$.

Following Box & Cox (1964), by using approximate asymptotic theory one can also obtain an approximate confidence interval for λ_j. The essential result is that a $100(1-\alpha)\%$ confidence interval for λ_j is defined by

$$2\{\mathscr{L}_{\max}(\hat{\lambda}_j) - \mathscr{L}_{\max}(\lambda_j)\} \leq \chi_1^2(\alpha),$$

where $\chi_\nu^2(\alpha)$ denotes the upper $100\alpha\%$ point of a chi-squared distribution with ν degrees of freedom.

The preceding discussion has been concerned with estimating power transformations of multiresponse data so as to improve marginal normality. Next, a method is described for choosing the transformations of Eq. 76 so as to enhance joint normality. To keep the computations simple, this description will be presented just in terms of a bivariate response situation. Thus the $n \times 2$ matrix $\mathbf{Y}' = ((y_{ij}))$, $i = 1, \ldots, n$; $j = 1, 2$, is the data matrix whose rows, $\mathbf{y}_i'$, are the bivariate observations, and it is assumed that after a transformation of the form in Eq. 76 the transformed data $(\mathbf{Y}^{(\boldsymbol{\lambda})})'$ may be statistically described by a *bivariate* normal density function with mean $\boldsymbol{\mu}'$ and covariance matrix $\boldsymbol{\Sigma}$.

Let $\boldsymbol{\Xi}' = \mathscr{E}[(\mathbf{Y}^{(\boldsymbol{\lambda})})'] = \mathbf{1} \cdot \boldsymbol{\mu}'$. [*Note*: For simplicity the sample is considered to be unstructured; however, the treatment for a structured sample with a general linear model specification is quite straightforward, requiring only that $\mathbf{X\Theta}$ be substituted for $\mathbf{1} \cdot \boldsymbol{\mu}'$.] If

$$\boldsymbol{\lambda} = \begin{pmatrix} \lambda_1 \\ \lambda_2 \end{pmatrix}$$

is the set of transformation parameters yielding bivariate normality, the density function of the original data, $\mathbf{Y}$, is

$$f(\mathbf{Y} \mid \boldsymbol{\mu}, \boldsymbol{\Sigma}, \boldsymbol{\lambda}) = |\boldsymbol{\Sigma}|^{-n/2}(2\pi)^{-n} \exp[-\tfrac{1}{2}\text{tr}\, \boldsymbol{\Sigma}^{-1}(\mathbf{Y}^{(\lambda)} - \boldsymbol{\Xi})(\mathbf{Y}^{(\lambda)} - \boldsymbol{\Xi})']J,$$

where J, the Jacobian of the transformation from $\mathbf{Y}^{(\boldsymbol{\lambda})}$ to $\mathbf{Y}$, is

$$\prod_{j=1}^{2} \prod_{i=1}^{n} y_{ij}^{\lambda_j - 1}.$$

Thus the log likelihood of $\boldsymbol{\mu}$, $\boldsymbol{\Sigma}$, and $\boldsymbol{\lambda}$ is given (aside from an additive constant) by

$$\mathscr{L}(\boldsymbol{\mu}, \boldsymbol{\Sigma}, \boldsymbol{\lambda} \mid \mathbf{Y}) = -\frac{n}{2} \ln |\boldsymbol{\Sigma}| - \tfrac{1}{2}\text{tr}\, \boldsymbol{\Sigma}^{-1}(\mathbf{Y}^{(\boldsymbol{\lambda})} - \boldsymbol{\Xi})(\mathbf{Y}^{(\boldsymbol{\lambda})} - \boldsymbol{\Xi})'$$
$$+ \sum_{j=1}^{2} \left[(\lambda_j - 1) \sum_{i=1}^{n} \ln y_{ij} \right].$$

For specified λ_1 and λ_2, the maximum likelihood estimates of $\boldsymbol{\mu}$ and $\boldsymbol{\Sigma}$ are given, respectively, by

$$\hat{\boldsymbol{\mu}} = \frac{1}{n} \mathbf{Y}^{(\lambda)} \cdot \mathbf{1},$$

and

$$\hat{\boldsymbol{\Sigma}} = \frac{1}{n}(\mathbf{Y}^{(\boldsymbol{\lambda})} - \hat{\boldsymbol{\Xi}})(\mathbf{Y}^{(\boldsymbol{\lambda})} - \hat{\boldsymbol{\Xi}})',$$

where $\hat{\boldsymbol{\Xi}}' = \mathbf{1} \cdot \hat{\boldsymbol{\mu}}'$. If these estimates are substituted in the above log-likelihood function, the resulting maximized function (up to an additive constant) is

$$\mathcal{L}_{\max}(\lambda_1, \lambda_2) = -\frac{n}{2}\ln|\hat{\boldsymbol{\Sigma}}| + \sum_{j=1}^{2}\left[(\lambda_j - 1)\sum_{i=1}^{n}\ln y_{ij}\right], \tag{78}$$

a function of two variables that may be computed and studied. The maximum likelihood estimates $\hat{\lambda}_1$ and $\hat{\lambda}_2$ may be obtained by numerically maximizing Eq. 78. Also an approximate $100(1-\alpha)\%$ confidence region for λ_1 and λ_2, obtained on the basis of asymptotic considerations, is

$$2\{\mathcal{L}_{\max}(\hat{\lambda}_1, \hat{\lambda}_2) - \mathcal{L}_{\max}(\lambda_1, \lambda_2)\} \le \chi_2^2(\alpha),$$

where $\chi_2^2(\alpha)$ is the upper $100\alpha\%$ of the chi-squared distribution with 2 degrees of freedom.

It is easy to see that Eq. 77 is the univariate version of the bivariate version in Eq. 78, and that both result from using the power transformations in Eq. 76 and a likelihood approach, except that Eq. 77 is the result of specifying marginal normality whereas Eq. 78 is a consequence of specifying bivariate normality. In fact, for the general case of p responses, if one were to start with the transformations in Eq. 76 and specify a p-variate normal distribution, $N[\boldsymbol{\mu}, \boldsymbol{\Sigma}]$, for the transformed observations, then, following the same arguments used in arriving at Eq. 78 for the bivariate case, one would obtain for the general case the following log-likelihood function of $\boldsymbol{\lambda} = (\lambda_1, \lambda_2, \ldots, \lambda_p)'$ after initial maximization with respect to $\boldsymbol{\mu}$ and $\boldsymbol{\Sigma}$:

$$\mathcal{L}_{\max}(\lambda_1, \lambda_2, \ldots, \lambda_p) = -\frac{n}{2}\ln|\hat{\boldsymbol{\Sigma}}| + \sum_{j=1}^{p}\left[(\lambda_j - 1)\sum_{i=1}^{n}\ln y_{ij}\right], \tag{79}$$

where y_{ij} is the ith observation on the (untransformed) jth response $(i = 1, \ldots, n;\ j = 1, 2, \ldots, p)$, and the $p \times p$ matrix

$$\hat{\boldsymbol{\Sigma}} = \frac{1}{n}(\mathbf{Y}^{(\boldsymbol{\lambda})} - \hat{\boldsymbol{\Xi}})(\mathbf{Y}^{(\boldsymbol{\lambda})} - \hat{\boldsymbol{\Xi}})',$$

$$\hat{\boldsymbol{\Xi}}' = \begin{cases} \dfrac{1}{n}\mathbf{1}\cdot\mathbf{1}'(\mathbf{Y}^{(\boldsymbol{\lambda})})' & \text{for an unstructured sample,} \\ \mathbf{X}\hat{\boldsymbol{\Theta}} & \text{for a general linear model specification.} \end{cases}$$

For this general p-response case, however, $\mathcal{L}_{\max}(\boldsymbol{\lambda})$ is a function of p variables, $\lambda_1, \lambda_2, \ldots, \lambda_p$, and thus the problem of studying and numerically maximizing it is more complex than in the bivariate case (see Chambers, 1973, for a discussion of optimization techniques). Formally, however, if $\hat{\lambda}_1, \ldots, \hat{\lambda}_p$ are the values that maximize $\mathcal{L}_{\max}(\lambda_1, \ldots, \lambda_p)$, an approximate confidence region for $(\lambda_1, \ldots, \lambda_p)$, analogous to the bivariate one mentioned above, is defined by

$$2\{\mathcal{L}_{\max}(\hat{\lambda}_1, \ldots, \hat{\lambda}_p) - \mathcal{L}_{\max}(\lambda_1, \ldots, \lambda_p)\} \leq \chi_p^2(\alpha),$$

where $\chi_p^2(\alpha)$ is the upper $100\alpha\%$ point of the chi-squared distribution with p degrees of freedom.

In certain situations, data may exhibit nonnormality in some but not all directions in the space of the original responses. One way of thinking about the two approaches discussed thus far is that the one directed toward improving marginal normality is concerned with p directions, one for each of the original coordinates, whereas the approach directed toward enhancing joint normality is concerned with all possible directions. The method to be described next is concerned with identifying directions (not necessarily confined to prespecified directions such as those of the coordinate axes) of possible nonnormality and then estimating a power transformation of the projections of the original observations onto these directions so as to improve normality along them. The specification of a direction will in general depend on several, and possibly all, coordinates, and hence the method no longer involves just transformations of each coordinate separately.

As before, let $\mathbf{Y}'$ denote the data matrix whose rows, $\mathbf{y}_i'$, $i = 1, \ldots, n$, are the multiresponse observations. With a general multivariate linear model specification, $\mathscr{E}(\mathbf{Y}') = \mathbf{X\Theta}$ (which includes the case of an unstructured sample by specifying $\mathbf{X}$ as an n-vector of unities, $\mathbf{1}$, and $\mathbf{\Theta}$ as the unknown mean vector, $\boldsymbol{\mu}'$), one can obtain the residual error covariance matrix, $\mathbf{S}_{\text{error}}$, defined in Eq. 66. For brevity of notation $\mathbf{S}_{\text{error}}$ will be denoted as $\mathbf{S}$ in the following discussion.

If $\mathbf{S}^{1/2}$ denotes the symmetric square root of $\mathbf{S}$, one can obtain the set of scaled residual vectors,

$$\mathbf{z}_i' = (\mathbf{y}_i' - \mathbf{x}_i' \cdot \hat{\mathbf{\Theta}})\mathbf{S}^{-1/2}, \qquad i = 1, \ldots, n,$$

where $\mathbf{x}_i'$ denotes the ith row of the design matrix $\mathbf{X}$. [*Note*: Once again, for an unstructured sample, $\mathbf{x}_i' \cdot \hat{\mathbf{\Theta}}$ will just be the sample mean vector, $\bar{\mathbf{y}}'$.] Any nonnormal characteristics of the observations $\mathbf{y}_i'$ will be reflected in corresponding (nonnormal) characteristics of the $\mathbf{z}_i'$, and the direction of any nonnormal clustering of points, if present, may perhaps be identified by studying a normalized weighted sum of the $\mathbf{z}_i'$:

$$\mathbf{d}_\alpha' = \frac{\sum_{i=1}^{n} w_i \mathbf{z}_i'}{\left\| \sum_{i=1}^{n} w_i \mathbf{z}_i \right\|}, \qquad w_i = \|\mathbf{z}_i\|^\alpha,$$

where $\|\mathbf{x}\|$ denotes the Euclidean norm, or length, of the vector $\mathbf{x}$, and α is a constant to be chosen.

The vector $\mathbf{d}'_\alpha$ provides a parametrization of directions in the z-space (and hence in the y-space of the original observations) in terms of the single parameter α. If $\alpha=-1$, $\mathbf{d}'_\alpha$ is a function only of the orientation of the $\mathbf{z}_i$'s, while if $\alpha=+1$, $\mathbf{d}'_\alpha$ becomes sensitive primarily to the observations, $\mathbf{y}'_i$, that are far from the mean $\bar{\mathbf{y}}'$. More generally, for $\alpha>0$ the vector $\mathbf{d}'_\alpha$ will tend to point toward any clustering of observations far from the mean, while for $\alpha<0$ the vector $\mathbf{d}'_\alpha$ will point in the direction of any abnormal clustering near the center of gravity of the data. If the scaled residuals are skewed in one direction, $\mathbf{d}'_\alpha$ will tend to point in that direction.

For a specified α, the vector $\mathbf{d}'_\alpha$ (chosen to be sensitive to particular types of nonnormal clusterings if any are present) corresponds to the vector $\mathbf{d}^{*\prime}_\alpha=\mathbf{d}'_\alpha\mathbf{S}^{1/2}$ in the space of the original observations. The projections of the original observations onto the unidimensional space specified by the "direction" $\mathbf{d}^{*\prime}_\alpha$ constitute a univariate sample, and one can estimate a power transformation to improve the normality of these projections by using the univariate technique of Box & Cox (1964) on the unidimensional "sample" of the projections. The effect of the transformation is to alter the data only in the direction $\mathbf{d}^{*\prime}_\alpha$.

The advantage of this method of enhancing directional normality is that the relatively small class of power transformations may be applied to very complex data. The procedure may be applied iteratively, using a different value of α at each stage so as to transform along a different direction. The computations for estimating transformations along each direction are univariate (in the sense that one is working only with the projections onto the unidimensional space specified by each direction), and this is an important pragmatic advantage of this approach.

As mentioned in the initial definition of the power transformation, $X\rightarrow X^{(\lambda)}$, a requirement for using this transformation is that the data be nonnegative since otherwise the transformed values may become imaginary for fractional values of λ. A simple way of conforming to this requirement is to shift all of the observations by an arbitrary amount to make them all nonnegative. A different way of handling the problem is to use the more general shifted-power class of transformations, $X\rightarrow(X+\xi)^{(\lambda)}$, where $(X+\xi)$ replaces X, and to treat ξ as an unknown parameter as well. For using the shifted-power transformations the requirement is that X not be smaller than $-\xi$, rather than that it be nonnegative. The main difficulties in using the shifted-power instead of the power transformation are that the computations become more complex and that the interpretation of the resulting estimates, $\hat{\xi}$ and $\hat{\lambda}$, may be complicated because of high correlations between them. Nevertheless, for the transformation approaches that involve only univariate computations (i.e., the schemes aimed at improving marginal normality and directional normality), it is not out of the question to use the shifted-power class of transformations. On the other hand, for the method aimed at improving joint normality, if one were to use the shifted-power transformation the log-likelihood function corresponding to Eq. 79 would be a function of $2p$ parameters, $\{\lambda_1, \xi_1, \lambda_2, \xi_2, \ldots, \lambda_p, \xi_p\}$, so that even for bivariate response data one would in general have to consider maximizing a function of four variables to obtain the required maximum likelihood estimates of the transformation parameters.

Example 23. The data consist of $50\,(=n)$ sets of bivariate normal deviates generated on a computer. Pairs of random standard normal deviates, (x_{1i}, x_{2i}), were transformed using the relationships

$$\left.\begin{aligned} y_{1i} &= x_{1i} \\ y_{2i} &= \rho x_{1i} + \sqrt{1-\rho^2}\,x_{2i} \end{aligned}\right\}, \qquad i = 1, 2, \ldots, 50,$$

to obtain the 50 samples, (y_{1i}, y_{2i}), from

$$N\left[\begin{pmatrix}0\\0\end{pmatrix}, \begin{pmatrix}1 & \rho\\ \rho & 1\end{pmatrix}\right].$$

To avoid negative values (so that power transformations could be employed), the mean vector was shifted sufficiently away from the origin by adding a constant vector (c, c) to each of the observations. A range of values for ρ was used to provide a basis for comparing the different approaches discussed above for transforming observations. For convenience in referring to the approaches, the method of Box and Cox (1964) applied to each variable separately so as to improve marginal normality is called Method I, the method for enhancing joint normality is termed Method II, and the one aimed at improving directional normality is designated as Method III.

Exhibit 23a shows the estimates of the transformation parameters obtained by the three approaches described earlier. The actual outputs of the analyses consist not only of the maximum likelihood estimates involved in each case but also, for Methods I and III, plots of the log-likelihood functions involved, together with the associated approximate confidence intervals, and for Method II a contour plot of the log-likelihood surface displayed with the approximate confidence sets for this case. To minimize the number of displays, only a few sample plots are included here.

Exhibit 23a. Monte Carlo normal data $(p=2, n=50)$

	I		II		III	
ρ	$\hat{\lambda}_1$	$\hat{\lambda}_2$	$\hat{\lambda}_1$	$\hat{\lambda}_2$	$\hat{\lambda}$	$d^{*\prime}_{1.0}$
0	0.896	0.957	0.878	0.984	0.688	−0.7, 0.7
0.1	0.896	1.021	0.892	1.009	0.811	−0.6, 0.8
0.3	0.896	1.085	0.890	1.035	0.714	−0.9, 0.4
0.5	0.896	1.041	0.882	1.011	0.728	−1, 0.2
0.75	0.896	0.810	0.887	0.887	0.725	−0.6, 0.8
0.8	0.896	0.745	0.886	0.852	0.729	−0.5, 0.8
0.9	0.896	0.614	0.884	0.782	0.735	−0.3, 0.9
0.95	0.896	0.596	0.884	0.769	0.719	−0.3, 0.9
0.975	0.896	0.642	0.883	0.784	0.737	−0.3, 0.9
0.999	0.896	0.833	0.884	0.859	0.715	−0.6, 0.8

Exhibit 23b. Plot of log-likelihood function with associated confidence intervals (Method I); mle of $\lambda = 0.596$

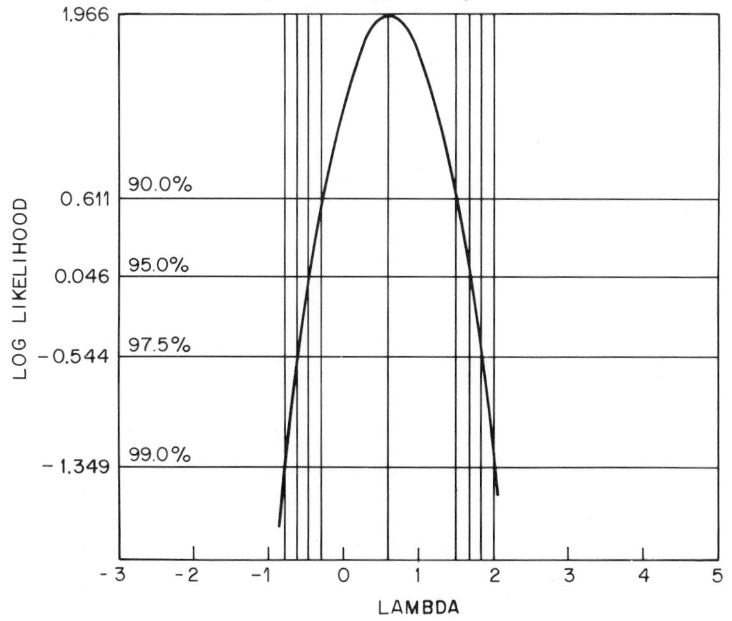

Over the range of 10 values of ρ shown in Exhibit 23*a*, it can be seen that the estimates of λ_1 and λ_2 obtained by Method I vary between 0.596 and 1.085. [*Note*: Because of the scheme used in generating the data, the sample of values of the first variable does not change as ρ changes and hence the estimate of λ_1 obtained by Method I remains the same for all ρ.] Moreover, every 95% confidence interval includes not only the "true" value of $\lambda = 1$ (since the original distributions are all normal) but also every other estimate of λ. Exhibit 23*b* shows a plot of the log-likelihood function of λ_2 when $\rho = 0.95$, the case in which Method I yielded the smallest (and farthest from 1) estimate of the transformation parameter.

Exhibit 23c. Contour plot of log-likelihood surface with associated confidence regions (Method II) for data with ρ=0.5; mle of λ'=(0.882, 1.011)

```
(*) ∈ 90% conf. set

(*&=) ∈ 95% conf. set

(*,=&X) ∈ 97.5% conf. set

(*,=,X&0) ∈ 99% conf. set


-0.118|
      |
      |
      |          0000000
      |        0X=====XX0
      |       0X=******=X0
      |      0X=********=X0
      |      0=**********=X0
      |      X=**********=X0
      |      X=**********=X0
 0.882|      X=**********=X0
      |      0=**********=X0
      |      0X*********=X0
      |       X=*******==00
      |       0X=*****=XXX0
      |        0X=====X00
      |          00XXX00
      |            000
      |
      |
 1.882|
      |____________________________________
       0.011          1.011         2.011
```

Exhibit 23d. Contour plot of log-likelihood surface with associated confidence regions (Method II) for data with ρ=0.95; mle of λ'=(0.884, 0.769)

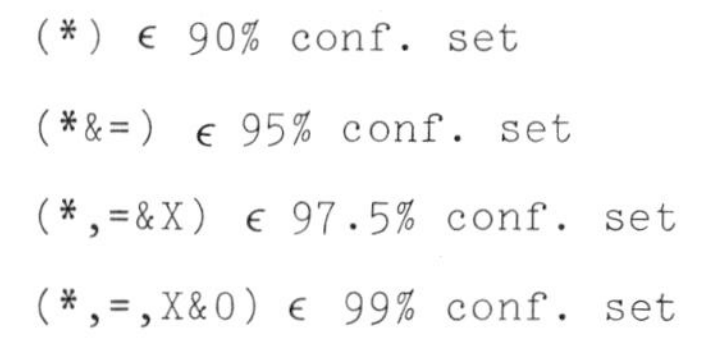

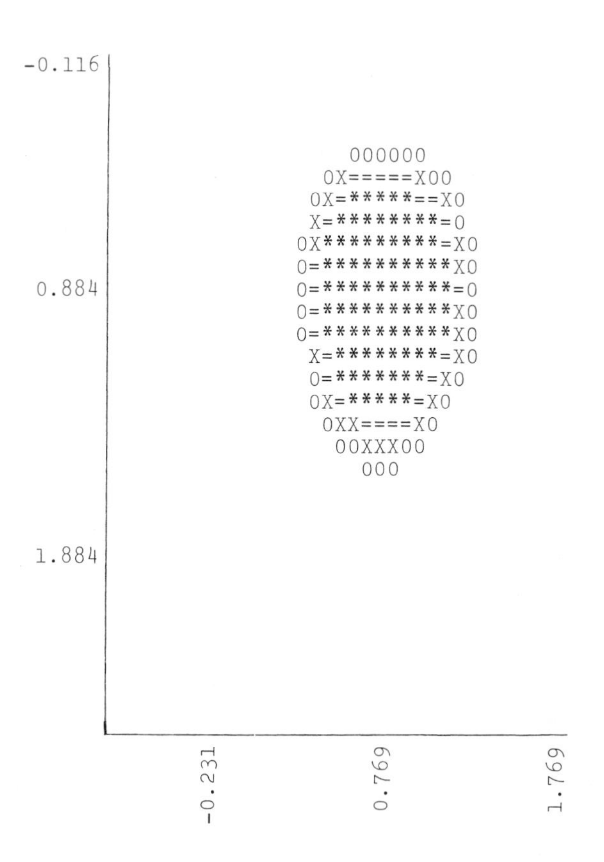

Method II yielded estimates of λ_1 and λ_2 that range between 0.878 and 1.035, and Exhibits 23*c* and *d* show the contour plots of the log-likelihood surfaces for the cases when $\rho = 0.5$ and $\rho = 0.95$. Even on the coarse grid used for generating these plots, there is some indication that the contours are tighter for the higher value of ρ.

A very interesting feature of the results in Exhibit 23*a* is the greater

Exhibits 23e,f. Scatter plots of data before and after transformation by Method III

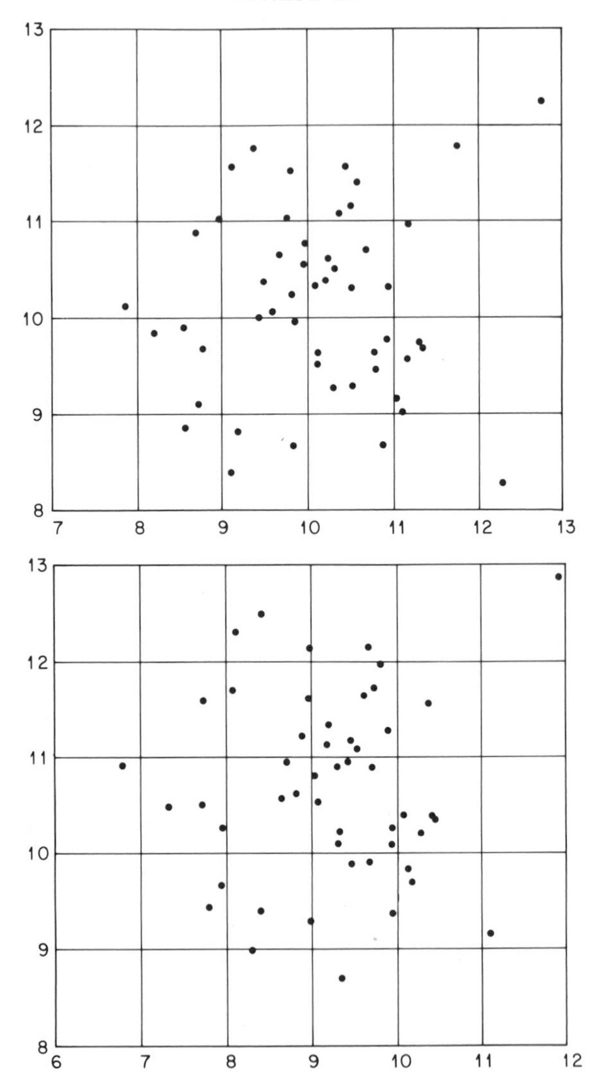

stability shown by the estimates obtained by Method II as compared to the ones yielded by Method I. The stability is particularly noticeable as ρ increases, although it is evident even for small values of ρ. This is encouraging in that the bivariate approach, i.e., seeking joint normality while still using coordinatewise transformations, is yielding "more" than the repeated application of the univariate approach with each variable separately. It is always legitimate to ask whether one gains anything significant by using a multivariate approach. In the present case it seems that a multivariate approach may be able to exploit the intercorrelations among the variables to advantage and lead to more stable estimates.

The results of applying Method III are also included in Exhibit 23*a*. The value of α used for obtaining the direction of possible nonnormality, $\mathbf{d}_\alpha^{*\prime}$, was 1. The estimate of λ as well as the direction, $\mathbf{d}_{1.0}^{*\prime}$, is shown. In this Monte Carlo example, the method appears to identify an arbitrary direction; and, as seen by the $\hat{\lambda}$ values and from the fact that all the 95% confidence intervals included the value 1, the transformation has not altered the data very much. This is also evident in Exhibits 23*e* and *f*, which show, respectively, the data before and after the transformation.

Example 24. To illustrate the use of the method for improving directional normality in a "nonnull" case, bivariate observations were generated for which the first coordinate was distributed lognormally whereas the second was distributed normally independent of the first. For these data, using $\alpha = 1$ in the directional method leads to identifying the direction of nonnormality as $\mathbf{d}_{1.0}^{*\prime} = (1, 0)$, as it should, and $\hat{\lambda} = -0.03$, which again is sufficiently close to 0, the value one would expect. Exhibits 24*a* and *b* show scatter plots of the data before and after transformation, and the achievements of the transformation are clear.

Exhibit 24a. Scatter plot of untransformed data

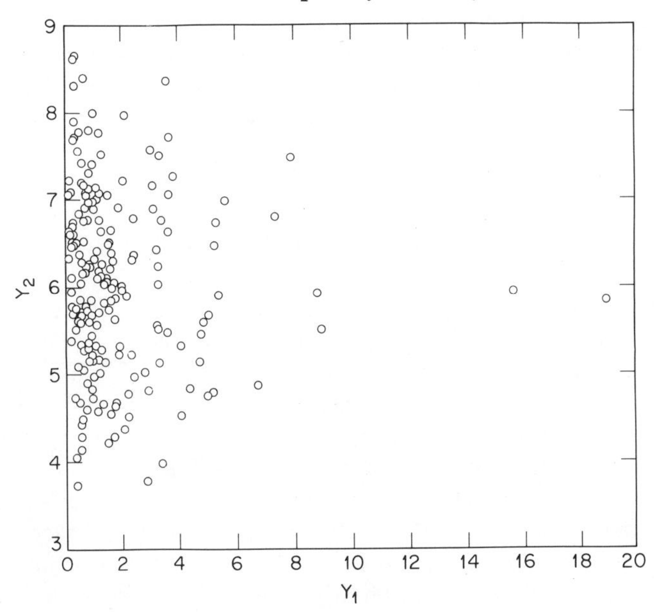

Exhibit 24b. Scatter plot of data transformed by Method III

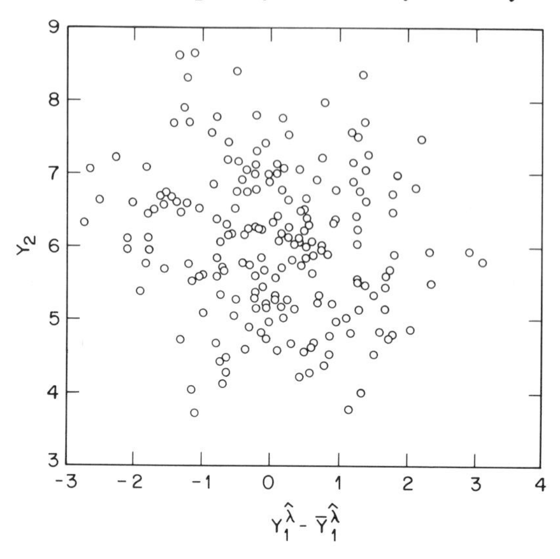

Other examples of the use of the data-based transformation methods discussed in this section will be given in Chapter 6 (see Example 37 in Section 6.3.1). A key point regarding the transformation methods described heretofore is that, although the objective in estimating a transformation is perhaps to improve normality, there is no guarantee that the resultant transformation will actually achieve adequate normality in any particular application. In other words, some kinds of nonnormality may not be ameliorated by relatively simple types of nonlinear transformations.

5.4. ASSESSMENT OF DISTRIBUTIONAL PROPERTIES

Statistical distributions play a useful role in modeling data. Both fitting and assessing the fit of various distributions to univariate data are common exercises. One reason for the interest in using appropriate distributions for modeling data is the feasibility of obtaining parsimonious representations of data in terms of the parameters (hopefully much fewer in number than the observations) of such distributions.

The variety of univariate distributional models is, of course, very rich, whereas this is not so in the multivariate situation. In fact, the multivariate normal distribution has been almost exclusively at the center of much of the development of multivariate methodology, and although other multivariate distributions have been proposed as alternative models, far less use has been made of these in practice (see Kendall, 1968). The lack of availability of a variety of alternatives is perhaps one explanation for the relative lack of emphasis, in the multiresponse situation as opposed to the univariate one, on assessing distributional properties and assumptions in the light of the data. Nevertheless, certain questions can be posed and solutions to them proposed, and the two subsections that follow are concerned with methods addressed to two sets of questions. Section 5.4.1 will discuss methods for evaluating the similarity of the marginal distributions of the responses, and Section 5.4.2 will describe techniques for assessing the normality of multiresponse data.

5.4.1. Methods for Evaluating Similarity of Marginal Distributions

Many of the theoretical multivariate distributions that have been proposed as bases or models for statistical analyses of data have the feature that the marginal distributions are either identical or common up to origin (or location) and/or scale parameters. For instance, in addition to the multivariate normal, the usual definitions of the multivariate t, F, and beta (or Dirichlet) distributions all incorporate this feature.

Many multivariate summaries (e.g., correlation coefficients or the covariance matrix) seem to depend, for sense and interpretability, on the degree of similarity of the marginal distributions of the components of a multiresponse observation. Also, to the extent that multivariate normality motivates many of the usual multivariate methods, a preliminary step for matching a body of data to such methods might be the assessment of the degree of "*commonality*" of the marginal distributions.

The problem to be considered here is the following: given a set of multiresponse observations which, for purposes of analysis, is viewed as a random sample from a single multivariate distribution, provide ways of assessing the degree of similarity or commonality of the marginal distributions of the components. Two types of approaches to this problem have been proposed by Gnanadesikan (1972) and will be described here. The first consists of informal graphical methods in the spirit of probability plotting techniques, while the second is based on the methods of Section 5.3 for developing data-based transformations to improve the normality of the observations.

One simple approach to the problem of assessing commonality is to ignore the multivariate nature of the observations and to study quantile-versus-quantile (Q-Q) probability plots (see Wilk and Gnanadesikan, 1968, and the brief description in Section 6.2) of the observations on the individual components separately, using a common distribution (e.g., univariate normal) as the standard for comparison. Although this method can often be useful, it is not parsimonious in some ways and in certain applications may lead to findings in the separate analyses that are difficult to integrate into a cohesive overall conclusion.

A second approach, which is more parsimonious than the above one, is to calculate the individual averages of the corresponding ordered observations for all of the components (i.e., average of smallest observation on each component, average of second smallest, etc.) and to make Q-Q probability plots of these averages. This is merely an adaptation, to the multiresponse case, of a method proposed by Laue and Morse (1968) for studying the assumed common distribution underlying several mutually independent, but comparable, sets of univariate data. Possible noncommensurability of the components in the multiresponse case may be handled by averaging the ordered observations after standardizing the individual components. This method, too, can be quite useful in some circumstances. However, like the first approach, it also uses an external standard distribution for comparison purposes; moreover, the averaging involved is likely to mask the differences among the marginal distributions, and thus, for the purpose of assessing the degree of similarity of marginal distributions, the method may not be sufficiently sensitive.

A third approach, which avoids the drawbacks of the first two while having its own limitations, is the joint plotting of component order statistics as now described. Let the rows of the $n \times p$ matrix, $\mathbf{Y}' = ((y_{ij}))$, $i = 1, \ldots, n$; $j = 1, \ldots, p$, denote the n p-dimensional observations. The jth column of $\mathbf{Y}'$ then consists of the n observations on the jth response, and one can order these observations to obtain

$$y_{[1j]} \leq y_{[2j]} \leq \cdots \leq y_{[nj]}$$

for each value of j separately. The first approach mentioned above consists in obtaining, for each standard distribution chosen, p Q-Q plots of these p sets of ordered observations. The second approach involves obtaining n averages, $\sum_{j=1}^{p} y_{[ij]}/p$ for $i = 1, \ldots, n$, and studying a single Q-Q plot of these for every chosen standard distribution. [*Note*: One version of the first approach which would avoid the need for choosing an external standard distribution would be to plot the n ordered observations on the jth response against the ordered values of the n averages defined in the second approach. For obtaining usefully stable averages, however, this would require a reasonably large value of p and hence a very much larger value of n, which might prove to be a severe requirement in some applications.]

For the third approach, a set of n sample "*multivariate order statistics*" is obtained by collecting together the corresponding ordered observations,

$$\mathbf{y}'_{[i]} = (y_{[i1]}, y_{[i2]}, \ldots, y_{[ip]}) \qquad \text{for } i = 1, 2, \ldots, n.$$

A plot of these n derived points in p-space is called a *component probability plot* (CPP for short). Actual graphical displays may be obtained for two- and three-dimensional projections of the n points, i.e., for subsets of sizes two and three from among the original p variates.

The motivation for this method of intercomparing the distributions of the components of the multivariate observation is that if, in the original coordinate system, the marginal distributions are the same up to origin and scale parameters, one may expect that the combined (i.e., multivariate) order statistics will conform to a linear configuration. Departures from linearity would indicate noncommonality of the marginal distributions. The procedure does not depend on any specific distributional assumptions and is addressed to the assessment of the composite hypothesis that the marginal distributions are the same up to origin and scale. A negative indication from this analysis may suggest the need for a nonlinear transformation on one or more of the coordinates.

In practice, the procedure may be particularly relevant and revelant when all the correlations among the variates are nonnegative. One implication of this is that two-dimensional CPP's are likely to be particularly useful (since a change of sign of one of the variables will accomplish this) and may be employed for assessing the similarity of marginal distributions of bivariate observations.

For the case of two variates, when there is no dependence between the variates the CPP is just a Q-Q probability plot since one is essentially plotting one set of empirical quantiles (sample order statistics) against another. Also, it is apparent that, when there is perfect positive correlation between the two variates, the two-dimensional CPP will be an exact linear configuration. In general, as the correlation decreases toward 0, the scatter about a linear configuration may be expected to increase. A useful supplement to the CPP is to fit a straight line to the scatter of the n points in p-space by minimizing the sum of squares of perpendicular deviations of the points from the line and to compute the value of the achieved minimum orthogonal sum of squares (MOSS). The algorithms for fitting the MOSS line and computing the MOSS value are simply linear principal components analysis ones (i.e., eigenanalysis on covariance matrices of the points plotted in the component probability plot).

Noncommensurability of the p components will introduce the usual difficulties of principal components analysis. Hence, both as a more reasonable graphical scaling technique and as a way of standardizing the above fitting procedure, one can define, display, and work with a *standardized component probability plot* (SCPP), which is a component probability plot whose coordinates have been standardized to have unit variance.

The crux of the graphical nature of the CPP or SCPP is the linearity of the configuration under null conditions and the departures from linearity otherwise. In two- or three-dimensional representations the picture is an adequate conveyor of information on conformity to linearity, but in higher-dimensional (and perhaps even in three-dimensional) space one needs some summary statistics to facilitate the assessment. The covariance and correlation matrices, $\mathbf{S}_0$ and $\mathbf{R}_0$, respectively, of the points in the CPP are natural starting points. Eigenvalues, and functions derived from them, of $\mathbf{S}_0$ and/or $\mathbf{R}_0$ may also be studied. Specifically, for instance, the MOSS associated with a p-dimensional CPP (SCPP) is the sum of the $(p-1)$ smallest eigenvalues of $\mathbf{S}_0$ ($\mathbf{R}_0$). For this and other summary statistics, it is useful to obtain some idea of their null distributions (i.e., distributions when the marginal distributions are the same up to origin and/or scale) so that some benchmarks will be available against which to compare observed values.

A different type of approach to the question of evaluating the similarity of marginal distributions can be based on the transformation techniques discussed in Section 5.3. The basic idea in this approach is, first, to transform the observations on each coordinate of a multivariate random variable so as to make the distributions of the transformed quantities more nearly the same; and, second, to intercompare the transformations, deciding that if they are in some sense identical or similar the original marginal distributions must have been equally similar. The choice of the class of transformations to be employed is an important issue. For present purposes only the power class of transformations (see Section 5.3) is considered. Specifically, one looks for a set of p transformation parameters, $\boldsymbol{\lambda}=(\lambda_1, \ldots, \lambda_p)'$, to transform the set of observations, $\mathbf{Y}'$, to $(\mathbf{Y}^{(\boldsymbol{\lambda})})'=((y_{ij}^{(\lambda_j)}))$, where

$$y_{ij}^{(\lambda_j)}=\frac{y_{ij}^{\lambda_j}-1}{\lambda_j} \qquad \text{for} \qquad \lambda_j\neq 0,$$

$$=\ln y_{ij} \qquad \text{for} \qquad \lambda_j=0,$$

$y_{ij}>0$; $i=1,\ldots,n$; $j=1,\ldots,p$. The objective of transforming the initial observations is to make the transformed observations have more nearly the same marginal distributions, and a natural choice for the common base distribution is the normal distribution. Hence the problem is to estimate the parameters, $\lambda_1, \ldots, \lambda_p$, from the data so as to enhance normality on the transformed scales (this is exactly the problem discussed earlier in Section 5.3) and then to develop methods for comparing the estimates, $\hat{\lambda}_1, \hat{\lambda}_2, \ldots, \hat{\lambda}_p$, to assess the reasonableness of assuming that they are all essentially estimates of a common parameter, λ. In this formulation, if indeed $\hat{\lambda}_1, \ldots, \hat{\lambda}_p$ turn out to be a cohesive set of estimates of a common parameter, it will not be unreasonable to conclude that the original marginal distributions are quite similar except possibly for differences in location and/or scale.

Corresponding to the methods of Section 5.3, there are two possibilities for specifying normality of the transformed scales, viz., improving marginal normality and enhancing joint normality. As discussed in Section 5.3, the method concerned with improving marginal normality would involve a consideration of the p log-likelihood functions, $\mathscr{L}_{\max}(\lambda_j)$, for $j = 1, \ldots, p$, defined in Eq. 77, and the associated maximum likelihood estimates, $\hat{\lambda}_1, \ldots, \hat{\lambda}_p$, as well as the p approximate confidence intervals for $\lambda_1, \lambda_2, \ldots$ and λ_p involved here (cf. the discussion in Section 5.3). As a procedure for assessing the similarity of marginal distributions, one can study plots of $\mathscr{L}_{\max}(\lambda_j)$ for $j = 1, \ldots, p$ on a single plot, or the p confidence intervals for $\lambda_1, \ldots, \lambda_p$, respectively, and infer the cohesiveness of the estimates. Thus, if the plots of $\mathscr{L}_{\max}(\lambda_j)$ overlap considerably, or, equivalently, if the confidence interval for λ_j includes not only $\hat{\lambda}_j$ but also $\hat{\lambda}_k$ for every $k \neq j$ and this happens for every j, one can conclude that $\hat{\lambda}_1, \ldots, \hat{\lambda}_p$ behave as if they are estimates of a common parameter, λ. [*Note:* For convenience in scaling the superimposed plots, it is desirable to plot the likelihood ratios, $L(\lambda_j)/L(\hat{\lambda}_j)$, where $L(\lambda_j) = \exp(\mathscr{L}_{\max})$, instead of the log-likelihood functions, $\mathscr{L}_{\max}$, since all the ratios have a maximum value of 1.]

Adopting the more explicitly multivariate approach, one would obtain the maximum likelihood estimates that enhance joint normality as the values $\hat{\lambda}_1, \ldots, \hat{\lambda}_p$ that maximize the log-likelihood function, $\mathscr{L}_{\max}(\lambda_1, \ldots, \lambda_p)$, defined in Eq. 79. A simple test of the significance of the hypothesis that $\lambda_1 = \lambda_2 = \cdots = \lambda_p = \lambda$, say, may be obtained by using the approximate asymptotic theory associated with the likelihood approach. In particular, the statistic

$$2\{\mathscr{L}_{\max}(\hat{\lambda}_1, \hat{\lambda}_2, \ldots, \hat{\lambda}_p) - \mathscr{L}_{\max}(\hat{\lambda}, \hat{\lambda}, \ldots, \hat{\lambda})\} \tag{80}$$

may be referred to the chi-squared distribution with $(p-1)$ degrees of freedom. The first term within the curly brackets in Eq. 80 is just the maximum value of $\mathscr{L}_{\max}(\lambda_1, \ldots, \lambda_p)$ of Eq. 79. The second term is the maximum value of $\mathscr{L}_{\max}(\lambda_1, \ldots, \lambda_p \mid \lambda_1 = \cdots = \lambda_p = \lambda)$, which may be defined by analogy with Eq. 79 just by replacing the $y_{ij}^{(\lambda_j)}$ by $y_{ij}^{(\lambda)}$, wherein a common value λ is used in place of the separate λ_j. The second term in Eq. 80, therefore, involves just a one-dimensional maximization, whereas the first term entails a p-dimensional maximization that may require considerable computational effort. (See Chambers, 1973, for a discussion of available numerical optimization algorithms.)

The computational effort involved in the transformation approach would increase considerably if the class of transformations were to be enlarged to include the shifted-power transformation [viz., with $(y_{ij} + \xi_j)$ in place of y_{ij}], which, among other advantages, would enable one to handle negative observations as well as positive ones. Apart from this important consideration, however, once again in principle the above approach can handle the shifted-power class of transformations.

Other classes of transformations, which remain simple and yet provide additional flexibility, need to be considered. There are, of course, limitations to the transformation approach, including general conceptual ones such as the possible nontransformability of some distributions by simple classes of transformations. Also, in some circumstances, it may be misleading to conclude that the marginal distributions are similar in shape just because the power transformations of the variables are essentially identical. This is illustrated in Example 27.

The two approaches to evaluating the similarity of marginal distributions have been applied to a variety of computer-generated two- and three-dimensional data. For instance, with bivariate normal data it was found repeatedly (and comfortingly) that both the graphical technique and the transformation test led to no striking or significant departures from null expectations (see Gnanadesikan, 1972, for a typical example of this sort). The performances of the techniques under nonnull conditions would, of course, be more interesting to study, and the next three examples illustrate specific aspects of the two approaches as they are revealed in the context of particular types of departures from the case of similar marginal distributions. For simplicity of discussion and display, the data in each of these examples are two-dimensional.

Example 25. The data for this example are a computer-generated sample of 100 bivariate observations in which one component has a standard normal distribution and the other an independent lognormal distribution, $\Lambda(0, 1)$, in the notation of Aitchison and Brown (1957, p. 7). All of the observations on the first coordinate were shifted to make them positive so as to allow the use of the simple power transformation.

Exhibit 25 shows the SCPP for this example. The MOSS value here is 0.16, which is about an order of magnitude larger than the typical values observed with bivariate normal data (see Example 1 of Gnanadesikan, 1972). The departure from linearity is striking and clearly suggests the extreme dissimilarity of the two marginal distributions.

Exhibit 25. SCPP of lognormal vs. normal

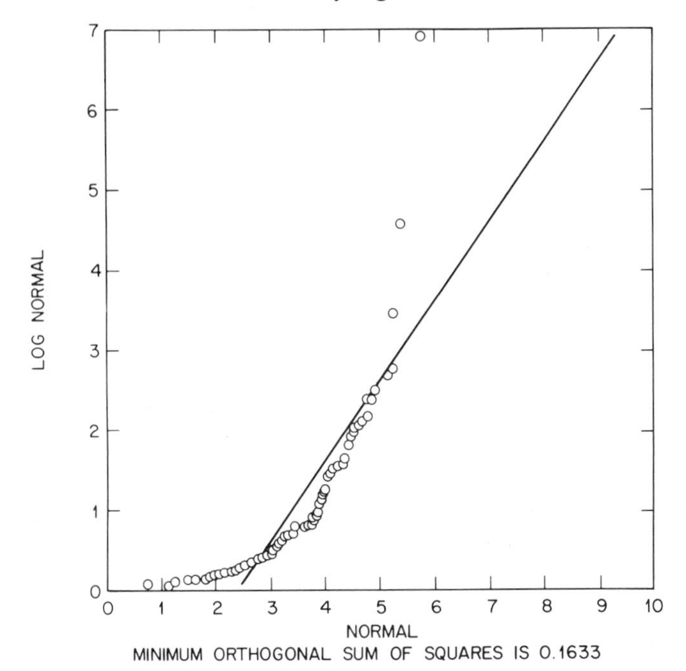

The value of the log-likelihood ratio test statistic defined in Eq. 80 turns out to be 16.06 in this example. The associated probability of exceedance in the $\chi^2_{(1)}$ distribution is 6×10^{-5}, indicating a highly significant departure from commonality.

Example 26. This example is based on a bivariate subset of trivariate data in which one component has a $\chi^2_{(2)}$ distribution, another has an independent $\chi^2_{(3)}$ distribution, and the third component is derived as the sum of the first two components, so that it has a $\chi^2_{(5)}$ distribution that is not independent of the first two distributions. The value of n is 50, and the subset chosen is the $[\chi^2_{(2)}, \chi^2_{(5)}]$ combination. Exhibit 26a shows the SCPP for this example, together with the fitted straight line and the associated MOSS value of 0.06. The systematically curved nature of the configuration on this plot would suggest dissimilarity of the marginal distributions.

On the other hand, the statistic defined in Eq. 80 turns out, in this example, to have the value 2.14, which is exceeded in a $\chi^2_{(1)}$ distribution

Exhibit 26a. SCPP of $\chi^2(2)$ vs. $\chi^2(5)$

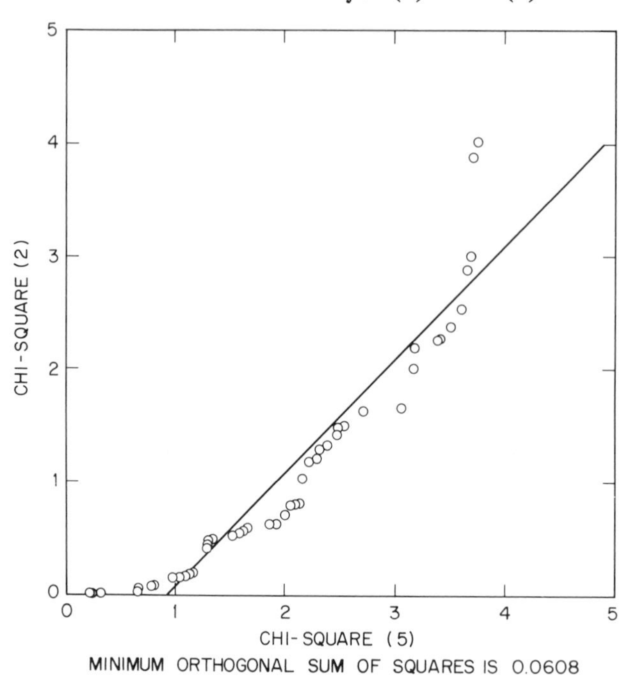

Exhibit 26b. Superimposed likelihood-ratio plots with associated confidence intervals

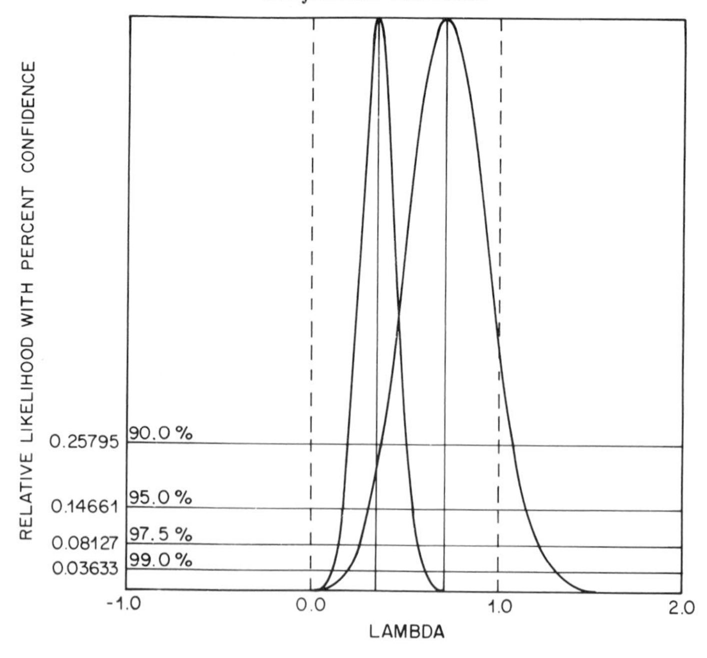

with a probability of 0.14, thus suggesting no highly significant departure from commonality. The estimated values of λ_1 and λ_2, in this bivariate transformation approach, are 0.41 and 0.69, respectively, while the estimate of λ (the hypothesized common value of λ_1 and λ_2) is 0.49. The corresponding univariate transformation approach leads to the estimates $\hat{\lambda}_1 = 0.33$ and $\hat{\lambda}_2 = 0.73$, and Exhibit 26*b* shows the superimposed plots of $L(\lambda_j)/L(\hat{\lambda}_j)$, $j = 1, 2$, with approximate confidence intervals for λ_1 and λ_2 also displayed on the figure. Although the univariate approach tends to pull the transformations of the two variables apart to a greater degree than does the bivariate approach, the overall indication from both approaches is of a moderate but not very strong difference in the two marginal distributions in this example.

In this example, therefore, the graphical display via the SCPP tends to be more revealing than the more formal test of significance based on the transformation approach. The next example brings out the same result even more forcefully.

Example 27. The bivariate data for this example consist of 100 observations simulated to be a random sample from the bivariate lognormal distribution, $\Lambda[\boldsymbol{\mu}, \boldsymbol{\Sigma}]$, where

$$\boldsymbol{\mu}' = (0, 5) \qquad \text{and} \qquad \boldsymbol{\Sigma} = \begin{pmatrix} 1 & 2.7 \\ & 9 \end{pmatrix}.$$

Exhibit 27*a* shows the SCPP for this example, and the departure from linearity is very striking. The MOSS value for this SCPP is 0.25.

The results of using the transformation approach, however, are totally unrevealing in this example. The bivariate approach leads to the estimates $\hat{\lambda}_1 = 0.02$, $\hat{\lambda}_2 = 0.04$, and $\hat{\lambda} = 0.03$, and the value of the log-likelihood ratio statistic is 0.07, with an associated exceedance probability of 0.79. The univariate transformation approach yields $\hat{\lambda}_1 = 0.03$ and $\hat{\lambda}_2 = 0.05$, and Exhibit 27*b* shows the superimposed plots of $L(\lambda_j)/L(\hat{\lambda}_j)$, $j = 1, 2$. The closeness of the estimates of λ_1 and λ_2 and the considerable overlapping of the curves in Exhibit 27*b* should be anticipated in this example, since, although the two lognormal distributions are distinctly different in shape (as judged by the difference between the diagonal elements of $\boldsymbol{\Sigma}$ above), the power transformation needed to transform both lognormal distributions to normal distributions is just the logarithmic one, i.e., the one corresponding to $\lambda = 0$. In fact, all the above estimates of λ_1 and λ_2 are statistically close to this zero value.

This example thus illustrates a limitation of the transformation approach in that the closeness of the transformations (within a class such as

Exhibit 27a. SCPP for bivariate lognormal data

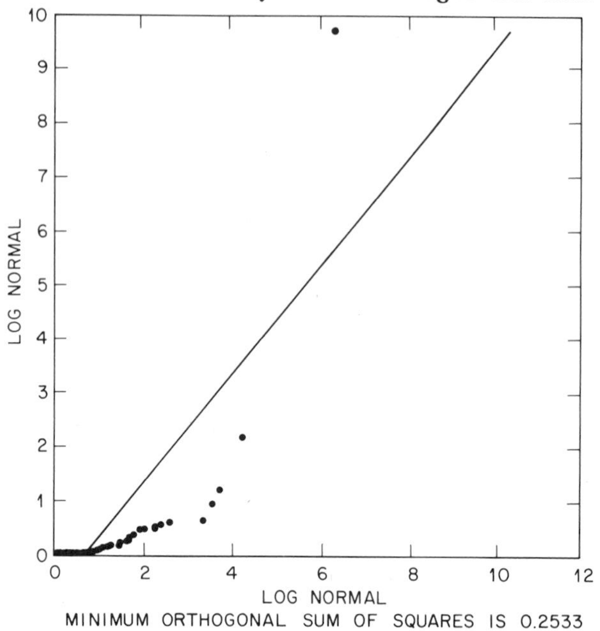

Exhibit 27b. Superimposed likelihood ratio plots with associated confidence intervals

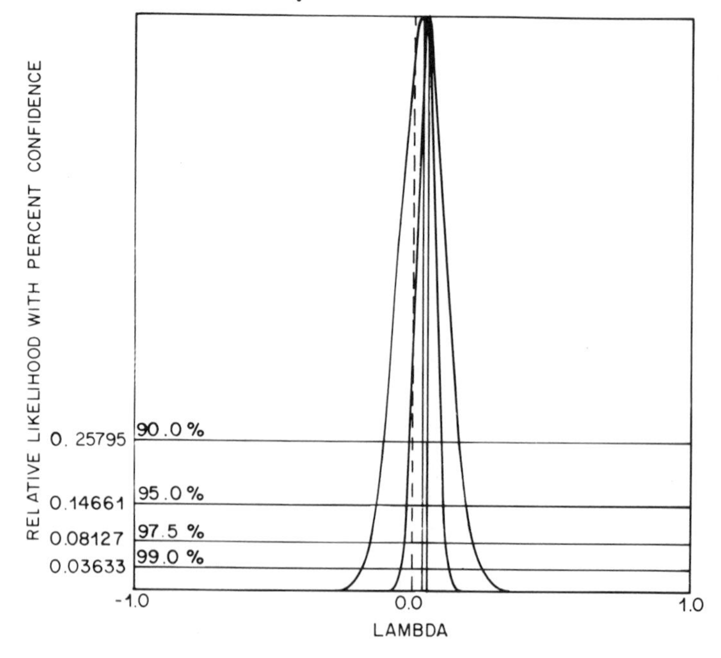

the power one) required to enhance normality of the marginal distributions is not a sufficient condition for similarity of the distributions of the untransformed variables.

5.4.2. Methods for Assessing Normality

The assumption of multivariate normality underlies much of the standard "classical" multivariate statistical methodology. The effects of departures from normality on the methods are not easily or clearly understood. Moreover, for analyzing multiresponse data, statistical methods that are robust to such distributional departures are still at a relatively early stage of development (see Section 5.2.3 and also Gnanadesikan and Kettenring, 1972; Devlin et al., 1975). Thus it would be useful to have procedures for verifying the reasonableness of assuming normality for a given body of multiresponse observations. If available, such a check would be helpful in guiding the subsequent analysis of the data, perhaps by suggesting the need for and nature of a transformation of the data to make them more nearly normally distributed, or perhaps by indicating appropriate modifications of the models and methods for analyzing the data.

Not only is there a paucity of multivariate nonnormal distributional models, but also most of the proposed alternative distributions (e.g., multivariate lognormal, exponential) are defined so as to have properties that are similar to those of the multivariate normal (e.g., that all marginal distributions belong to the same class). Real data will, of course, not necessarily conform to such specialized forms of multivariate nonnormality.

With multiresponse data it is clear that the possibilities for departure from joint normality are indeed many and varied. One implication of this is the need for a variety of techniques with differing sensitivities to the different types of departures; seeking a single best method would seem to be neither pragmatically sensible nor necessary. Developing several techniques and enabling an accumulation of experience with, and insight into, their properties is a crucial first step. Aitkin (1972), Andrews et al. (1973), and Malkovich and Afifi (1973) have proposed different methods for assessing normality, and the discussion in this subsection draws heavily from the work of Andrews and his colleagues.

One way of seeing the need for a variety of techniques in the multivariate case is in terms of the degree of commitment one wishes to make to the coordinate system for the multiresponse observations. (See also the discussion of this issue in Section 5.2.3 regarding robust estimates of multivariate location.) At one extreme is the situation in which interest is completely confined to the observed coordinates. In this case the marginal distributions of each of the observed variables and conditional distributions of certain of these, given certain others, will be the objects of interest. On the other hand, the interest may lie in the original coordinates as well as all possible orthogonal transformations of them, and here summaries (such as Euclidean distance) that remain invariant under orthogonal transformations will be the ones of interest. More generally, the class of all nonsingular linear transformations of the observed variables may be the one of interest, and then affine invariance will guide the analysis. Aside from linear transformations, one may sometimes be willing to make simple nonlinear transformations (perhaps of each coordinate separately) so as to be able to use simple models and techniques. In this case the methods used should reflect an awareness of this degree of flexibility and should attempt to incorporate it statistically. Much of the formal theory of multivariate analysis has been concerned solely with affine invariance, thus limiting the class of available procedures. The present subsection will consider techniques that are applicable to situations with different degrees of commitment to the observed coordinate system, including the classical one requiring affine invariance.

Another important issue with multivariate techniques is that, although some complexity of the methods is to be expected, they should, if possible, be kept computationally economical. The feasibility of extensive computing, made easily accessible by modern computers, does not imply that every technique is economically tenable. One objective used in developing the methods to be described below was that, computationally, they be reasonably economic and efficient.

The methods for assessing normality to be discussed here may be grouped under the following headings; (i) univariate techniques for evaluating marginal normality; (ii) multivariate techniques for evaluating joint normality; and (iii) other procedures based on unidimensional views of the multiresponse data. As mentioned earlier (see also Section 5.3), performing an initial transformation on the data and then using "standard" methods of analysis constitute a prevalent and often useful approach in analyzing data. Hence, as a general approach under each of the three categories of methods mentioned above, the assessment of normality may be made by inquiring about the need for a transformation. However, an approach that is not explicitly dependent on data-based transformations is also possible. Techniques of both types are discussed below.

Evaluating Marginal Normality. In practice, a single overall multivariate analysis of data is seldom sufficient or adequate by itself, and almost always it needs to be augmented by analyses of subsets of the responses, including univariate analyses of each of the original variables. Although marginal normality does not imply joint normality, the presence of many types of nonnormality is often reflected in the marginal distributions as well. Hence a natural, simple, and preliminary step in evaluating the normality of multiresponse data is to study the reasonableness of marginal normality for the observations on each of the variables. For this purpose one can use a variety of well-known tests for univariate normality, some of which are described next.

Perhaps the most classical method of evaluating the normality of the univariate observations $X_1, \ldots, X_n$ is by means of the well-known skewness and kurtosis coefficients:

$$\sqrt{b_1} = \frac{\sqrt{n} \sum_{i=1}^{n} (X_i - \bar{X})^3}{\left\{ \sum_{i=1}^{n} (X_i - \bar{X})^2 \right\}^{3/2}},$$

$$b_2 = \frac{n \sum_{i=1}^{n} (X_i - \bar{X})^4}{\left\{ \sum_{i=1}^{n} (X_i - \bar{X})^2 \right\}^{2}}.$$

Tables of approximate 5% and 1% points of these two statistics may be found in Pearson and Hartley (1966, pp. 207–208).

D'Agostino and Pearson (1973) have proposed improved schemes for using $\sqrt{b_1}$ and b_2 to test normality rather than employing these coefficients directly. Specifically, they provide (i) graphs (based on extensive computer simulations) for calculating the empirical probability integral of b_2 (under the null hypothesis of sampling from a normal) for a specified sample size n ($20 \le n \le 200$), and (ii) a table for calculating a standardized normal equivalent deviate $X(\sqrt{b_1})$ corresponding to $\sqrt{b_1}$—the table gives values of δ and $1/\lambda$ for use in the definition, $X(\sqrt{b_1}) = \delta \sinh^{-1}(\sqrt{b_1}/\lambda)$, for values of $n = 8(1)50(2)100(5)250(10)500(20)1000$.

Given an observed couplet of values $\sqrt{b_{1,0}}$ and $b_{2,0}$ derived from a sample of n_0 observations, one may use the graphs of the empirical probability integral of b_2 to obtain a value of the cumulative probability, $P(b_{2,0} | n_0) = P(b_2 \le b_{2,0} | n = n_0)$, and then the equivalent standard normal deviate, $X(b_{2,0})$, corresponding to this probability. Also, the table of values of δ and $1/\lambda$ can be used to calculate $X(\sqrt{b_{1,0}})$. These standard normal deviates, $X(\sqrt{b_{1,0}})$ and $X(b_2)$, may be utilized individually for testing skewness and kurtosis departures, and, in addition, they can be combined into a single omnibus test statistic,

$$\chi^2_{(2)} = X^2(\sqrt{b_{1,0}}) + X^2(b_{2,0}),$$

which can be referred to a chi-squared distribution with 2 degrees of freedom. D'Agostino and Pearson (1973) also suggest a second omnibus test based on tail probabilities rather than the equivalent normal deviates, but they state that the two tests are likely to produce very similar results.

Shapiro and Wilk (1965) have suggested a different omnibus test for normality that has appealing power properties including generally good sensitivity to a wide variety of alternatives to the normal (see Shapiro et al., 1968). The statistic proposed for assessing the univariate normality of $X_1, \ldots, X_n$ is

$$W = \frac{\left(\sum_{i=1}^{n} a_i X_{(i)}\right)^2}{\sum_{i=1}^{n} (X_i - \bar{X})^2},$$

where $X_{(1)} \leq X_{(2)} \leq \cdots \leq X_{(n)}$ denote the ordered observations, and the unit-length vector $\mathbf{a}' = (a_1, \ldots, a_n)$ is defined in terms of the vector of expected values, $\mathbf{m}'$, of standard normal order statistics and their covariance matrix, $\mathbf{V}$, as

$$\mathbf{a}' = \frac{\mathbf{m}'\mathbf{V}^{-1}}{\|\mathbf{m}'\mathbf{V}^{-1}\|}.$$

The numerator of W is, except for a multiplicative constant, the square of the best linear unbiased estimate of the standard deviation from the order statistics of a sample assumed to be from a normal population (see Sarhan and Greenberg, 1956, Section 10C), and the denominator is, of course, $(n-1)$ times the usual unbiased estimate of the variance. Shapiro and Wilk (1965) provide tables of values of the coefficients $\{a_i\}$ for $n = 2(1)50$.

Small values of W correspond to departure from normality, and percentage points are given by Shapiro and Wilk for $n = 3(1)50$.

For handling $n > 50$ without extensive tabulation of coefficients (or percentage points), D'Agostino (1971) has proposed an alternate test statistic,

$$D = \frac{T}{n^{3/2}\left\{\sum_{i=1}^{n} (X_i - \bar{X})^2\right\}^{1/2}},$$

where

$$T = \sum_{i=1}^{n} \left[i - \frac{n+1}{2}\right] X_{(i)}$$

is essentially Gini's mean difference and also, except for the multiplicative constant $2\sqrt{\pi}/n(n-1)$, the estimator of the standard deviation of a normal distribution proposed by Downton (1966). Thus D is a constant times the ratio of two estimates of the standard deviation, and both large and small deviations from its expected value correspond to departures from normality. D'Agostino (1971) gives a brief table of percentage points of a standardized version of D for sample sizes up to 1000.

For moderately large samples another simple test for normality has been proposed by Andrews et al. (1972). The test is based on the normalized gaps,

$$g_i = \frac{X_{(i+1)} - X_{(i)}}{m_{i+1} - m_i}, \qquad i = 1, \ldots, (n-1),$$

where $\mathbf{m}' = (m_1, \ldots, m_n)$, as before, is the vector of expected values of standard normal order statistics.

If the distribution of X is normal with mean μ and variance σ^2, the g_i will be approximately independently and exponentially distributed with scale parameter σ. Under an alternative, the configuration of the ordered observations may be expected to depart from the m_i with a corresponding effect on the configuration of the g_i. One approach for studying relatively smooth departures from the null configuration of the g_i proceeds via an examination of means of adjacent g_i. Specifically, one can compute sums of the first quarter, the middle half, and the last quarter of the g_i:

$$S_L = \sum_{i=1}^{[(n-1)/4]} g_i, \qquad S_M = \sum_{i=[(n-1)/4]}^{[3(n-1)/4]} g_i, \qquad S_U = \sum_{i=[3(n-1)/4]}^{n-1} g_i.$$

Let n_1 be the number of normalized gaps involved in S_L and S_U, and n_2 the number involved in S_M, so that $2n_1 + n_2 = (n-1)$. Then, under null conditions,

$$g_L = \frac{1}{n_1} S_L \qquad \text{and} \qquad g_U = \frac{1}{n_1} S_U$$

will have mean σ and variance σ^2/n_1, while

$$g_M = \frac{1}{n_2} S_M$$

has mean σ and variance σ^2/n_2.

On the basis of the approximate exponential distribution of the normalized gaps, the ratios $r_L = g_L/g_M$ and $r_U = g_U/g_M$ will each have an F distribution with degrees of freedom $2n_1$ and $2n_2$. Thus, for large n, r_L and r_U may each be treated as approximately normal with mean 1 and variance $(1/n_1+1/n_2)$. Also, a test statistic (distributed approximately as chi-squared with 2 degrees of freedom) which will tend to be more omnibus by combining the information in r_L and r_U is the quadratic form in (r_L-1) and (r_U-1) with compounding matrix $n_1\mathbf{I}-[n_1^2/(2n_1+n_2)]\mathbf{J}$, where $\mathbf{I}$ is the identity matrix and $\mathbf{J}$ is a matrix of unities. With $n_2=2n_1$, this $\chi^2_{(2)}$ statistic is

$$q=\frac{n_1}{4}\{3(r_L-1)^2-2(r_L-1)(r_U-1)+3(r_U-1)^2\}.$$

Thus, three statistics, r_L, r_U, q, together with approximate significance levels, may be calculated. The statistics r_L and r_U may be useful in interpreting and acting on significant nonnormality detected by the more omnibus q.

In addition to the direct tests for univariate normality discussed thus far, one can inquire into tests based on transforming the data. One such test can be developed in conjunction with the method proposed by Box and Cox (1964) for estimating shifted-power transformations,

$$X\to X^{(\xi,\lambda)}=\begin{cases}[(X+\xi)^\lambda-1]/\lambda & \text{for } \lambda\neq 0,\\ \ln(X+\xi) & \text{for } \lambda=0.\end{cases}$$

The estimation problem and an approach of Box and Cox to it was discussed in Section 5.3, where the detailed development was presented for the case in which ξ, the shift parameter, is taken to be 0. For the purpose of deriving an associated test for univariate normality, using both the shift and power parameters would seem to be more advantageous than using just the power parameter, λ. Basically, λ appears to be sensitive to skewness, whereas ξ seems to respond to kurtosis and heavy-tailedness. Also, in the univariate situation, including ξ implies a two-parameter effort in computational aspects, and this is not too difficult to handle.

Thus, if $X_1, X_2, \ldots, X_n$ are univariate observations, which are to be transformed by a shifted-power transformation of the above form so as to improve normality on the transformed scale, then, following Box and Cox (1964) and essentially the same steps as outlined in Section 5.3, one can obtain a log-likelihood function (initially maximized with respect to the mean and the variance for given ξ and λ) quite analogous to the one in Eq. 77 of Section 5.3:

$$\mathcal{L}_{\max}(\xi,\lambda)=-\frac{n}{2}\ln\hat{\sigma}^2+(\lambda-1)\sum_{i=1}^{n}\ln(X_i+\xi),$$

where $\hat{\sigma}^2$, a function of both ξ and λ, is the maximum likelihood estimate of the variance of the presumed normal distribution of the transformed observations, e.g.,

$$\hat{\sigma}^2 = \frac{1}{n}\sum_{i=1}^{n}[X_i^{(\xi,\lambda)} - \bar{X}^{(\xi,\lambda)}]^2, \qquad \bar{X}^{(\xi,\lambda)} = \frac{1}{n}\sum_{i=1}^{n} X_i^{(\xi,\lambda)}$$

for an unstructured sample. The above log-likelihood function may be maximized to obtain the maximum likelihood estimates, $\hat{\xi}$ and $\hat{\lambda}$, and approximate asymptotic theory yields a $100(1-\alpha)\%$ confidence region for ξ and λ, defined by

$$2\{\mathcal{L}_{\max}(\hat{\xi}, \hat{\lambda}) - \mathcal{L}_{\max}(\xi, \lambda)\} \leq \chi_2^2(\alpha),$$

where $\chi_2^2(\alpha)$ denotes the upper $100\alpha\%$ point of a chi-squared distribution with 2 degrees of freedom. A simple transformation-related procedure for assessing the normality of the distribution of X consists in not rejecting (at a $100\alpha\%$ level of significance) the hypothesis of normality if the above confidence region overlaps with the line $\lambda = 1$. A related, more stringent "likelihood-ratio test" would consist of comparing the value of $2\{\mathcal{L}_{\max}(\hat{\xi}, \hat{\lambda}) - \mathcal{L}_{\max}(\hat{\xi}, 1)\}$ to a chi-squared distribution with 1 degree of freedom. [Note that $\mathcal{L}_{\max}(\xi, 1)$ is independent of ξ so that any value, including $\hat{\xi}$, maximizes $\mathcal{L}_{\max}(\xi, 1)$ as a function of ξ.]

When the observations on the variable X are structured (i.e., some design or regression structure underlies the observations), Andrews (1971) has proposed exact procedures (confidence regions as well as tests) for formal inferences regarding ξ and λ, and one can use these in place of the approximate procedures described above. In many applications the conclusions from using the exact procedures are not likely to be markedly different from those arrived at by the approximate methods.

The preceding discussion has been concerned with formal tests of significance for detecting departures from univariate normality. For data-analytic purposes, plotting on normal probability paper or making a normal Q-Q (quantile-versus-quantile) probability plot (cf. Section 6.2) is often a very useful method of assessing the univariate normality of observations. The technique consists in plotting the ordered observation, $X_{(i)}$, against the quantile, q_i, of the standard normal distribution corresponding to the cumulative probability $(i-\frac{1}{2})/n$ (or $i/n+1$ or similar fraction) for $i = 1, \ldots, n$. [*Note*: $q_i = \Phi^{-1}(p_i)$, i.e., q_i is defined by the equation,

$$\int_{-\infty}^{q_i} \frac{1}{\sqrt{2\pi}} \exp(-\tfrac{1}{2}t^2)\, dt = p_i,$$

where $p_i = (i-\frac{1}{2})/n$ or similar fraction.] A linear configuration on such a plot would correspond to adequate normality of the observations, while systematic and subtle departures from normality would be indicated by deviations from linearity.

Although a normal probability plot does not provide a single-statistic-based formal test, as a graphical tool it conveys a great deal more information about the configuration of the observations than any single summary statistic is likely to do. In fact, one motivation for the W statistic of Shapiro and Wilk (1965) mentioned earlier is that it provides a comparison of the square of the slope of a normal probability plot of the observations against the usual estimate of variance and hence is directed towards assessing the linearity of such a plot. Devising tests directed toward detecting specific departures from linearity (e.g., quadratic or cubic) would be natural extensions of the W test. The normal probability plot is, however, likely to be a valuable supplement to any single test procedure.

A graphical display of the normalized gaps is also possible. A plot of g_i versus i, for $i=1,\ldots,(n-1)$, should appear as a random horizontal scatter revealing no systematic patterns or extremely deviant observations, provided that the original data are reasonably normally distributed. Under several nonnormal alternatives, the g_i have expected values that deviate smoothly but noticeably in the tails, and this will show up as deviations from horizontality at the left and right ends of the plot of g_i versus i. To reduce the "noisy" appearance, some smoothing of such a plot may be helpful. Exponential probability plots of the normalized gaps [i.e., a plot of the ith ordered value, $g_{(i)}$, versus the "corresponding quantile," viz., the quantile for a fraction such as $(i-\frac{1}{2})/n$, for the exponential distribution] can also be made and studied. Another variant is to make a normal probability plot of the cube roots of the normalized gaps.

Evaluating Joint Normality. In practice, except for rare or pathological examples, the presence of joint nonnormality is likely to be detected quite often by methods directed at studying the marginal normality of the observations on each variable. However, there is a need for tests that explicitly exploit the multivariate nature of the data in order, it is hoped, to yield greater sensitivity. Some methods addressed to this need are discussed next.

Classical goodness-of-fit tests, such as the chi-squared and Kolmogorov-Smirnov tests, would be possibilities for use in testing for multivariate normality. However, the drawbacks of these tests in univariate circumstances (e.g., choice of the number and boundaries of cells for the chi-squared test) are likely to be magnified for the multivariate case, and this may be part of the reason for the noticeable lack of use of these procedures with multivariate data. Also, Weiss (1958) and Anderson (1966) have suggested tests based on local densities of the observations, but, perhaps because of the difficulty of the computations involved, neither of these has seen wide application.

A relatively simple test, called the *nearest distance test*, has been proposed by Andrews et al. (1972) for testing joint normality. In this test nearest neighbor distances for each point are transformed through a series of steps to standard normal deviates. Under the null hypothesis these transformed distances are independent of the coordinates of points from which they are measured. This independence may be tested by multiple regression techniques.

The first step in the procedure is to transform the data to the unit hypercube, using the sample version of a transformation discussed by Rosenblatt (1952). One way of implementing the transformation is to initially transform the observations, using the sample mean vector and covariance matrix so as to make the transformed data have zero mean and identity covariance matrix. Then one applies the probability integral transformation to each "observation" on each coordinate separately, using the standard normal distribution as the null basis for the probability integral transformation. [*Note*: The degree of nonuniformity in small samples, resulting from using the univariate probability integral transformation with estimated values of the parameters substituted for the parameters, has been studied by David and Johnson (1948).] For adequately large sample sizes, it is perhaps not unreasonable to expect the data, if they conform to the null hypothesis of joint normality, to be transformed to the unit hypercube by this means. Also, for large sample sizes (>50 when p is small), the occurrence of points in disjoint parts of this space may be usefully approximated by independent Poisson events.

For each point $\mathbf{x}_i$ in this hypercube, a nearest neighbor distance may be calculated by using the metric

$$d(\mathbf{x}_i, \mathbf{x}_j) = \max\{\min[|x_{ki} - x_{kj}|, ||x_{ki} - x_{kj}| - 1|]\}.$$

[*Note:* To avoid boundary effects, the metric "wraps around" opposite faces of the unit hypercube. Other ways of handling this problem may also be worth considering.] Other metrics, such as the Euclidean one, may also be used. However, with moderate-sized samples, many distances have to be calculated, and the above metric is relatively inexpensive to compute. It seems well suited to algorithms that make use of sorted arrays of each coordinate.

The volume of the set enclosed by a distance d from the point $\mathbf{x}_i$,

$$\{\mathbf{x}_j : d(\mathbf{x}_i, \mathbf{x}_j) \leq d\},$$

is given by

$$V(d) = (2d)^p.$$

Since the points are assumed to be uniformly distributed in the space, the variable $V(d)$, where d is the distance to the nearest neighbor, has an exponential distribution, and

$$P[V(d) \leq V(d_i)] = 1 - \exp\{-\lambda V(d_i)\}.$$

Conditionally, given that $d \leq d_0$, the probability

$$p(d_i) = P[V(d) \leq V(d_i) \mid d_i \leq d_0] = \frac{1 - \exp\{-\lambda V(d_i)\}}{1 - \exp\{-\lambda V(d_0)\}}.$$

To this probability there corresponds a standard normal deviate,

$$w_i = \Phi^{-1}\{p(d_i)\}.$$

If the w_i are calculated from disjoint parts of the unit p-cube, they should not show any dependence on $\mathbf{x}_i$, the coordinates of the center from which nearest neighbors are measured. Such dependence may be tested by examining the regression sum of squares associated with fitting to w a quadratic surface in the elements of $\mathbf{x}$. Under the null hypothesis this regression sum of squares has a chi-squared distribution with $(p+1)(p+2)/2$ degrees of freedom. Using this distribution, one may readily assess the significance level associated with the observed regression sum of squares. If only a first-order (i.e., linear in elements of $\mathbf{x}$) model is used, the degrees of freedom are $(p+1)$.

For the $n \times p$ multiresponse data matrix $\mathbf{Y}'$, whose rows $\mathbf{y}_i'$ $(i = 1, \ldots, n)$ are taken for simplicity of discussion to constitute an unstructured sample, the computations involved in the nearest distance test are outlined by the following steps:

1. Compute the sample mean vector, $\bar{\mathbf{y}}$, and covariance matrix, $\mathbf{S}$; obtain the scaled residuals, $\mathbf{z}_i = \mathbf{S}^{-1/2}(\mathbf{y}_i - \bar{\mathbf{y}})$; and, if z_{ij} denotes the jth element of $\mathbf{z}_i$, compute the standard normal probability integral value, $x_{ij} = \Phi(z_{ij})$, $i = 1, \ldots, n$; $j = 1, \ldots, p$. Let $\mathbf{x}_i$ denote the p-dimensional vector whose jth element is x_{ij}.
2. Calculate the distances

$$d(i, i') = \max_k[\min\{|x_{ki} - x_{ki'}|, \, ||x_{ki} - x_{ki'}| - 1|\}]$$

and

$$d_{\min}(i) = \min_{i' \neq i} d(i, i').$$

3. For each point $\mathbf{x}_i$, *if* $d_{\min}(i) < 1/2n^{1/p}$ and *if* $d(i, i') > 1/2n^{1/p}$, $i' < i$, calculate

$$w_i = \Phi^{-1}\left[\frac{1-\exp\{-n[2d_{\min}(i)]^p\}}{1-\exp\{-1\}}\right].$$

4. For the $\mathbf{z}_i$ used in step 3, regress w_i on 1, $x_{i1}, \ldots, x_{ip}, x_{i1}^2, \ldots, x_{ip}^2$, $x_{i1}x_{i2}, \ldots, x_{i(p-1)}x_{ip}$; i.e., fit the quadratic relationship $\mathscr{E}(w) = \beta_0 + \beta_1 x_1 + \cdots + \beta_p x_p + \beta_{11}x_1^2 + \cdots + \beta_{pp}x_p^2 + \beta_{12}x_1x_2 + \cdots + \beta_{(p-1)p}x_{p-1}x_p$, using the n' [$\leq n$ and, it is hoped, $>(p+1)(p+2)/2$] points that survive step 3, and thus obtain a regression sum of squares with $(p+1)(p+2)/2$ degrees of freedom.

5. Compare the obtained value of the regression sum of squares to a chi-squared distribution with $(p+1)(p+2)/2$ degrees of freedom, rejecting joint normality for large values of the regression sum of squares.

Just as the univariate transformation approach of Box and Cox (1964) was utilized to obtain a test of marginal normality, the transformation approach of Andrews et al. (1971) directed toward enhancing the joint normality of multiresponse data (see Section 5.3) may be used for providing a transformation-related test of multivariate normality. The essential idea in a transformation-related approach is that evidence suggesting that a nonlinear transformation is required to significantly improve joint normality is considered as evidence that the untransformed data are nonnormal. (See, however, the discussion near the end of this subsection regarding a limitation of this formulation.)

For present purposes, even when p is not larger than 2, in order to keep the computational effort down and also be able to display some of the analyses graphically, the transformations actually employed are just power transformations of each variable separately, viz., $Y_j^{\lambda_j}$, with no shift parameters involved. (See, however, some of the earlier comments and the discussion of Example 29 for possible limitations imposed by not including shift parameters.)

For the power family, the linear transformation $\boldsymbol{\lambda} = (\lambda_1, \ldots, \lambda_p)' = \mathbf{1}$ is the only transformation consistent with the hypothesis that the data are normally distributed. A likelihood-ratio test of the hypothesis $\boldsymbol{\lambda} = \mathbf{1}$ may be based on the asymptotically approximate $\chi^2_{(p)}$ distribution of

$$2\{\mathscr{L}_{\max}(\hat{\boldsymbol{\lambda}}) - \mathscr{L}_{\max}(\mathbf{1})\},$$

where $\mathscr{L}_{\max}(\boldsymbol{\lambda})$ is the log-likelihood function defined in Eq. 79 of Section 5.3, and $\hat{\boldsymbol{\lambda}}$ is the value of $\boldsymbol{\lambda}$ that maximizes $\mathscr{L}_{\max}(\boldsymbol{\lambda})$. This $\chi^2_{(p)}$ distribution may be used to obtain both a significance level, α, associated with the observed $\hat{\boldsymbol{\lambda}}$ and a confidence set for $\boldsymbol{\lambda}$. In that the estimation method discussed earlier in Section 5.3 is built into this procedure, it not only indicates when data are nonnormal—which we may be willing to grant for many large samples—but also suggests data transformations that may be used to enhance normality.

The discussion heretofore of methods for assessing joint normality was oriented toward numerical rather than graphical techniques. For evaluating univariate normality, normal probability plots were mentioned as having particular appeal as a graphical aid in analyzing data. For evaluating joint normality, Andrews et al. (1973) have suggested an informal graphical procedure that utilizes a *radius-and-angles* representation of multiresponse data.

The first step in conceptualizing the method, in the simple context of an unstructured sample, is to obtain the scaled residuals

$$\mathbf{z}_i = \mathbf{S}^{-1/2}(\mathbf{y}_i - \bar{\mathbf{y}}), \qquad i = 1, \ldots, n,$$

which were defined and used in the nearest distance test discussed earlier. Under the null hypothesis the scaled residuals are approximately spherically symmetrically distributed. The squared radii, or squared lengths of the $\mathbf{z}_i$,

$$r_i^2 = \mathbf{z}_i'\mathbf{z}_i = (\mathbf{y}_i - \mathbf{y})'\mathbf{S}^{-1}(\mathbf{y}_i - \mathbf{y}),$$

will have approximately a chi-squared distribution with 2 degrees of freedom in the bivariate case (and p degrees of freedom in the p-variate case). Also, in the bivariate case the angle θ_i that $\mathbf{z}_i$ makes with, say, the abscissa direction will be approximately uniformly distributed over $(0, 2\pi)$. All quantities, viz., the r_i^2's and the θ_i's, will be approximately independent for large n. The dependence enters, among other routes, via the estimates of the mean and the covariance matrix, and it is hoped that for adequately large samples this dependence will have no serious effects. A further comment which may be in order is that the exact *marginal* distribution of r_i^2 is known to be a constant multiple of a beta rather than a chi-squared distribution; but again, even for moderate samples (i.e., $n = 20$ or 25 in the bivariate case), the difference between using the beta and the chi-squared approximation appears to be insignificant (see Gnanadesikan and Kettering, 1972).

The properties mentioned above suggest that summaries in terms of radii and angles may be useful for assessing joint normality. Indeed some authors (e.g., Healy, 1968; Kessel and Fukunaga, 1972) have suggested procedures based purely on the squared radii. The simple graphical procedures to be described next are based on both radii and angles.

In the bivariate case the procedure is to make a $\chi^2_{(2)}$ probability plot of the r_i^2 and a uniform probability plot of the normalized form of θ_i, viz., $\theta_i^* = \theta_i/2\pi$. (See Section 6.2 for a brief discussion of probability plots.)

Specifically, the n squared radii, r_i^2 $(i=1, \ldots, n)$, are ordered in magnitude, and the ith-ordered value is plotted against the quantile of a $\chi^2_{(2)}$ distribution corresponding to a cumulative probability of $(i-\frac{1}{2})/n$, for $i=1, \ldots, n$. Also, the n values of the normalized angles, θ_i^* $(i=1, \ldots, n)$, are ordered, and the ith-ordered value is plotted against $(i-\frac{1}{2})/n$, for $i=1, \ldots, n$. If the data conform statistically to the null hypothesis of bivariate normality, the configurations on these two probability plots should be reasonably linear. Departures from linearity on either or both of the plots would indicate specific types of departure from null conditions. [*Note:* The origin on the plot of the θ_i^* is arbitrary. Also, the θ_i^* that correspond to large r_i^2 may be more statistically stable than those with very small r_i^2, and therefore one may wish to "trim" the observations with the smallest values of r_i^2 and to study an appropriate uniform probability plot of the θ_i^* only for the remaining observations.]

For bivariate data one can also combine the information in the radii and angles in a single two-dimensional display. Let u_i denote the probability integral transformation of r_i^2 based on a $\chi^2_{(2)}$ distribution of the latter, i.e., $u_i = P\{\chi^2_{(2)} \le r_i^2\}$ for $i=1, \ldots, n$. Then a plot of the n points whose coordinates are (u_i, θ_i^*), $i=1, \ldots, n$, may be made. Under the null hypothesis one would expect to get a uniform scatter of points on the unit square. Nonuniformity of scatter, or indication of any relationship between the two coordinates in the plot, would suggest departures from the null hypothesis. [*Note:* Formal tests for uniformity can also be made; however, the main value and appeal of the procedure is its graphical character.]

For higher-dimensional data (say, p-dimensional with $p>2$), the radius-and-angles representation will yield, for each observation, a squared radius (which now will be the squared length of the p-dimensional scaled residual corresponding to the observation) and $(p-1)$ independent angles. The next step is to suggest probability plots (which will be reasonably linear if the original data are sufficiently multivariate normally distributed) for the n squared radii values and for the n values of each of the $(p-1)$ independent angles, yielding p probability plots in all. The appropriate plot for the squared radii is a $\chi^2_{(p)}$ probability plot, and a uniform probability plot over $(0, 2\pi)$ is still the appropriate procedure for one of the $(p-1)$ angles. The remaining $(p-2)$ angles involved, however, have distributions whose densities are proportional to $\sin^{p-1-j}\theta_j$ $(0 \le \theta_j \le \pi,\ j=1,2, \ldots, p-2)$, so that the appropriate probability plot for these angles is a plot of the n ordered values of the jth angle against the n corresponding quantiles of this distribution. Since under null assumptions the squared radius and the $(p-1)$ angles would again be approximately mutually independent, bivariate scatter plots on unit squares of such things as the probability integral transforms of the squared radii and of the angles may be made and studied for departures from uniformity.

Mardia (1970) has proposed a large-sample test for multivariate normality based on measures of multivariate skewness and kurtosis. The measure of multivariate skewness suggested by him is

$$b_{1,p} = \sum_{a,b,c} \sum_{a',b',c'} s^{aa'} s^{bb'} s^{cc'} m_{111}^{(abc)} m_{111}^{(a'b'c')},$$

where

$$\mathbf{S}^{-1} = ((s^{ij}))$$

and

$$m_{111}^{(jkl)} = \frac{1}{n} \sum_{i=1}^{n} (y_{ij} - \bar{y}_j)(y_{ik} - \bar{y}_k)(y_{il} - \bar{y}_l), \qquad j, k, l = 1, \ldots, p.$$

The large-sample test for joint normality based on $b_{1,p}$ is to refer the statistic $A = nb_{1,p}/6$ to a chi-squared distribution with $p(p+1)(p+2)/6$. In the bivariate case the above expression simplifies to

$$\begin{aligned} b_{1,2} = (1-r^2)^{-3}\{g_{30}^2 + g_{03}^2 + 3(1+2r^2)(g_{12}^2 + g_{21}^2) - 2r^3 g_{30} g_{03} \\ + 6r[g_{30}(rg_{12} - g_{21}) + g_{03}(rg_{21} - g_{12}) - (2+r^2) g_{12} g_{21}]\}, \end{aligned}$$

where r is the sample correlation coefficient, and

$$g_{us} = \frac{\sum_{i=1}^{n} (y_{i1} - \bar{y}_1)^u (y_{i2} - \bar{y}_2)^s}{\left[\sqrt{\sum_i (y_{i1} - \bar{y}_1)^2}\right]^u \left[\sqrt{\sum_i (y_{i2} - \bar{y}_2)^2}\right]^s}$$

is the standardized (u, s)th sample bivariate moment. The statistic $A = nb_{1,2}/6$ would be referred to a $\chi^2_{(4)}$ distribution for testing skewness departures from bivariate normality.

The multivariate kurtosis measure proposed by Mardia (1970) is the arithmetic mean of the squares of the Mahalanobis generalized distances of the observations from the sample mean, i.e.,

$$b_{2,p} = \frac{1}{n} \sum_{i=1}^{n} \{(\mathbf{y}_i - \bar{\mathbf{y}})' \mathbf{S}^{-1} (\mathbf{y}_i - \bar{\mathbf{y}})\}^2 = \frac{1}{n} \sum_{i=1}^{n} r_i^4,$$

where the r_i^2's are the squared radii discussed earlier. The proposed large-sample test for kurtosis departures from joint normality is to compare $b_{2,p}$ to a normal distribution with mean $p(p+2)$ and variance

$8p(p+2)/n$. In other words, the statistic

$$B = \frac{b_{2,p} - p(p+2)}{[8p(p+2)/n]^{1/2}}$$

is to be compared against the percentage points of the standard normal distribution. Limited investigation of the normality of B, using simulated bivariate normal samples, suggests that one would need extremely large samples for the normal approximation to be adequate and that, even for moderately large n, the distribution of B can be positively skewed. Hence, in small and moderately large samples, the exact significance level associated with the above test may be quite different from the assumed nominal level.

Tests Based on Unidimensional Views. One attractive property of tests for marginal normality is that the computational effort involved increases only linearly with p, the dimensionality of the data. It is therefore not inappropriate to examine the possibility of using various unidimensional views of the data in addition to just the marginal variables. A study of the squared radii by themselves, as proposed by Healy (1968) and by Kessell and Fukunaga (1972), is one example. Investigating the degree to which the regression of each variable on all the others is linear (a property of the multivariate normal) is another example of using a collection of unidimensional views of the data.

Another obvious class of techniques to seek is based on the characterization of the multivariate normal distribution in terms of univariate normality of all linear combinations of the variables. Tests of multivariate normality that look at "all possible" unidimensional projections and utilize the union-intersection principle of Roy (1953) have received some attention (see Aitkin, 1972; Malkovich and Afifi, 1973). The computational efforts involved in some of these tests, however, tend to be prohibitive. A different scheme, based on looking at unidimensional projections of the multivariate data along specified, rather than "all possible," directions is described next.

Marginal analysis of each of the original variables considers the projections of the data onto each of the coordinate axes separately. Other one-dimensional projections may also be considered. It is of some interest to use the projections that are likely to exhibit certain types of marked nonnormality.

One approach to this problem is to look at projections of the data along directions that are in part determined by the data, but also in part chosen to be sensitive to particular types of nonnormality. The work of

Andrews et al. (1971) described in Section 5.3 in the context of estimating transformations to enhance directional normality provides a contact point for the testing problem of present concern.

From the discussion in Section 5.3, it will be recalled that the method consists in first obtaining the projections of the observations onto the unidimensional space specified by the direction vector $\mathbf{d}_\alpha^{*\prime}$, which has been chosen to be sensitive to particular types of nonnormality by appropriately specifying a value for α. Then, since these projections constitute a univariate sample, they may be studied by any of the univariate procedures (described earlier in the context of evaluating marginal normality) for detecting departures from univariate normality. For instance, the D'Agostino and Pearson (1973) test, the Shapiro-Wilk test, the shifted-power transformation test, and a normal probability plot are all candidates for use.

Because of the data-dependent, as well as certain other, aspects of the approach the significance levels associated with the formal tests are probably not formally applicable when used with the univariate "sample" of the projections. However, they do provide useful benchmarks for measuring the nonnormality along particular directions (employing different values of α would enable one to look in many directions) in the space of the original variables. If this measure is not significant, there is some hope that subsequent methods of analysis will behave as expected. If, on the other hand, this measure is highly significant, a further transformation may make the subsequent analysis more meaningful. Since the transformation test derives from the estimation technique described in Section 5.3 for enhancing directional normality, it provides an indication of what transformation will ameliorate the abnormalities when the data are viewed in specified directions.

The various methods for assessing normality described in this subsection are applied to specific sets of data in the three examples discussed next. The examples, which for simplicity are limited to bivariate observations, are based both on computer-simulated (Examples 28 and 29) and "real" (Example 30) data. The examples involving computer-simulated data are useful because the departure from normality is known since it is part of the data-generation process. The two such examples included here are extracted from a larger set studied by Andrews et al. (1972), who discuss a greater variety of nonnull (i.e., nonnormal) data.

The scheme involved in generating the computer data was to start with observations on two independent standard normal variables, X_1 and X_2, then to transform the observations on each of these variables separately to yield observations on two independently distributed variables, Z_1 and Z_2, with a specified (but same for Z_1 and Z_2) nonnormal distribution, and finally to combine the variables Z_1 and Z_2 to form correlated variables, Y_1 and Y_2, by using the linear transformation

$$\begin{pmatrix} Z_1 \\ Z_2 \end{pmatrix} \rightarrow \begin{cases} Y_1 = Z_1, \\ Y_2 = \rho Z_1 + \sqrt{1-\rho^2}\, Z_2. \end{cases} \tag{81}$$

The correlation coefficient between Y_1 and Y_2 would thus be ρ. A different scheme for generating correlated bivariate nonnormal distributions is also discussed and used by Andrews et al. (1972, 1973), but Examples 28 and 29 apply only the scheme just described.

Example 28. This mildly nonnormal example involves 100 observations from a bivariate correlated $\chi^2_{(10)}$ distribution. Two independent $\chi^2_{(10)}$ (i.e., Z_1 and Z_2 of Eq. 81 were $\chi^2_{(10)}$ variables) samples were taken and then correlated as in Eq. 81 with $\rho = 0.9$. The two variables have marginal distributions that are relatively close to normal, the second being "expected" to be more nearly normal than the first.

The first analyses to be performed on the data were addressed to assessing the univariate normality of each variable separately. For instance, the two-parameter family of transformations

$$y \rightarrow (y + \xi)^{\lambda}$$

yielded parameter estimates and a likelihood-ratio test for each marginal variable. These results are summarized in Exhibit 28*a*. This test gives strong evidence of nonnormality of both marginal distributions.

The skewness and kurtosis measures, $\sqrt{b_1}$, b_2, were calculated for both marginal variables, and the results are recorded in Exhibit 28*b*. There is some statistical evidence of skewness in the distributions of both variables but not of kurtosis.

Exhibit 28a. Results of Box-Cox transformation test

	Variable	
	Y_1	Y_2
Parameter estimates		
$\hat{\xi}$	−0.032	−5.429
$\hat{\lambda}$	0.382	0.604
Log likelihood-ratio value	5.724	7.639
Approximate significance level	0.0007	0.0001

Exhibit 28b. Marginal skewness and kurtosis

Measure		Y_1	Y_2
Skewness	$\sqrt{b_1}$	0.699	0.395
Approximate significance level		<0.01	<0.05
Kurtosis	b_2	3.20	2.784
Approximate significance level		Not sig.	Not sig.
Omnibus D'Agostino-Pearson test	$\chi^2_{(2)}$	8.382	2.850
Significance level		≃0.015	Not sig.

The Shapiro-Wilk test, applied to the marginal distributions of these data, yielded values of W of 0.954 and 0.965 for the two variables. Without precise tables of percentage points for the present sample size, it is difficult to conclude anything regarding statistical significance other than that the first value (viz., 0.954) is "possibly" mildly statistically significant. The D'Agostino test, when applied to the two variables, led to values of the D statistic both of which had significance levels greater than 0.2.

For applying the gaps test, the standardized spacings or gaps were calculated for both marginal distributions. In both cases the difference between left and right tail lengths was manifested by gaps shorter on the left and longer on the right. The values of $(r_L - 1)$ and $(r_U - 1)$ for each variable, together with approximate significance levels, are recorded in Exhibit 28*c*. The combined statistic, q, was also computed, and its value, together with the approximate significance level, is also shown in the exhibit. Both variables show a specific skewness departure from normality in that the tails of the distributions appear to be short on the left and long

Exhibit 28c. Standardized gaps test

		Variable 1	Variable 2
Left-hand gaps	$r_L - 1$	−0.267	−0.325
Approximate significance level		$2\Phi(-1.08)$ ≃0.28	$2\Phi(-1.31)$ ≃0.19
Right-hand gaps	$r_U - 1$	0.351	0.027
Approximate significance level		$2\Phi(-1.42)$ ≃0.16	$2\Phi(-0.11)$ Not sig.
Combined statistic	q	4.626	1.951
Approximate significance level		0.1	0.4

on the right. However, these departures are not extremely statistically significant as measured by the formal gaps test.

Exhibits 28*d* and *e* (see page 180) are normal probability plots of the data for the first and second variables, respectively. The departure from normality is quite striking in Exhibit 28*d*. In Exhibit 28*e* departure from normality in the second variable, although not as striking, can be detected in the gentle curvature away from the hypothesized linear configuration.

The marginal techniques have indicated with differing degrees of strength the apparent nonnormality in the marginal distributions for this example. The results of applying the techniques for assessing joint normality are described next.

Exhibit 28*f* (see page 181) is a scatter plot of the bivariate data. The bivariate transformation technique described earlier for assessing bivariate normality was applied, and two transformation parameters, λ_1, λ_2, were estimated by maximum likelihood. The asymptotic properties of the likelihood ratios yielded an approximate test of the null hypothesis $\lambda_1 = \lambda_2 = 1$. The results of this procedure are summarized in Exhibit 28*g* (see page 181). From this test there is some, though not very strong, evidence of nonnormality.

The nearest distance test did not yield significant results. The significance level was about 0.4. This test seems to have relatively low power against smooth departures from normality, as is the case in the present example.

Next, the techniques based on radii and angles were applied. Exhibit 28*h* (see page 182) is a scatter plot of the radius-and-angle reparametrization of these data. Under the null hypothesis the plotted points have a uniform distribution on the unit square. Departures from this null hypothesis are quite apparent in this plot in that several cells are empty and also several horizontal and some vertical strips (e.g., the ones marked with arrows) are sparse in points relative to other strips. Exhibit 28*i* (see page 183) is a $\chi^2_{(2)}$ probability plot of the squared radii for this example. This plot appears to be reasonably linear, exhibiting no marked departures of the squared radii from null expectations.

Exhibit 28*j* (see page 184) is a uniform probability plot of the normalized angles. Under the null hypothesis these normalized angles should have a uniform distribution. This plot, however, appears quite irregular, especially at the upper end. A chi-squared goodness-of-fit test based on 10 equal cells yields a statistic of 25.4 with a corresponding significance level of ~ 0.002.

The multivariate skewness and kurtosis tests proposed by Mardia (1970) were also used with these data. Whereas the skewness statistic revealed a striking departure from bivariate normality, the kurtosis statistic was not statistically significant.

Exhibits 28d,e. Normal probability plots for the two variables

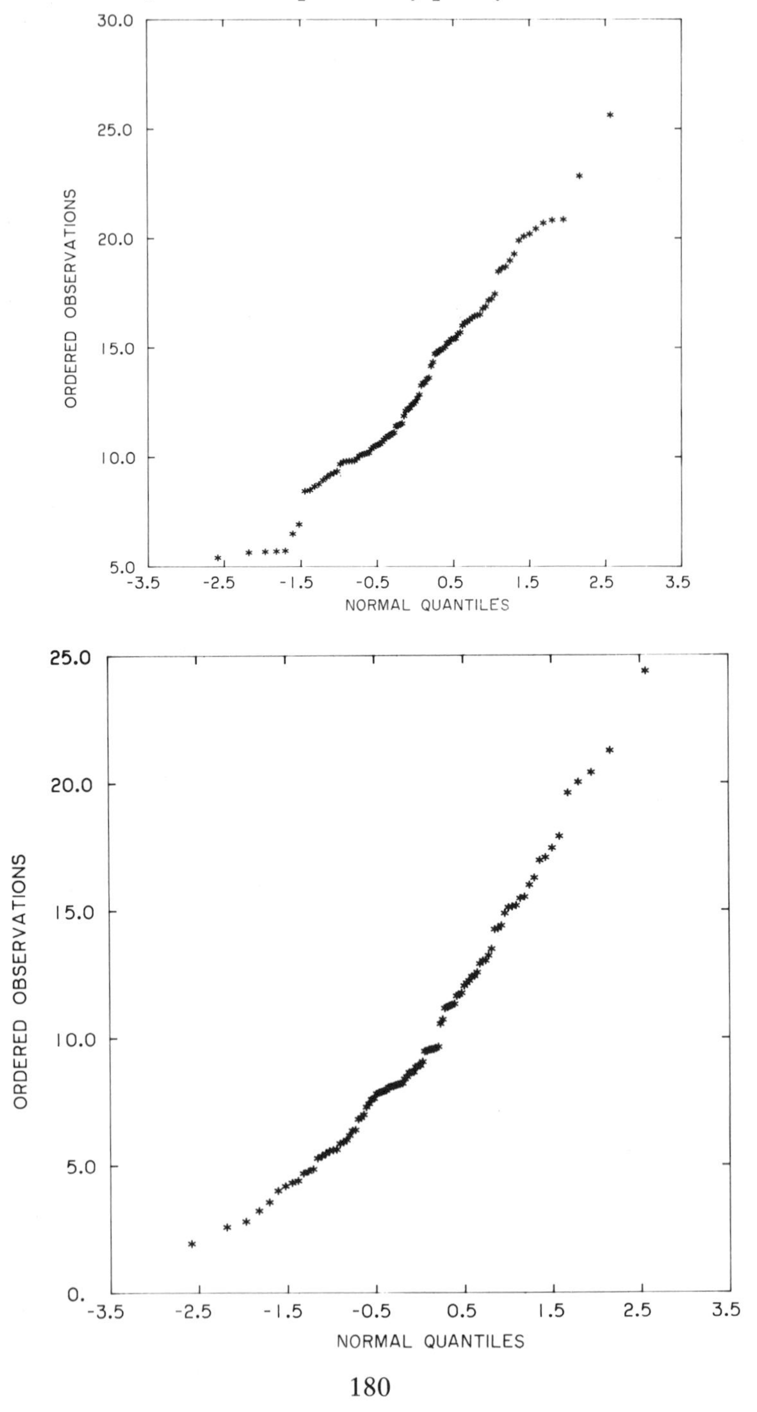

Exhibit 28f. Scatter plot of data

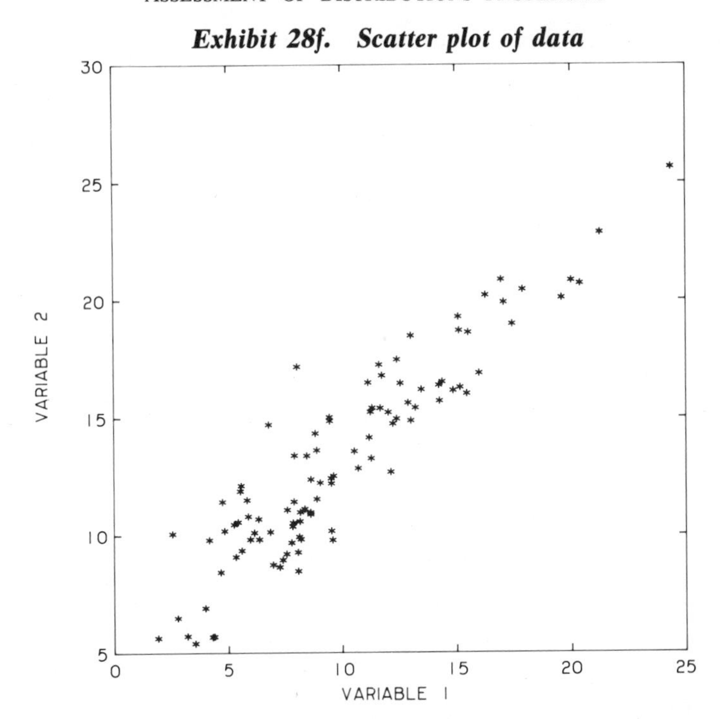

Finally, the technique of testing for directional normality was applied to the data in this example. This procedure (with $\alpha = 1.0$) selected a direction $\mathbf{d}_{1.0}^{*\prime} = (0.788, 0.616)$ and investigated the projections of the data on this one-dimensional subspace. The univariate shifted-power transformation procedure was then applied, with the results summarized in Exhibit 28k (see page 184). In the direction chosen, the data exhibit extreme nonnormality, much more marked than either of the marginal variables (cf. Exhibit 28a). Exhibit 28l (see page 185) shows a normal probability plot of the projections onto the unidimensional space specified by $\mathbf{d}_{1.0}^{*}$, and the departure from linearity here is just as striking as the one in Exhibit 28d.

Exhibit 28m (see page 185) summarizes the results of applying the various techniques to this example. Many of the techniques indicate

Exhibit 28g. Bivariate power transformation test

$\hat{\lambda}_1$	0.937
$\hat{\lambda}_2$	0.706
$\mathscr{L}_{\max}(\hat{\lambda}_1, \hat{\lambda}_2) - \mathscr{L}_{\max}(1, 1)$	2.8
Approximate significance level	0.061

Exhibit 28h. Scatter plot of normalized angles vs. probability integral transform of squared radii

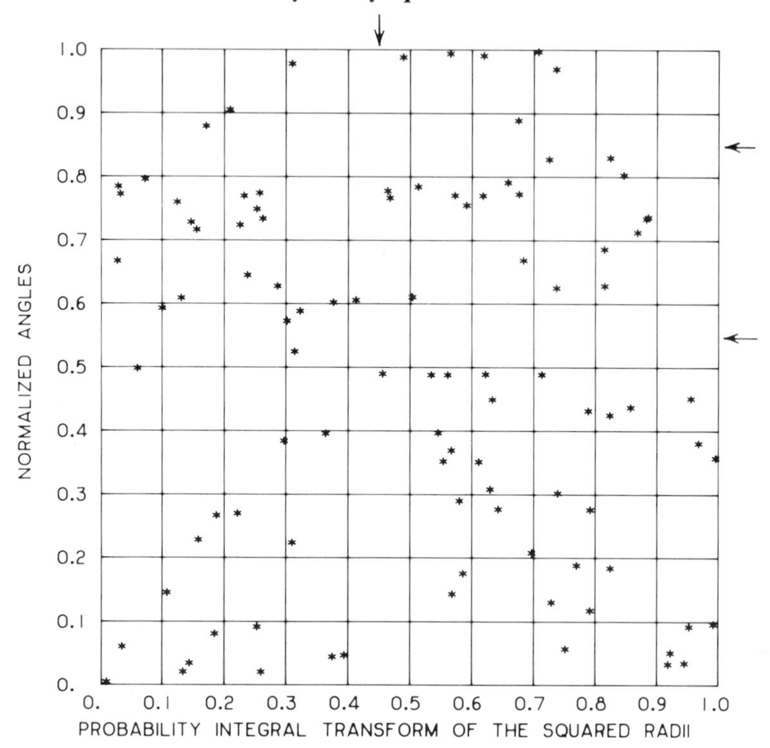

significant departures from multivariate normality. The exceptions include the D'Agostino test and the univariate gaps test for marginal normality, and the nearest distance test for bivariate normality. An important aspect of this example is the discovery of which methods did not detect the sort of departure incorporated in the data.

Example 29. The data for this example are 100 points from a correlated Laplace distribution, correlated by using Eq. 81 with $\rho = 0.9$. The distribution is long tailed but quite symmetric. Only the results of using the techniques for assessing joint and directional normality are described here.

The bivariate transformation procedure was applied to these data after initially shifting the observations to make them all lie in the first quadrant, and the results are summarized in Exhibit 29*a* (see page 186). The transformation utilized here involved only power parameters and no shift parameters. For this reason, one would expect sensitivity to skewness but not to long-tailedness in the presence of symmetry. Including shift parameters in the transformations of the variables would most probably remedy the situation, but the computational effort required would be substantially higher. At any rate, the nonsignificant result in Exhibit 29*a* is at least interpretable.

The nearest distance test also failed to detect any significant departure from normality in this case—the observed significance level was 0.7. On the other hand, the multivariate skewness and kurtosis tests revealed significant skewness and kurtosis departures; observed levels were 0.0055 and $<10^{-4}$, respectively. Examination of a scatter plot of the data suggested that it is quite reasonable to reject bivariate normality on grounds of both skewness and kurtosis.

The plotting procedures for radii and angles also proved useful once again. Exhibits 29*b–d* (see page 186–187) show the combined scatter and marginal probability plots of radii and angles for these data. The long-tailedness of the data is clearly evident in the $\chi^2_{(2)}$ probability plot of the

Exhibit 28i. Chi-squared (df=2) probability plot of squared radii

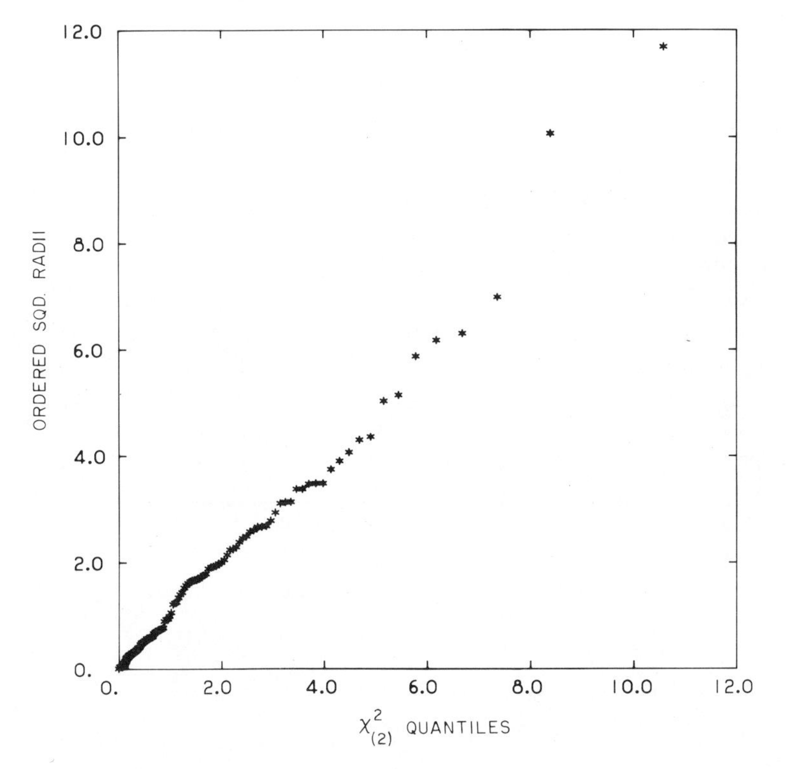

Exhibit 28j. Uniform probability plot of normalized angles

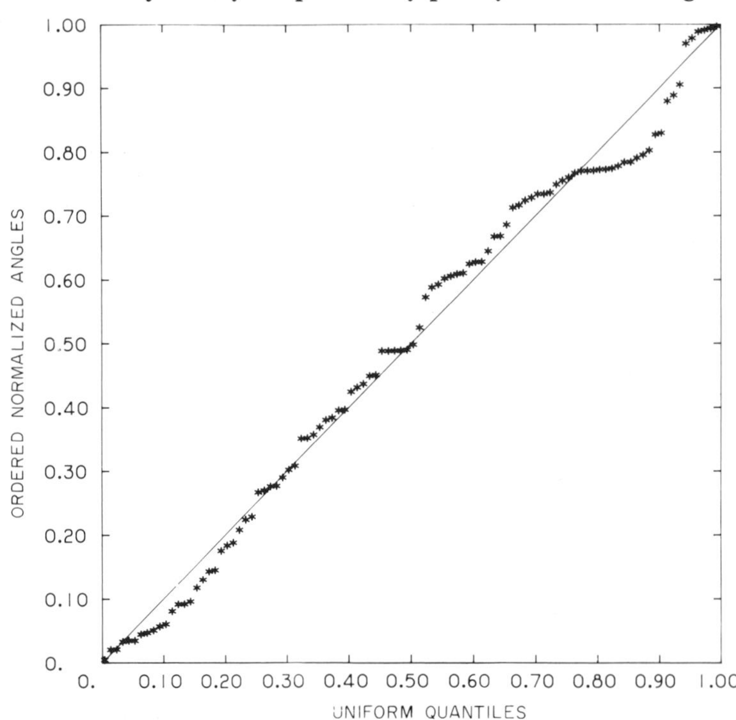

squared radii (Exhibit 29*c* see page 187). Some evidence of lack of spherical symmetry of the scaled residuals is provided by the uniform probability plot of the angles (Exhibit 29*d*).

Lastly, as a means of studying directional normality in this example, projections of the data were explored by employing directions $\mathbf{d}_{\alpha}^{*\prime}$ for a range of values of α, viz., $\alpha = -1, -0.5, -0.1, 0.1, 0.5, 1$. Exhibit 29*e* (see page 188) gives for each value of α the resulting direction, $\mathbf{d}_{\alpha}^{*\prime}$, together with the results of the univariate shifted-power transformation test procedure applied to the projections on the direction involved. The

Exhibit 28k. Directional normality test

Shift parameter estimate	$\hat{\xi}$	-5.66
Power parameter estimate	$\hat{\lambda}$	0.53
$\mathscr{L}_{\max}(\hat{\xi}, \hat{\lambda})$		-168.2
$\mathscr{L}_{\max}(\hat{\xi}, 1)$		-178.3
Approximate significance level		0.00001

Exhibit 28l. Normal probability plot of projections onto direction of nonnormality

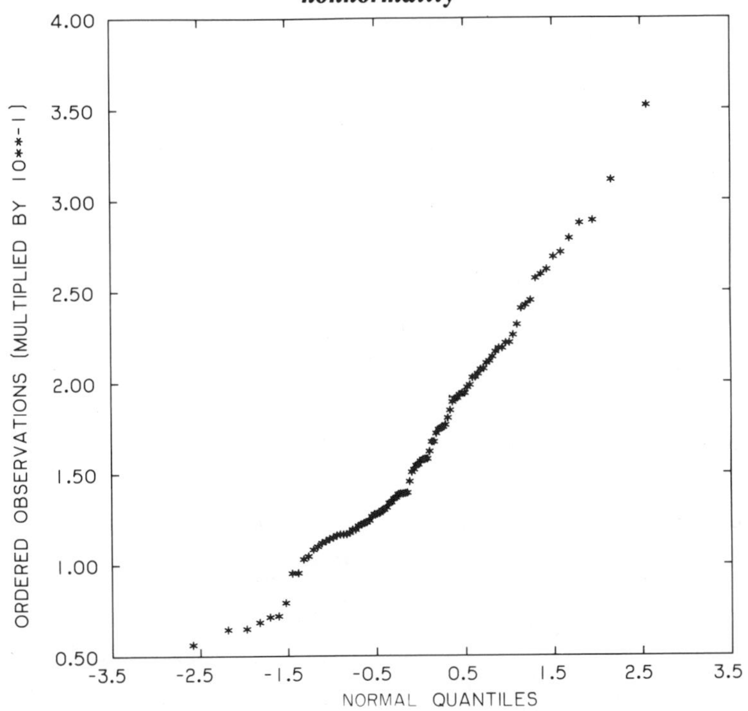

Exhibit 28m. Example 28 summary

Technique	Significance Level	
Marginal		
Marginal Box-Cox	0.0007	0.0001
Skewness	<0.01	<0.05
Kurtosis	Not sig.	Not sig.
D'Agostino-Pearson	≃0.015	Not sig.
Shapiro-Wilk	?	?
D'Agostino	>0.2	>0.2
Univariate gaps	~0.2	
Marginal probability plots	Some evidence of nonnormality	
Joint		
Scatter plot	Some evidence of nonnormality	
Bivariate transformation	0.06	
Nearest distance	0.4	
Radius and angles	Good evidence of nonnormality	
Mardia's tests	0.0014($b_{1,2}$)	$b_{2,2}$ not sig.
Directional transformation	0.00001	

Exhibit 29a. Bivariate power transformation test

Estimates of transformation parameters	
$\hat{\lambda}_1$	1.22
$\hat{\lambda}_2$	1.20
$\mathcal{L}_{max}(\hat{\lambda}_1, \hat{\lambda}_2) - \mathcal{L}_{max}(1, 1)$	0.91
Approximate significance level	0.40

directions determined by using $\alpha < 0$, being sensitive to the center of the data, do indicate more significant departures, and this is not very surprising in view of the difference between the densities of the Laplace and the normal in the center.

In summary, as expected, this symmetric nonnormality, which is an important though not sufficiently extreme departure, was not clearly detected by some procedures. The results for this example are summarized in Exhibit 29*f*. (See page 188)

Exhibit 29b. Scatter plot of normalized angles vs. probability integral transform of squared radii

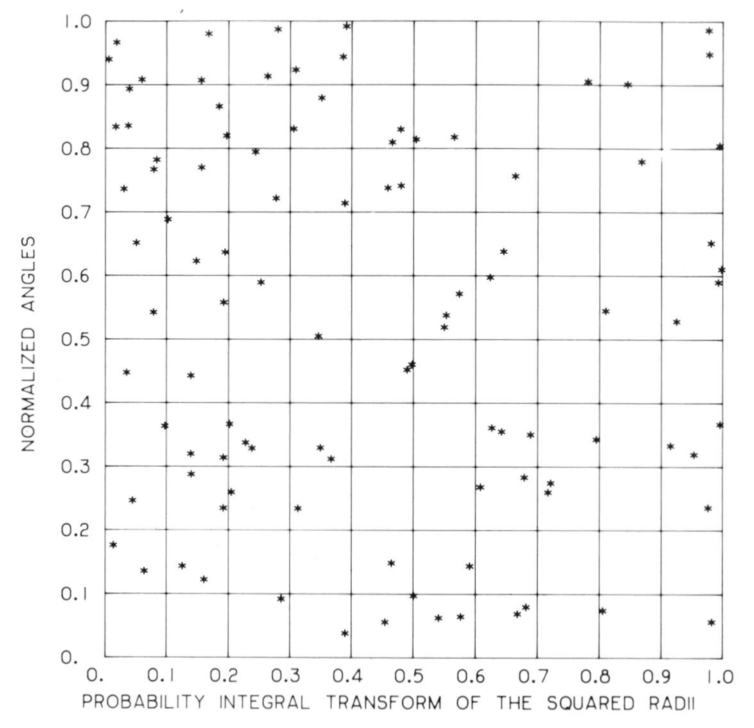

Exhibit 29c. Chi-squared (df=2) probability plot of squared radii

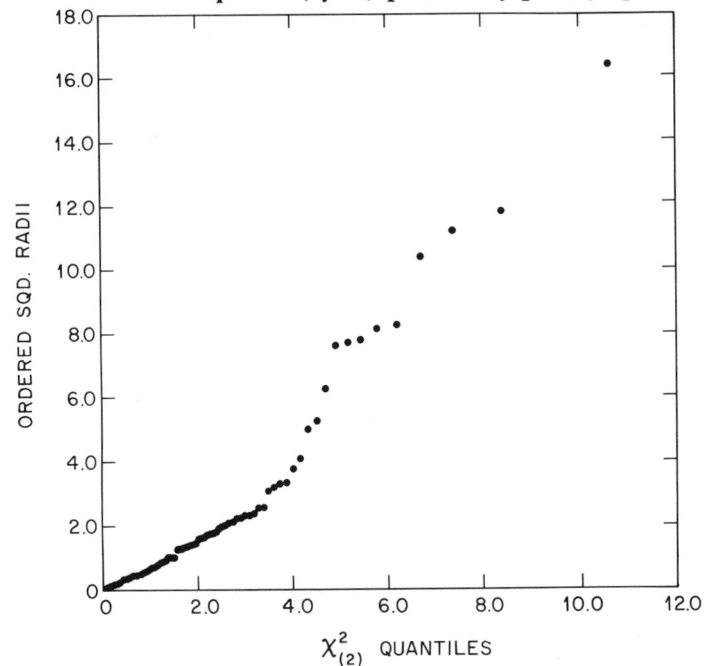

Exhibit 29d. Uniform probability plot of normalized angles

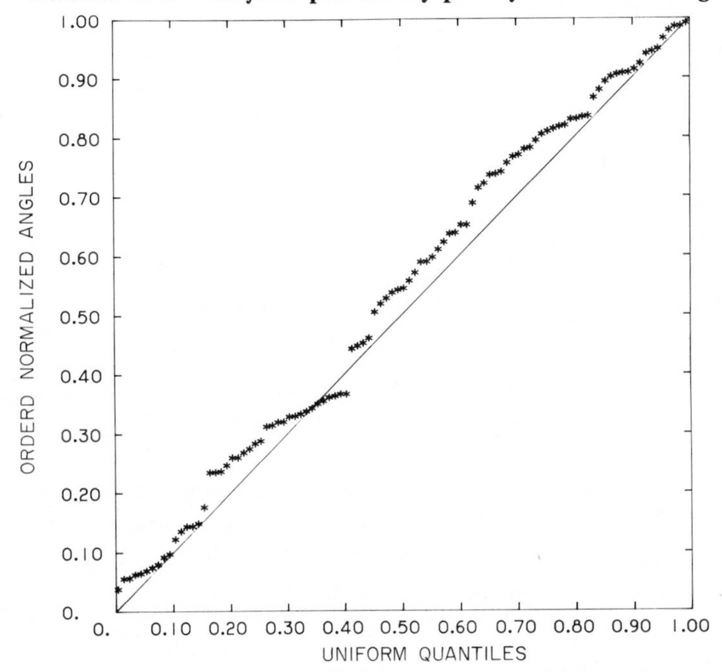

Exhibit 29e. Directional normality tests

α	$\mathbf{d}_\alpha^{*\prime}$	$\hat{\xi}$	$\hat{\lambda}$	$\mathcal{L}_{\max}(\hat{\xi}, \hat{\lambda}) - \mathcal{L}_{\max}(\hat{\xi}, 1)$	Approximate Significance Level
1.0	−0.65, −0.76	23.2	−1.0	3.2	0.011
0.5	−0.63, −0.78	24.5	−1.0	3.2	0.011
0.1	−0.59, −0.81	24.5	−1.0	3.1	0.013
−0.1	0.56, 0.83	10.0	1.8	4.1	0.004
−0.5	0.48, 0.88	9.7	1.8	4.0	0.005
−1.0	0.31, 0.95	8.9	1.7	3.9	0.005

Exhibit 29f. Example 29 summary

Technique	Approximate Significance Level
Bivariate power transformation	0.4
Nearest distance	0.7
Mardia's tests	$0.0055(b_{1,2})$, $<10^{-4}(b_{2,2})$
Radius-and-angles decomposition	Indication of departures from normality
Directional normality	<0.015 for $\alpha>0$, $\simeq 0.005$ for $\alpha<0$

Example 30. Since real data may not conform to any prespecified type of nonnormality of the kinds reflected in Examples 28 and 29, it is instructive to apply the techniques for assessing normality to observations that are not simulated. Thus the data for this example, which is taken from Standard and Poor's COMPUSTAT tape, consist of observed values of debt ratio and the dividends/price ratio for each of 94 utilities for the year 1969. Exhibit 30*a* (see page 189) is a scatter plot of the observations, and departures from normality are evident even in this simple plot.

The test proposed by D'Agostino (1971) was applied to both marginal distributions and did not indicate any strikingly significant departures from normality. The results of estimating a shifted-power transformation of each variable by the methods of Box and Cox (1964), and of applying the associated likelihood-ratio test of univariate normality to each variable separately, are summarized in Exhibit 30*b* (see page 189). The transformation-based test indicates a highly significant departure from normality for the distribution of values of the dividends/price ratio.

Exhibits 30*c* (see page 190) and 30*d* (see page 191) are normal probability plots of debt ratio and dividends/price ratio, respectively. Both plots exhibit noticeable deviations from the null straight line config-

Exhibit 30a. Scatter plot of dividends/price vs. debt ratio for 94 utilities in 1969

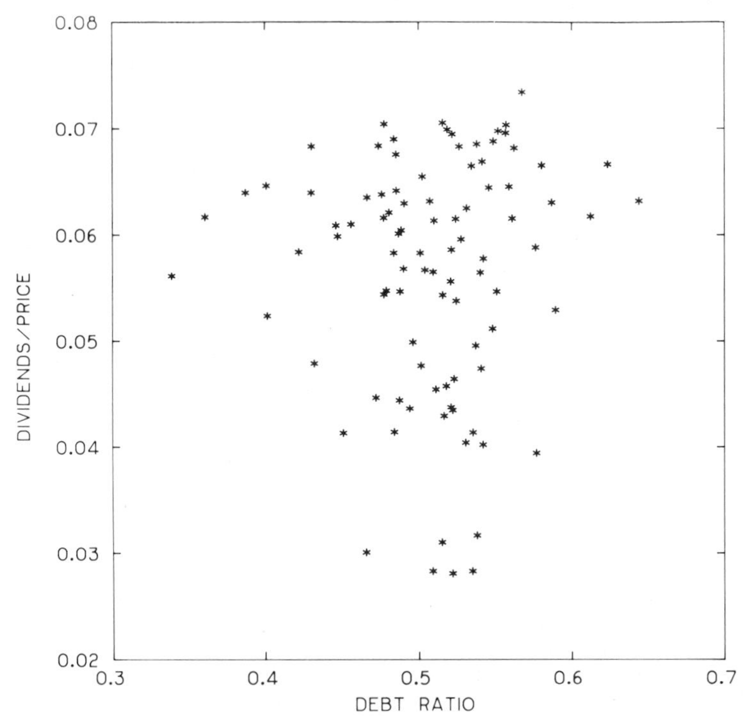

uration to be expected for normality. Exhibit 30*d* with marked curvature indicates an abnormally short upper tail of the distribution of the dividends/price ratio. Also, six observations in the lower tail are distinctly separated from the rest of the data.

The bivariate transformation procedure also detected significant non-normality. Exhibit 30*e* presents the results of this procedure. Some

Exhibit 30b. Results of Box-Cox transformation test

	Debt Ratio	Dividends/Price
Parameter estimates		
$\hat{\xi}$	−0.127	0.195
$\hat{\lambda}$	1.980	10.873
Log likelihood-ratio value	1.826	7.854
Approximate significance level	$0.05 \leq p < 0.058$	$0.00006 \leq p \leq 0.00008$

Exhibit 30c. Normal probability plot for debt ratio

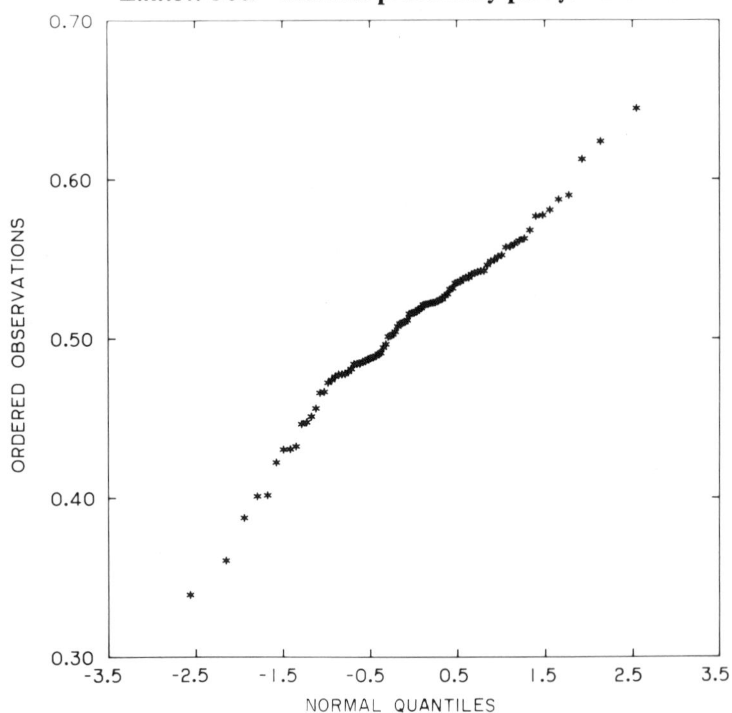

evidence of nonnormality appears in both variables, as indicated by the values of $\hat{\lambda}_1$ and $\hat{\lambda}_2$.

The nearest distance test was used with both linear and quadratic regressions as the basis for studying the dependence of transformed nearest neighbor distances on the location of the point from which the distances were measured. The significance levels of these two regression tests were 0.031 and 0.093, respectively. In this example, therefore, this test provides some indication, although not very strong evidence of nonnormality.

Exhibit 30*f* (see page 192) is a scatter plot of the radii-and-angles decomposition. Exhibit 30*g* (see page 193) is a $\chi^2_{(2)}$ probability plot of the squared radii, and Exhibit 30*h* (see page 194) is the uniform probability plot of the normalized angles. The nonuniform scatter in Exhibit 30*f*, and especially the sparseness in several contiguous blocks, indicate departures from bivariate normality. Exhibit 30*g* shows peculiarities in the upper tail of the distribution of the squared radii, Exhibit 30*h* also manifests some departures from spherical symmetry in the distribution of the scaled residuals.

Exhibit 30d. Normal probability plot for dividends/price

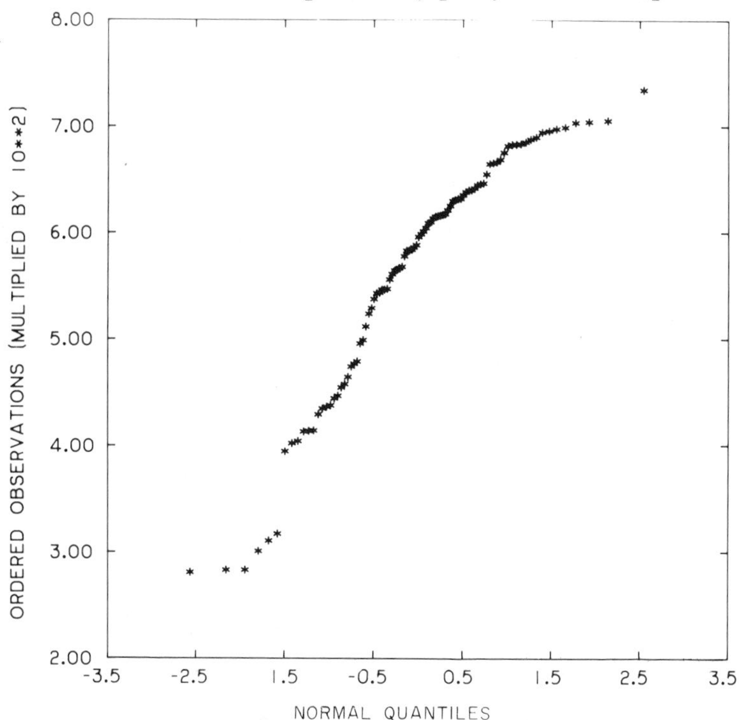

The directional normality procedure also suggested a somewhat, but not strikingly, significant departure from normality. Exhibit 30*i* (see page 194) presents the results of the directional normality test for the case of $\alpha = 1.0$. Here $\mathbf{d}_{1.0}^*$ is clearly influenced heavily by the second variable (dividends/price ratio). This is not too surprising in the light of the more striking nonnormality of the second variable, as revealed by the tests for marginal normality discussed earlier.

The results for this example are summarized in Exhibit 30*j* (see page 195).

In this subsection various techniques have been described for assessing the normality of the distribution of multiresponse data. Many of the new

Exhibit 30e. Bivariate power transformation test

$\hat{\lambda}_1$	2.719
$\hat{\lambda}_2$	2.375
$\mathcal{L}_{\max}(\hat{\lambda}_1, \hat{\lambda}_2) - \mathcal{L}_{\max}(1, 1)$	8.227
Approximate significance level	0.00027

Exhibit 30f. Scatter plot of normalized angles vs. probability integral transform of squared radii

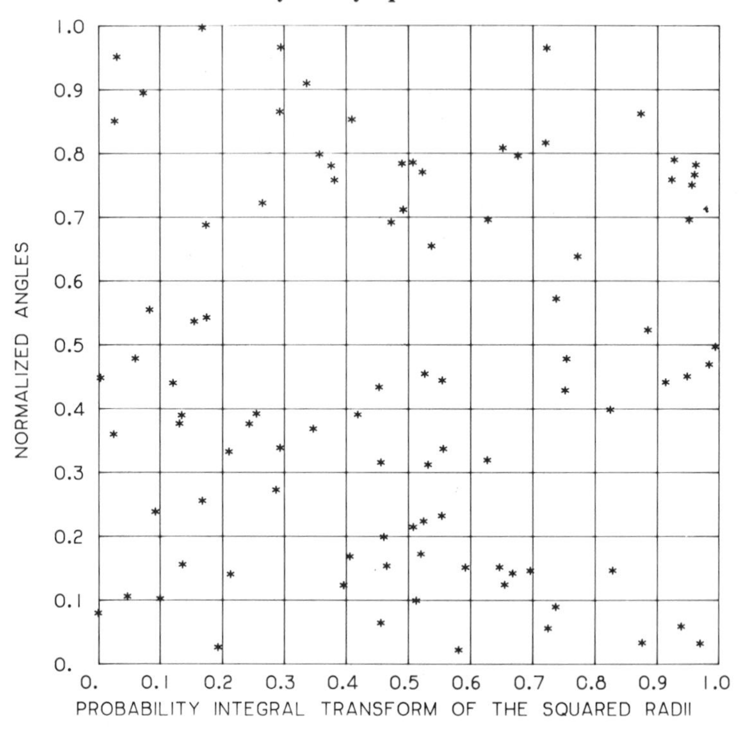

techniques (e.g., the nearest distance test, the radius-and-angles plots) need further theoretical investigation, as well as practical use and exposure. On the theoretical front some refinement of the distributional approximations involved in some of the procedures (e.g., the univariate gaps test and the multivariate nearest distance test) may be in order. Also, a better understanding is needed of issues such as the nonuniqueness of some of the preliminary transformations of the data (e.g., the Rosenblatt, 1952, transformation mentioned in connection with the nearest distance test) and the effects of using the sample mean vector and covariance matrix in place of the corresponding population quantities in some of the methods.

More work, of course, is needed to promote understanding of the relative sensitivities of the different procedures. This is necessary, not for picking an optimal test for normality, but for general guidance in interpreting the results in specific applications of these techniques.

General indications concerning the newer techniques are that the transformation-related methods and the plotting procedures based on the radius-and-angles representation of multivariate data appear to be promising tools for data analysis. The transformation-related methods have appeal above and beyond serving as tests of significance because of the fact that estimates of the transformation (admittedly within some class such as the shifted-power one) are included as an integral part of the method and are likely to be very useful in the next step of the analysis. Here too, however, enlarging the class of transformations to include additional types would be useful for practitioners. Also, a limitation of the transformation-related approach to testing normality should be noted. Since there is no guarantee that a specific member of a class of transformations, such as the shifted-power class, will necessarily achieve normality, the evidence, as provided by the transformation test, that no transformation is required cannot be taken entirely at face value as adequate support for normality. Specifically, if one were to transform a set of data by the techniques described in Section 5.3 and treat the resulting transformed data as input to the same transformation techniques, one would

Exhibit 30g. Chi-squared (df=2) probability plot of squared radii

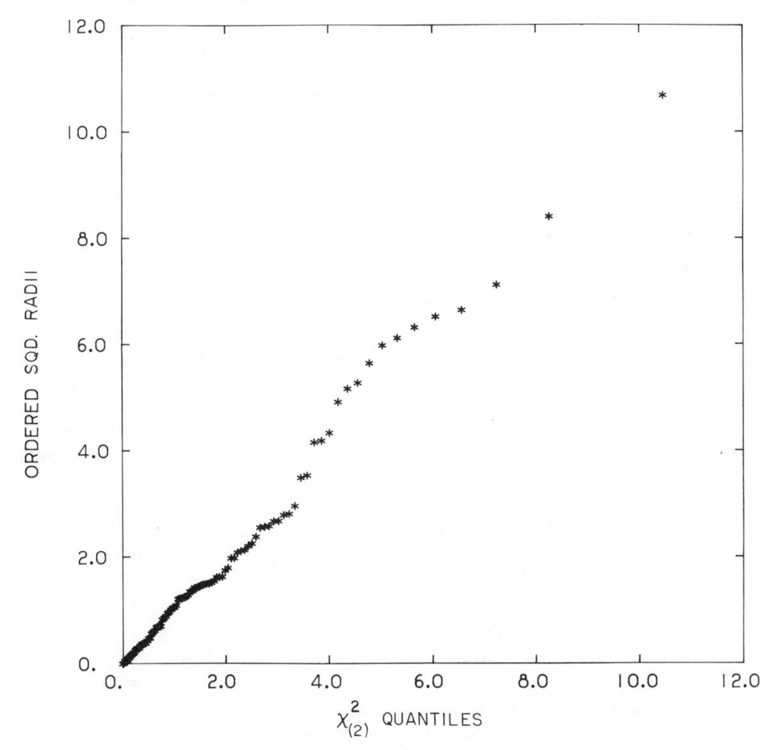

Exhibit 30h. Uniform probability plot of normalized angles

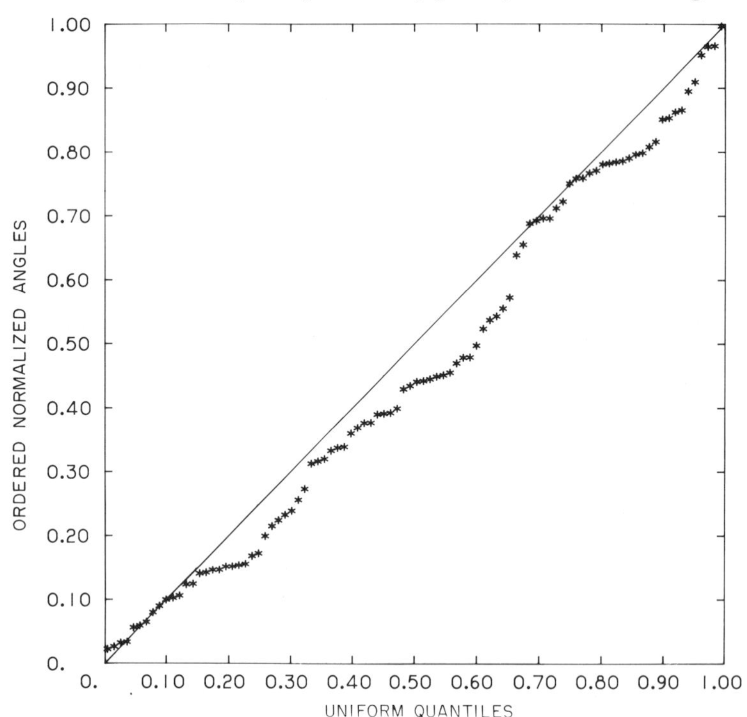

necessarily get indications that no further transformation (within the class considered) was required, but this would be just an artifact of iterating the transformation technique. The "direct" techniques (i.e., those not related to transformations) for assessing normality do not suffer from such a limitation.

The plotting procedures associated with the radius-and-angles decompositions have particular appeal as informal but informative graphical aids for data analysis. Additional graphical methods, especially directed toward assessing the normality of high-dimensional data, would indeed be

Exhibit 30i. Directional normality

$\mathbf{d}^{*\prime}_{1.0} = (-0.462, -0.887)$

Shift parameter estimate	$\hat{\xi}$	4.415
Power parameter estimate	$\hat{\lambda}$	−10.703
$\mathcal{L}_{\max}(\hat{\xi}, \hat{\lambda}) - \mathcal{L}_{\max}(\hat{\xi}, 1)$		2.487
Approximate significance level		0.026

Exhibit 30j. Example 30 summary

Technique	Approximate Significance Level	
Marginal Box-Cox	0.06	0.00008
Marginal probability plots	Good evidence of marginal nonnormality	
Bivariate power transformation	0.0003	
Radii and angles	Good evidence of joint nonnormality	
Directional normality	0.03	

worth developing (see the discussion of one such tool in Section 6.2). Such graphical techniques often exemplify the significant value of a statistical tool which may have been designed for one purpose but turns out to have a variety of additional applications. Thus, for instance, the usefulness of the techniques considered in this subsection for assessing distributional normality is greatly enhanced by their possible utility in detecting additional data anomalies such as outliers.

REFERENCES

Section 5.1 Anderson (1958), Puri & Sen (1971), Rao (1965), Roy (1957).

Section 5.2.1 Anderson (1958), Chambers (1974), Golub & Reinsch (1970), Pillai & Jayachandran (1967), Roy & Bose (1953), Roy et al. (1971), Scheffé (1953, 1959), Stein (1956), Wilkinson (1970).

Section 5.2.2 Anderson (1958, 1969), Jöreskog (1973), Lindley (1972), Rao (1965), Roy & Gnanadesikan (1957, 1962), Srivastava (1966), Stein (1956, 1965).

Section 5.2.3 Andrews (1973, 1974), Andrews et al. (1972), Bickel (1964), Devlin et al. (1974), Gentleman (1965), Gnanadesikan & Kettenring (1972), Hampel (1968), Huber (1964, 1970, 1972, 1973), Johnson & Leone (1964), Mallows (1973), McLaughlin & Tukey (1961), Mood (1941), Teichroew (1956), Tukey (1960), Tukey & McLaughlin (1963), Wilk & Gnanadesikan (1964), Wilk et al. (1962).

Section 5.3 Andrews et al. (1971), Box & Cox (1964), Chambers (1973), Cox (1970, 1972), Moore & Tukey (1954), Puri & Sen (1971), Tukey (1957).

Section 5.4 Kendall (1968).

Section 5.4.1 Aitchison & Brown (1957), Chambers (1973), Gnanadesikan (1972), Laue & Morse (1968), Wilk & Gnanadesikan (1968).

Section 5.4.2 Aitkin (1972), Anderson (1966), Andrews (1971), Andrews, Gnanadesikan & Warner (1971, 1972, 1973), Box & Cox (1964), D'Agostino (1971), D'Agostino & Pearson (1973), David & Johnson (1948), Devlin et al. (1975), Downton (1966), Gnanadesikan & Kettenring (1972), Healy (1968), Kessell & Fukunaga (1972), Malkovich & Afifi (1973), Mardia (1970), Pearson & Hartley (1966), Rosenblatt (1952), Roy (1953), Sarhan & Greenberg (1956), Shapiro & Wilk (1965), Shapiro et al. (1968), Weiss (1958).

CHAPTER 6

Summarization and Exposure

6.1. GENERAL

The main function of statistical data analysis is to extract and explicate the informational content of a body of data. The processes of description and communication of the information involve *summarization*, perhaps in terms of a statistic (e.g., a correlation coefficient) which may be undergirded by some reasonably tightly specified model, or perhaps in terms of a simple plot (e.g., a scatter plot). In addition to the well-recognized traditional role of summarization, however, the meaningful exercise of processes of data analysis requires *exposure*, i.e., the presentation of analyses so as to facilitate the detection of not only anticipated but also unexpected characteristics of the data. For instance, an x-y scatter plot of data is a pictorial representation that is useful not only for interpreting the computed value of the correlation coefficient for that body of data (see also the scatter plots described in Section 6.4.2) but also for indicating the adequacy of assuming linearity of the relationship between x and y. A more substantial example of the twin-pronged process of summarization and exposure is fitting a straight line to y versus x data and then studying the residuals in a variety of ways, especially through different plots of them, such as plots of residuals against observed values of x and of y, perhaps against values of relevant extraneous variables such as time, and also probability plots of the residuals. Although the fitting provides summarization, the study of the residuals is often crucial in exposing the inadequacies of various assumptions that underlie the fitting procedure (e.g., constancy of variance).

Pedagogy, publications, and, more generally, the codification of statistical theory and methods have been concerned almost exclusively with formal procedures such as tests of hypotheses, confidence region estimation, and various optimality criteria and associated methodology. Even when the concern has been with developing methods for summarization, with a clear awareness of the possible inadequacies or inappropriateness of certain "standard" assumptions, the goal of summarization has often been somewhat artificially separated from that of exposure. For instance, much of the work on robust estimation, while usefully concerned with summary statistics that are not unduly influenced by a small fraction of possibly deviant observations, has adopted the formal and familiar framework of point estimation with its criteria of bias, efficiency, etc., and relatively little attention has been paid to the exposure value of such things as the residuals obtained from using the robust estimates in place of the standard estimates.

The manifold theories of statistical inference that have been advanced as the focal points for "unifying" statistics have only relatively recently (Cox, 1973; Cox and Hinkley, 1974) been considered and carefully scrutinized in terms of their relevance for and relationship to the needs of applications of statistics. No single formal theory of statistical inference seems able to encompass and completely subsume the flexible and interactive processes involved in summarization and exposure. There are, however, less formal techniques that are perhaps not in the mainstream of any formal statistical theory but nevertheless are useful tools of informative inference directed toward the dual objectives of summarization and exposure.

The treatment in this book has attempted, even when a problem has been formulated fairly narrowly (e.g., tests for commonality of marginal distributions), to combine formal methods where available with informal procedures for revealing the relevant information in multiresponse data. A feature common to most of the informal procedures is their graphical nature. In the following sections of this chapter, some general problem areas of summarization and exposure are distinguished (and inevitably these overlap to some degree the concerns of the earlier chapters), and some techniques of relevance to these problems are discussed. The emphasis throughout this chapter is on graphical methods.

6.2. STUDY OF AN UNSTRUCTURED MULTIRESPONSE SAMPLE

One is often interested in examining a body of data *as if* it were an unstructured collection or sample, and many of the techniques discussed in the earlier chapters of this book have been described in the context of analyzing an unstructured sample (e.g., linear and generalized principal components analysis in Chapter 2, robust estimates of location and dispersion in Chapter 5, and the assessment of distributional properties in

Chapter 5). With uniresponse data several graphical and semigraphical techniques are available for analyzing an unstructured sample. Some examples of such techniques are stem-and-leaf displays and box plots (see Chapters 1 and 5 of Tukey, 1970) and the more familiar histogram, empirical cumulative distribution function (or ecdf, which may be defined as a plot of the ith ordered observation against $(i-\frac{1}{2})/n$), and the class of techniques loosely called *probability plotting methods* (see Wilk and Gnanadesikan, 1968).

Wilk and Gnanadesikan (1968) describe two basic types of probability plots, called *P-P* and *Q-Q* plots, respectively. Figure 8 may be used for defining the two types. In comparing two distribution functions, a plot of points whose coordinates are the cumulative probabilities $\{p_x(q), p_y(q)\}$ for different values of q is a *P-P* plot, while a plot of the points whose coordinates are the quantiles $\{q_x(p), q_y(p)\}$ for different values of p is a *Q-Q* plot. For conceptual convenience both of the distribution functions displayed in Figure 8 are shown as smooth curves, but this is not an essential requirement in that one or both of the distribution functions involved can be a step function or an ecdf. In fact, the usual form of the comparison is one in which an ecdf for a body of univariate data is compared to a specified (or theoretical) distribution function, i.e., a step function is compared to a continuous one. Also, *Q-Q* probability plots tend to be more widely used than *P-P* probability plots. Perhaps one

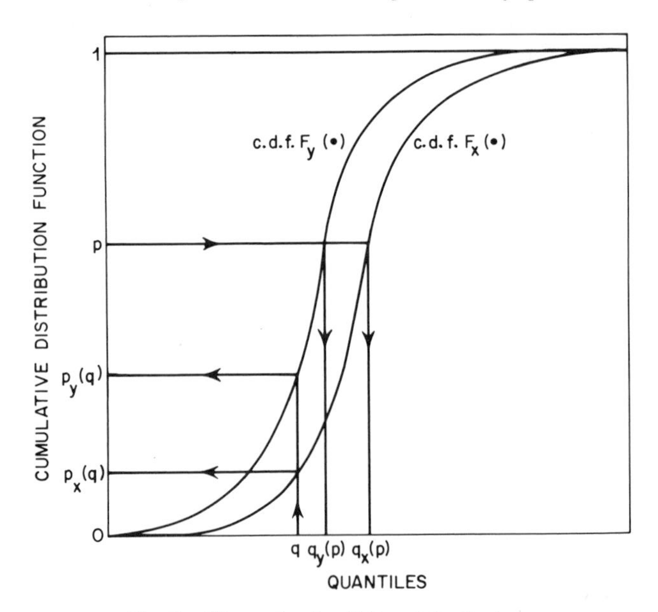

Fig. 8. Illustration for *P-P* and *Q-Q* plots.

reason for this is a property of linear invariance possessed by Q-Q but not P-P plots, viz., when the two distributions involved in the comparison are possibly different only in location and/or scale, the configuration on the Q-Q plot will still be linear (with a nonzero intercept if there is a difference in location, and/or a slope different from unity if there is a difference in scale), whereas the configuration on a P-P plot will in general be *necessarily* linear (with zero intercept and unit slope) only if the two distributions are identical in all respects, including location and scale.

At any rate the following is a canonical description of a Q-Q probability plot in its most widely used form, wherein an ecdf of an unstructured sample, $x_1, \ldots, x_n$, of size n is to be compared with a hypothesized standardized (i.e., origin or location parameter is zero and scale parameter is unity) distribution function $F(x; \boldsymbol{\theta})$ (where the parameters $\boldsymbol{\theta}$, which do not include origin or location and/or scale parameters, have specified values): if $x_{(1)} \le x_{(2)} \le \cdots \le x_{(n)}$ are the ordered observations, plot the n points $\{\tilde{x}_i, x_{(i)}\}$, $i = 1, \ldots, n$, where $\tilde{x}_i$ is the quantile of the distribution function F corresponding to a cumulative probability p_i $[(i-\alpha)/(n-2\alpha+1)$ with $\alpha = \frac{1}{2}, \frac{1}{3}$, or 0 as some of the possible choices], i.e., $\tilde{x}_i$ is defined by $F(\tilde{x}_i; \boldsymbol{\theta}) = p_i$. The well-known use of normal probability paper for plotting data (which was mentioned, for instance, in Section 5.4.2 in connection with evaluating marginal normality) is an example of the making of a Q-Q plot with F taken as the distribution function Φ of the standard normal distribution. Other examples of specification of F are a chi-squared distribution with a specified degree of freedom, a gamma distribution with a specified shape parameter, and a beta distribution with values specified for both of its shape parameters. (See Wilk et al., 1962a; Gnanadesikan et al., 1967.)

For uniresponse observations, in addition to ecdf's, probability plots, and the other displays mentioned above, there are some simple graphical displays for aiding in the assessment of symmetry of the data distribution. From the viewpoint of multiresponse data analysis, symmetry of the marginal distributions of each response is not an unreasonable requisite for the meaningful use of several summary statistics such as correlation coefficients and covariance matrices. If the raw data are quite asymmetric, a preliminary transformation of the observations (perhaps by the methods of Section 5.3) to enhance symmetry will often be a sensible first step before the subsequent univariate or multivariate analyses that may be performed on the transformed data.

The ecdf itself is often a good means of studying symmetry. However, other plots specifically useful for investigating possible asymmetry in data can also be made. For instance, if $x_{(1)} \le x_{(2)} \le \cdots \le x_{(n-1)} \le x_{(n)}$

denote the ordered observations, Wilk and Gnanadesikan (1968) have suggested plotting the points whose coordinates are $\{x_{(1)}, x_{(n)}\}$, $\{x_{(2)}, x_{(n-1)}\}$, etc. If the observations are symmetric around a center of symmetry $x = b$, such a plot should look reasonably linear with intercept approximately equal to $2b$ and slope approximately equal to -1. Departures from such a linear configuration will indicate the type of asymmetry present in the data. For instance, an upward bow to the plot will indicate a longer upper tail; a downward bow, a longer lower tail. In a variant of this plot for assessing symmetry the points plotted have coordinate values that are deviations from the median of the observations, i.e., the points plotted are $\{x_M - x_{(1)}, x_{(n)} - x_M\}$, $\{x_M - x_{(2)}, x_{(n-1)} - x_M\}$, etc., where x_M denotes the median. For symmetric data the configuration on such a plot will, therefore, be linear with zero intercept and unit slope, and departures from such a "null" configuration can be easily appreciated and interpreted. Another plotting procedure for studying symmetry, proposed by Tukey (cf. Wilk and Gnanadesikan, 1968), consists in plotting the points whose coordinates are, respectively, differences and sums of the symmetrically situated pairs of ordered observations $x_{(i)}$ and $x_{(n-i+1)}$, i.e., the plotted points are $\{x_{(n)} - x_{(1)}, x_{(1)} + x_{(n)}\}$, $\{x_{(n-1)} - x_{(2)}, x_{(2)} + x_{(n-1)}\}$, etc. This is a scheme for "tilting" the plots so that the "null" configuration becomes a horizontal linear one, and departures of the data from symmetry will be indicated by deviations from horizontality. The next example illustrates the use of these three graphical methods for assessing symmetry.

Example 31. The observations are maximum daily ozone measurements (in ppm) as observed at a particular air monitoring site in New Jersey during certain months in 1973. Exhibit 31*a* (see page 201) shows a plot of the symmetrically situated ordered observations $\{x_{(i)}, x_{(n-i+1)}\}$, $i = 1, \ldots, [n/2]$, and the upward bow of the plot suggests a positively skewed distribution for the observations. Exhibit 31*b* (see page 202) shows a plot of the deviations from the median, viz., $\{x_M - x_{(i)}, x_{(n-i+1)} - x_M\}$ for $i = 1, \ldots, [n/2]$, and the departure from the line of zero intercept and unit slope is quite strikingly indicative of a positively skewed distribution. Exhibit 31*c* (see page 203) shows the plot suggested by Tukey; the systematic deviation from a horizontal linear configuration is evident here too.

In an attempt to improve the symmetry of the data in this example, square roots of the observations were taken, and Exhibit 31*d* (see page 204) shows a plot of the deviations from the median for the transformed data. A comparison with Exhibit 31*b* reveals the clear accomplishment of the square-root transformation in symmetrizing the data. The lognormal distribution has often been used as the model for ambient air quality

Exhibit 31a. Plot of upper vs. lower half of the sample for ozone data

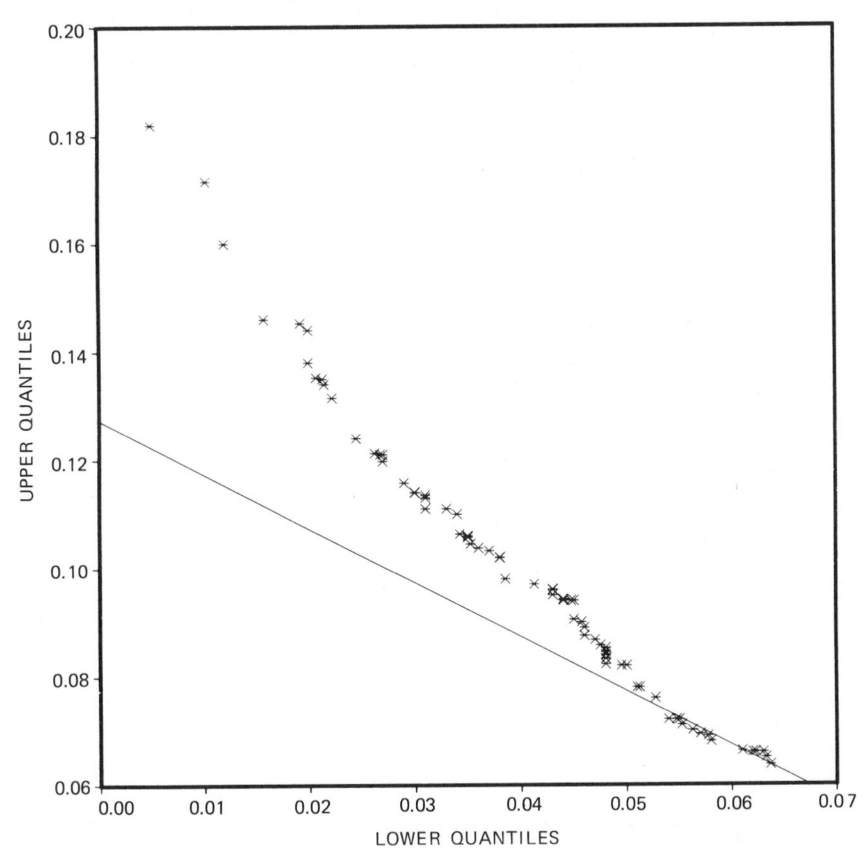

measurement data (see e.g., Zimmer & Larsen, 1965). For the present data Exhibit 31*e* (see page 205) shows a plot of deviations from the median when logarithms of the observations are used, and a comparison of Exhibits 31*b*, *d*, and *e* reveals that the square-root transformation is better for symmetrizing the data than the logarithmic transformation, which results in a negatively skewed distribution for the data in this example. More extensive evidence in favor of the square-root transformation in connection with ambient air quality data is contained in the work of Cleveland et al. (1975).

For multiresponse data there does not seem to be any extension of uniresponse *Q-Q* probability plotting, perhaps because no unique (or even generally useful) way of defining quantiles is available. Even more basically, it is only recently (see Hartigan, 1973) that convenient graphical

Exhibit 31b. Plot of deviations of upper quantiles from median vs. deviations of lower quantiles from median for ozone data

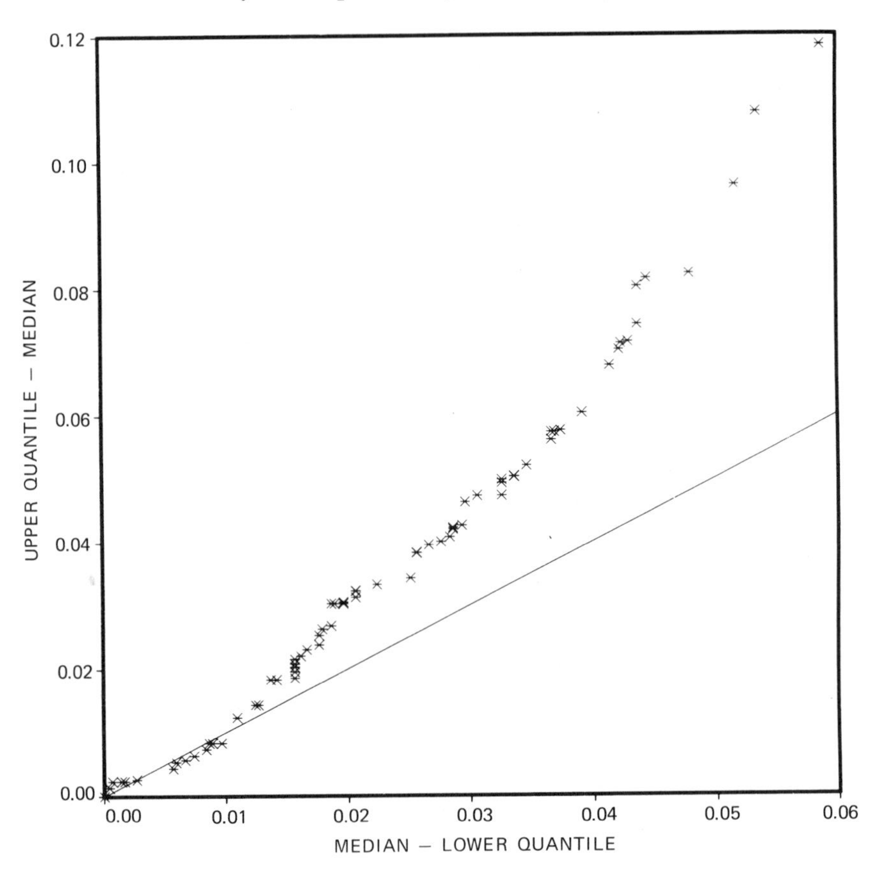

representations of multivariate (especially with $p \geq 3$) histograms have been proposed.

Nevertheless some things *can* be done to provide insights into multiresponse data configurations, and a few graphical techniques (some of which have been described and used in earlier chapters) are worth explicit mention here.

1. Two- and three-dimensional scatter plots of bivariate and trivariate subsets of the original data can be useful for studying cohesiveness, separation within the sample, possible outliers, and general shape.

Devlin et al. (1975) have suggested a way of augmenting the pictorial value of two-dimensional scatter plots for judging the effects of individual observations on a correlation coefficient computed from the points exhibited in a scatter plot. The suggestion is to display on the scatter plot the contours of a so-called influence function (see Hampel, 1968, 1974) of the correlation coefficient so as to enable one not only to gain an overall appreciation of the strength of the correlation but also to gauge how much the computed value of the correlation coefficient can be altered (inflated or deflated) by the omission of individual observations. An example of a scatter plot with superimposed influence function contours is given in Section 6.4.2.

With the availability of interactive graphical display facilities, one could sweep through a series of two-dimensional projections in addition to those onto the original coordinate planes and gain a good appreciation of

Exhibit 31c. Plot of sum of and difference between upper and lower quantiles for ozone data

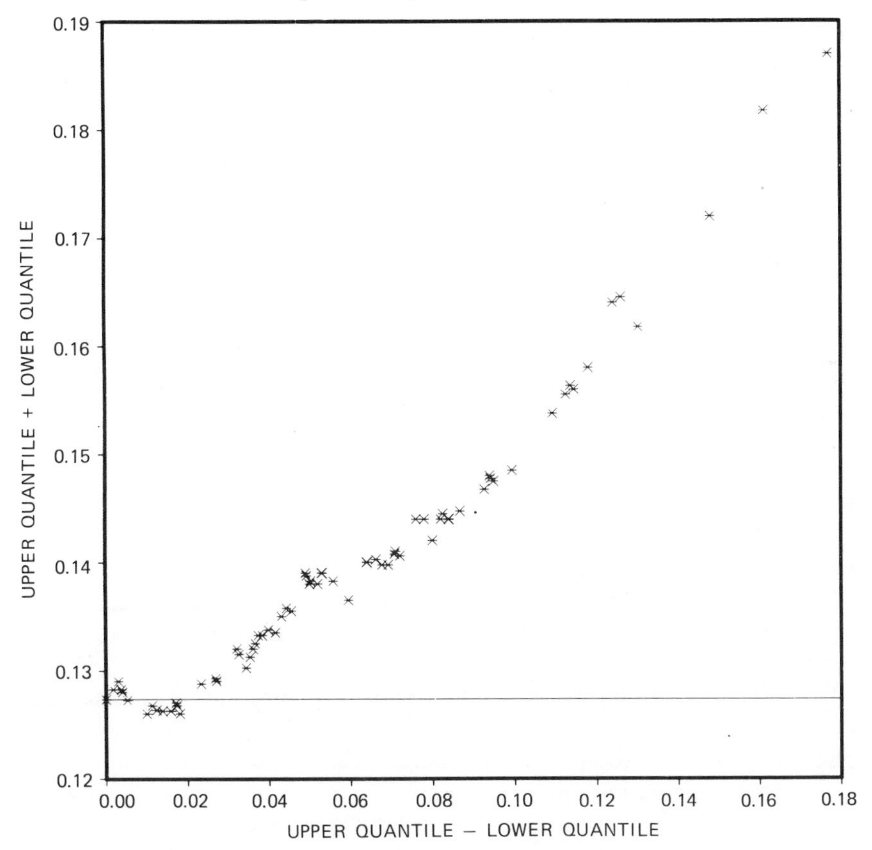

Exhibit 31d. Plot of deviations of upper quantiles from median vs. deviations of lower quantiles from median for square roots of ozone data

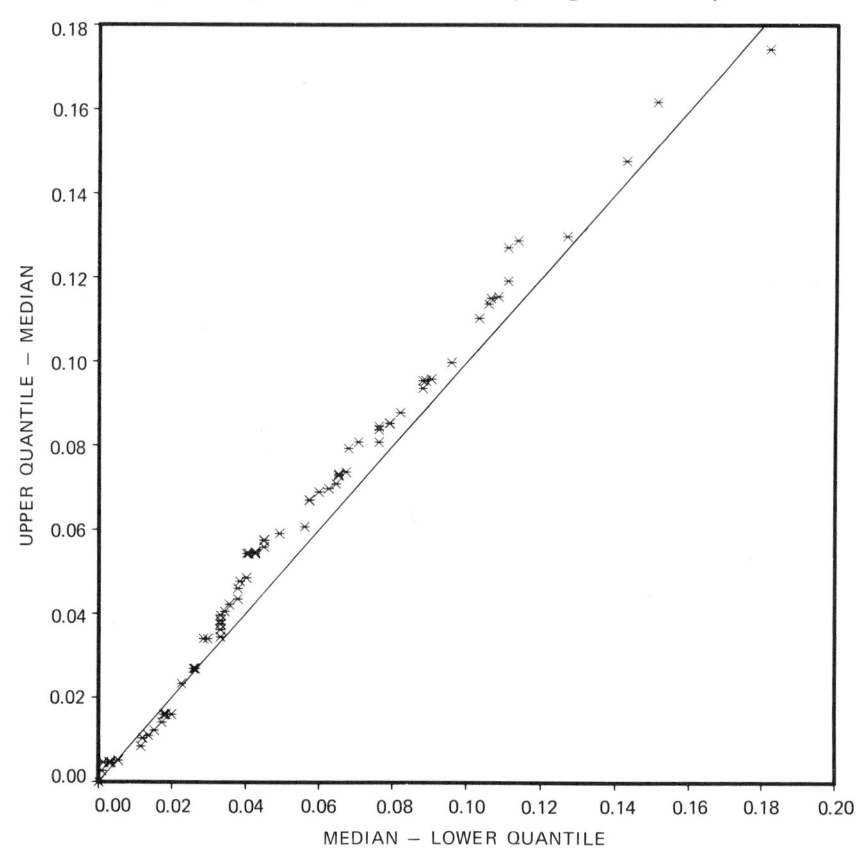

the structure of high-dimensional data. (See Fisherkeller et al. 1974, for a very interesting example of such an interactive display system.)

2. Probability plots of the observations on each response separately will generally be useful, in conjunction with other multivariate analyses. A natural base or starting point for such plotting will often be the normal distribution. Such plots may indicate the desirability of marginal transformations or of more appropriate and insightful kinds of probability plots.

3. Scatter plots and/or probability plots of projections onto eigenvectors from either linear or generalized principal components analysis can also be made and studied with benefit in many cases.

4. Joint evaluation of the eigenvalues of a sample covariance or correlation matrix is a problem often associated with the analysis of an

unstructured multiresponse sample. Such evaluations may, for instance, provide the key to possible reduction of dimensionality. The adequate assessment of specific sample eigenvalue results is not, however, an elementary task. The fact is that, even for large samples from spherical multivariate distributions, the eigenvalues may exhibit substantial variability (see Example 13 of Gnanadesikan and Wilk, 1969). For this problem of assessing a collection of eigenvalues, a Q-Q type of probability plot has been proposed (Gnanadesikan, 1968, 1973). The idea is to plot the ordered eigenvalues against their expected (or some other typical) values, determined by using the null assumption of sampling from a standard spherical normal distribution. The plotting positions (i.e., the "expected" values) may be determined by either a simple or a more

Exhibit 31e. Plot of deviations of upper quantiles from median vs. deviations of lower quantiles from median for logarithms of ozone data

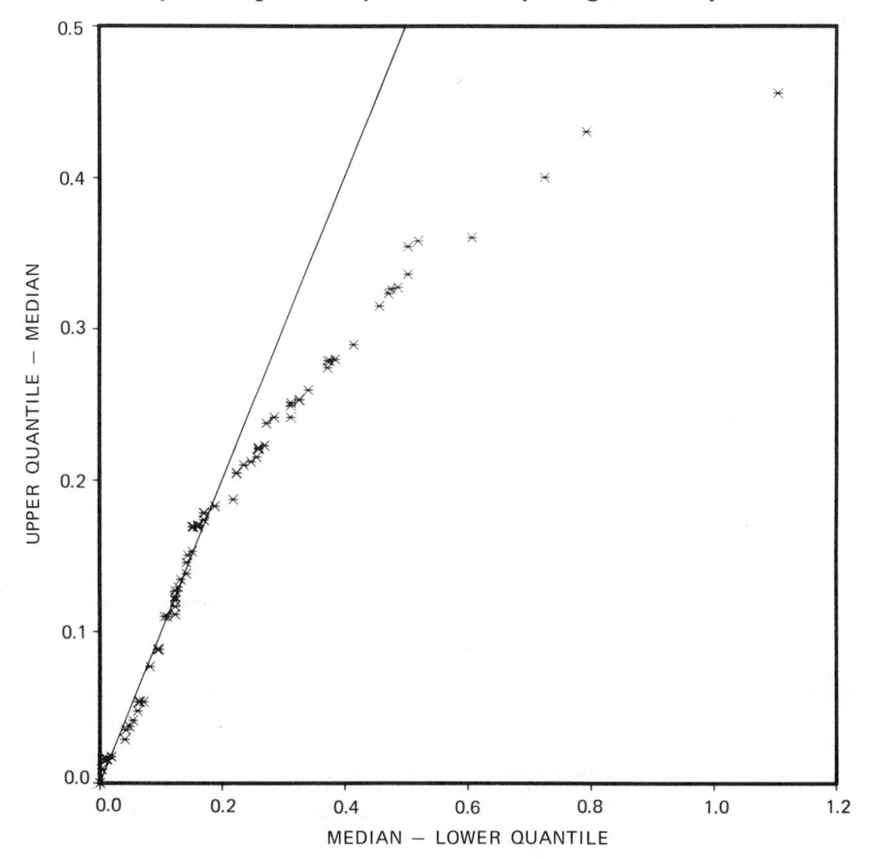

sophisticated and efficient (Hastings, 1970) Monte Carlo approach. The work of Stein (1969) and of Mallows and Wachter (1970) on asymptotic configurations of eigenvalues of Wishart matrices also provides a basis for plotting the ordered eigenvalues of a covariance (albeit not a correlation) matrix against a set of corresponding quantiles of a particular distribution.

Example 32. For this example 25 random deviates from a 10-dimensional normal distribution were generated. The underlying dispersion of the computer-generated data was much larger along two of the coordinates than along the other eight, so that the variability observed in the 10-dimensional sample would be expected to be confined largely to a two-dimensional subspace.

Exhibit 32 shows a plot of the 10 ordered eigenvalues of the sample covariance matrix against simple Monte Carlo estimates of their respective expected values under sampling from a 10-dimensional standard spherical normal distribution. The two largest eigenvalues clearly deviate from the configuration indicated by the smaller eight eigenvalues. Replotting (i.e., redetermination of the plotting positions on the basis of sampling from an eight-dimensional standard spherical normal) the smaller eight eigenvalues would be useful for studying the cohesiveness of and/or groupings among them.

This Q-Q type of graphical analysis of eigenvalues is useful not only for isolating large eigenvalues but also for identifying cases in which the

Exhibit 32. "Q–Q" plot of eigenvalues

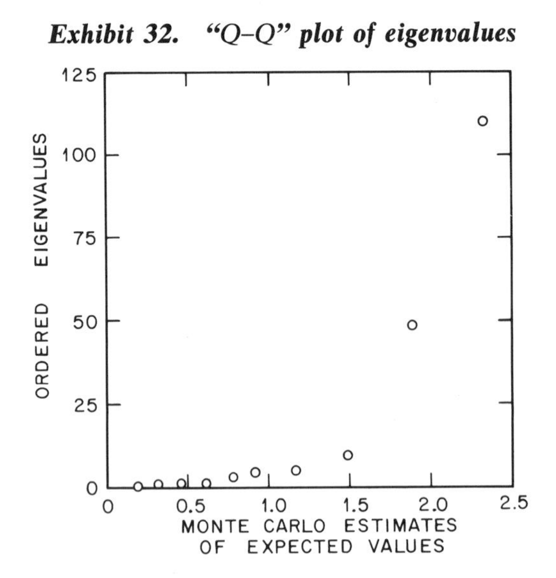

overall space of the responses is decomposable into subspaces within each of which the dispersion of the points is essentially spherical; this will be indicated by a plot that has several linear pieces with differing slopes.

5. The use of one, or preferably several, distance functions to convert the multiresponse data to single numbers, followed by the probability plotting of these numbers, can be very effective. A common useful class of distance functions is that of positive semidefinite quadratic forms, $\mathbf{x}'\mathbf{A}\mathbf{x}$, where both the vector, $\mathbf{x}$, and the matrix, $\mathbf{A}$, may be some functions of the multiresponse observations themselves. Example 7 discussed in Section 2.4, as well as Examples 36–38 described in Section 6.3, illustrate the idea involved here.

6. The CPP and SCPP techniques of component probability plotting described in Section 5.4.1, and the plotting procedures associated with the radius-and-angles decomposition discussed in Section 5.4.2, are additional examples of graphical methods that are useful in studying the distributional characteristics of multiresponse data.

7. Lastly, a technique proposed by Andrews (1972) and certain ramifications of it constitute a relatively recent and promising development in the graphical display of high-dimensional data. The rest of this subsection is devoted to a discussion and illustration of this class of displays.

The essential idea in Andrews's proposal is to map each multiresponse observation into a function, $f(t)$, of a single variable, t, by defining $f(t)$ as a linear combination of orthonormal functions in t with the coefficients in the linear combination being the observed values of the responses. For instance, given the p-dimensional observations, $\mathbf{y}_i = (y_{i1}, y_{i2}, \ldots, y_{ip})'$, $i = 1, \ldots, n$, one can map each observation, $\mathbf{y}_i$, into

$$\begin{aligned} f_{\mathbf{y}_i}(t) &= f_i(t) \\ &= y_{i1}a_1(t) + y_{i2}a_2(t) + \cdots + y_{ip}a_p(t) \\ &= \mathbf{y}_i'\mathbf{a}_t, \qquad i = 1, \ldots, n, \end{aligned} \tag{82}$$

where the functions $\{a_1(t), a_2(t), \ldots, a_p(t)\}$ are orthonormal in an interval, say, $0 \le t \le 1$. Specifically, Andrews (1972) suggests the set of functions

$$\begin{aligned} \mathbf{a}_t' &= \{a_1(t), a_2(t), \ldots\} \\ &= \left\{\frac{1}{\sqrt{2}}, \sin t, \cos t, \sin 2t, \cos 2t, \ldots\right\}, \end{aligned} \tag{83}$$

which are orthonormal on $(-\pi, +\pi)$. [*Note*: Simply by taking $2\pi t$ in place of t in Andrews's definition one would obtain a set of functions orthonormal on $(0, 1)$ instead of $(-\pi, +\pi)$.]

The n functions, $f_1(t)$, $f_2(t)$, ..., $f_n(t)$, may then be plotted simultaneously against values of t in the permissible range, e.g., (0, 1) or $(-\pi, +\pi)$. Thus the initial multiresponse observations, which are n points in p-space, will now appear as n curves in a two-dimensional display whose ordinate corresponds to the function value and whose abscissa is the range of values of t. At a specific value of t, say $t=t_0$, $f_i(t_0)$ is the length of the projection of the ith observation, $\mathbf{y}_i$, onto the vector (or one-dimensional subspace) $\mathbf{a}'_{t_0}=\{a_1(t_0), a_2(t_0), \ldots, a_p(t_0)\}$. Thus, on the Andrews *function plot*, at a specific value of t one is looking at the lengths of the projections of each of the n observations onto a specific one-dimensional subspace of the original p-space, and, as one scans the plot across several values of t, one is looking at a collection of several such one-dimensional views. An equivalent algebraic way of thinking about the function plot is that at each value of t one is looking at a specific linear combination of the p responses, and thus across different values of t one is looking at several different linear combinations.

Andrews (1972) has established various statistical properties of these function plots. For instance, since the definition of the functions in Eq. 82 is linear in the p variables, the technique preserves the mean in the sense that, if $\bar{\mathbf{y}}$ denotes the mean vector of the observations, then

$$f_{\bar{\mathbf{y}}}(t)=\frac{1}{n}\sum_{i=1}^{n} f_i(t),$$

so that the centroid of the observations will correspond to an "average curve" on the function plot. Another property of the function plot is that it preserves distances in a certain sense. Specifically, as a consequence of the orthonormality of the functions $\{a_j(t)\}$, $j=1, \ldots, p$, the squared distance between the pair of functions $f_i(t)$ and $f_l(t)$, defined as

$$\int_{\substack{\text{over the}\\ \text{total range}\\ \text{of } t}} [f_i(t)-f_l(t)]^2\,dt,$$

is just proportional to the squared Euclidean distance between $\mathbf{y}_i$ and $\mathbf{y}_l$ in the p-space of the original observations. This property enables one to think of close curves on the function plot as corresponding to close data points (at least as judged by Euclidean distance, which may not itself be a statistically appropriate measure of distance for certain kinds of multiresponse data, as discussed in Section 4.2.1) in the p-space of the responses.

Yet another property of the plot is that, provided the p responses have equal variances and are mutually uncorrelated, the variance across values of t in the function plot is essentially constant. This is so because $\mathbf{a}'_t$ as defined by Eq. 83, for instance, is of constant length ($=\sqrt{p/2}$) when p is odd, and when p is even and large it is of approximately constant length since its length is between $\sqrt{(p-1)/2}$ and $\sqrt{(p+1)/2}$. Consequently, if the p responses have a common variance σ^2 and, furthermore, are mutually uncorrelated, it follows from Eq. 82 that the variance σ_f^2 of $f(t)$ (defined with $\mathbf{a}'_t$ as in Eq. 83) is $\sigma^2 p/2$ when p is odd and lies between $\sigma^2(p-1)/2$ and $\sigma^2(p+1)/2$ when p is even. The requirement of a common variance for, and no intercorrelations among, the responses is, however, not only unrealistic but also self-defeating, in that if this were so the case for a multivariate approach to analyzing the data would not be very cogent. In practice, two different ways of moving the data toward meeting the requirement for constancy of variance of the function plot are (i) rotating the data to standardized principal component coordinates, and (ii) standardizing the variables initially (e.g., by scaling the observations on a response by either the standard deviation or the interquartile range) without any attempts to uncorrelate the data.

From the definition of $f_i(t)$ in Eq. 82, it is clear that the choice of the specific elements of $\mathbf{a}'_t$ to associate with each of the variables can be important. For instance, the suggestion in Eq. 83 would associate $1/\sqrt{2}$ with the first variable, sin t with the second, and so on. A different permutation of the coefficients would, of course, lead to weighting the variables differently. One suggestion for using a specific ordering of coefficients such as the one in Eq. 83 is to take the variables in the order of their importance; however, such an ordering by importance may not always be feasible, and as a general rule it may be advisable to try a few different permutations of the coefficients with a given set of variables. Since the appearance of the function plot is not invariant under permutations of the coefficients, the use of different permutations may lead to different insights into the data and thus prove to be valuable.

Ideally, as t varies across its total range of values, the values assumed by the vector $\mathbf{a}'_t$ will "cover" the sphere in p dimensions systematically and thoroughly so that no interesting unidimensional views (or linear combinations) are neglected. This seems to be too much to expect or require, however, even for moderately large p, especially if the set of coefficients is prespecified and not based on indications from the data. For providing a more complete coverage of the sphere and also for including the case of assigning equal weights to the p variables, a suggestion due to Tukey is the choice

$$\mathbf{a}'_t = (\cos t, \cos\sqrt{2}t, \cos\sqrt{3}t, \cos\sqrt{5}t, \ldots), \qquad 0 \le t \le k\pi, \tag{84}$$

for an appropriate value of k. Normalization of $\mathbf{a}'_t$ to constant length would seem advisable for comparisons across different values of t. At

$t = 0$, the weights for the variables are all equal. [*Note*: The lack of orthogonality among the elements of $\mathbf{a}_t'$ in Eq. 84 would imply that interpreting closeness among the curves directly in terms of closeness of the original p-dimensional observations would not be as easy as it would be in the case of Andrews's original suggestion for $\mathbf{a}_t'$, viz., Eq. 83.]

A different issue in using the plotting scheme as proposed initially by Andrews is its use in the situation where one has a very large number of multivariate observations. Since each observation is mapped into a curve, a routine Andrews plot with a very large number of curves would tend to look quite messy and not particularly revealing of anything but global aspects (e.g., clearly separated clusters or outliers) of the configuration of the data. For this case when n is large, an adaptation of the Andrews plot is, however, feasible and appears to be quite useful for studying the configurational and distributional aspects of multivariate data. The essential idea in the adaptation is to plot, for each point in a specified grid of values of t, only selected quantiles or percentage points (e.g., median, upper, and lower quartiles) of the distribution of the n values of f and, in addition, perhaps plot selected individual observations such as extreme values. The appearance and appreciation of such a *quantile contour plot*, or indeed of any of the versions mentioned above, can sometimes be improved by plotting an internally standardized set of values (such as deviations of f from its median divided by the interquartile range) rather than the values of f itself.

In addition to issues of choice of $\mathbf{a}_t$, that are shared by both function and quantile contour plots, the latter also involve the choice of quantiles for display purposes. As a general rule, plotting the median and quartiles (or, in the standardized version, centering the quantile contour plot at the median and scaling by the interquartile range) is useful. With regard to choosing specific quantiles beyond the quartiles, however, flexibility in the light of the specific application is in order. Thus, choosing the upper and lower 10%, 5%, and 1% quantiles is one possibility, while the upper and lower $12\frac{1}{2}$%, $6\frac{1}{4}$%, and $3\frac{1}{8}$% quantiles (i.e., equally spaced in probability) constitute another useful choice.

Since, in general, the function representing a specific multiresponse observation need not correspond to a particular quantile for all values of t [i.e., for instance, if $f_i(t_1)$ is the median value of the function at $t = t_1$ and $f_j(t_2)$ is the median value at $t = t_2$, it is not necessarily true that $i = j$], the quantile contours will not enable one to study the behavior of specific observations but will only aid in assimilating the general distributional aspects of the high-dimensional data. The functions corresponding to specific observations that are of particular interest can, of course, be displayed on a quantile contour plot, provided that the number of such observations is not so large as to interfere with appreciation of the plot as a whole.

Function plots and quantile contour plots are useful devices for detecting clusters and/or outliers. In view of the properties mentioned earlier, on an Andrews function plot the curves corresponding to the multiresponse observations in a cluster would cohere together, and distinct clusters (or outliers) would be indicated by clear separations among the curves (or sets of them). On a quantile contour plot the existence of strong clusters may be revealed by a disproportionate squeezing together of particular quantiles at some values of t. For instance, clustering that is revealed by multimodality of the distribution of the projections along the vector corresponding to a specific value of t will tend to show up as such a squeezing together of certain of the quantiles at that value of t since the multimodality implies that for a small change in some quantile values (perhaps usually the "outer" quantiles) there will be a large change in the corresponding cumulative probability values.

Also, the quantile contour plot may be useful for studying the shape and more subtle configurational aspects of high-dimensional data distributions. Symmetry (as revealed by the appearances of the contours of pairs of upper and lower quantiles, especially the outer ones) is most easily appreciated. More specifically, if the data distribution is essentially spherical, one would expect to see approximately equal spacings between any specified pair of quantiles across the entire plot. The existence of high intercorrelations among the responses, which will tend to induce "ellipsoidal" types of configurations for the data, is likely to be revealed by an approximately proportionate squeezing together of almost all the quantiles for some values of t. [*Note*: An interesting use of function and/or quantile contour plots, as a consequence of this property, occurs in the context of multiple regression analysis. When one suspects difficulties caused by possible multicollinearity in the data, a plot of this type for the observations (either just on the independent variables or on both the dependent and independent variables) may be useful for identifying the singularities, as well as the essentially linearly independent combinations of the variables—the singularities would correspond to directions of (or linear combinations with) zero variance, which would be revealed by all the curves (or all quantiles) going through a single point at one or more values of t, while directions in (or linear combinations for) which curves (or quantiles) spread out considerably would be useful for picking the essentially linearly independent combinations.]

With quantile contour plots, as a check on the normality of the distribution of the data, one can compare the ratios of the observed spacings between specified pairs of quantiles with the values of such ratios for the normal distribution. Thus, for instance, the spacing between the 10% quantile and the median is approximately twice (1.9, more precisely) the spacing between the quartile and the median for a normal distribution; and if this relationship is not adequately satisfied for one or more values of t by the three quantiles involved, one will have reason to question the normality of the distributions of the linear combinations of the variables corresponding to these values of t, and hence also to question the joint normality of the distribution of the initial observations. Since the values of t spanned in a quantile contour plot do not generally yield *all possible* linear combinations of the original variables, indications of reasonable conformity to normality for every value of t in the grid chosen for a quantile contour plot, although not equivalent to a confirmation of joint normality, will nevertheless be useful evidence for deciding to use methods based on normality assumptions for further analyses of the data. Also, if normality is singularly inapplicable for only a few values of t, one may be able to transform the data initially so as to improve directional normality (see Section 5.3) along just these directions without altering the data in other directions, and then use standard methods with the transformed data. At any rate the quantile contour plot at least provides an informal basis for verifying normality. [*Note*: Since several comparisons of ratios of spacings between quantiles may be involved, one may wish to automate this process and have the computer not only do the plotting but also provide printout flagging situations in which the departures from normality are sufficiently striking.]

A considerably different problem, which can be motivated in terms of the function and quantile contour plots, is that of choosing a "typical" multiresponse observation. The sample mean vector, $\bar{\mathbf{y}}$, and the more robust estimators of location discussed in Section 5.2.3 are examples of statistics that summarize one typical aspect of multiresponse data, viz., overall location. Even in the context of location estimation, for some applications one may be interested in choosing an actual observation as a typical value instead of a summary statistic. One approach to this problem is to choose as a typical observation the one whose representation as a curve on a function plot is closest (in some specified metric) to the set of median values on the corresponding quantile contour plot. In fact, this idea, in addition to its use in developing a location estimate, may be worth investigating as a means of defining multivariate order statistics and quantiles.

Example 33. This example, taken from Andrews (1972), pertains to a discriminant analysis described by Ashton et al. (1957) of eight measurements ($p = 8$) on teeth of different "races" of men and apes, so as to aid in the classification of some fossils on the basis of their measurements with respect to the same eight characteristics used for the men and the apes. Nine groups—three "races" of men and six groups (three types × two sexes) of apes were involved, so that there were eight CRIMCOORDS (cf. Section 4.2) in the discriminant analysis.

Exhibit 33*a* (see page 214) shows the coordinates of the nine group centroids in the eight-dimensional space of the CRIMCOORDS and also the representations for the six fossils in this space. Exhibit 33*b* (see page 215) shows a graphical representation, obtained by Ashton et al. (1957), of the group centroids as well as the fossils in the space of the first two CRIMCOORDS; also included in this picture are approximate 90% confidence regions for the locations of the nine groups in the two-dimensional CRIMCOORDS space. (See the discussion in Section 4.2 pertaining to the methodological details of such graphical displays.) In Exhibit 33*b* the coordinates of the centers of the circles and of the points that correspond to the fossils are the values shown in Exhibit 33*a* for just the first two CRIMCOORDS. Thus an inspection of the values in Exhibit 33*a* or of the visual portrayal in Exhibit 33*b* shows the clear separation (especially on the first CRIMCOORD) of the three "races" of men from the six groups of apes. Ashton et al. (1957) went further, on the basis of Exhibit 33*b*, to conclude that the fossil *Proconsul africanus* is very much like a chimpanzee, whereas the other five fossils are more akin to the "races" of men.

Andrews (1972), on the other hand, studied the eight-dimensional data in Exhibit 33*a* by means of his function plots and came to interesting but different conclusions regarding the classification of the fossils. Specifically, Exhibit 33*c* (See page 215) shows the function plot obtained by Andrews (1972) for the nine group centroids given in Exhibit 33*a*. The choice of $\mathbf{a}_t'$ in this plot is the one in Eq. 83 with $1/\sqrt{2}$ being associated with the first CRIMCOORD, $\sin t$ with the second, and so on. [*Note*: The CRIMCOORDS would have equal variances and be uncorrelated, *under the usual assumptions* of discriminant analysis, even if the original variables did not have these properties.] In Exhibit 33*c* the curves for the three human groups are quite well separated from the curves of the six groups of apes, and, among the apes, the chimpanzees stand out from the other two types. Also, the two sexes within each ape group tend to have relatively closely spaced curves, especially in the left part of the plot.

In addition, at specific values of t the separations among the groups are very pronounced in relation to the separations within the groups. For instance, at t_2 and t_4 the human groups are very cohesive and more clearly distinguished from the values for the curves for any of the apes. At t_1 and t_3 the chimpanzees seem to stand out more clearly from the

Exhibit 33a. Results of discriminant analysis of fossil data (Ashton et al., 1957; Andrews, 1972)

PERMANENT FIRST LOWER PREMOLAR

means of groups in the space of CRIMCOORDS

A.	West African	-8.09	+0.49	+0.18	+0.75	-0.06	-0.04	+0.04	+0.03
B.	British	-9.37	-0.68	-0.44	-0.37	+0.37	+0.02	-0.01	+0.05
C.	Australian aboriginal	-8.87	+1.44	+0.36	-0.34	-0.29	-0.02	-0.01	-0.05
D.	gorilla: male	+6.28	+2.89	+0.43	-0.03	+0.10	-0.14	+0.07	+0.08
E.	female	+4.82	+1.52	+0.71	-0.06	+0.25	+0.15	-0.07	-0.10
F.	orang-outang: male	+5.11	+1.61	-0.72	+0.04	-0.17	+0.13	+0.03	+0.05
G.	female	+3.60	+0.28	-1.05	+0.01	-0.03	-0.11	-0.11	-0.08
H.	chimpanzee: male	+3.46	-3.37	+0.33	-0.32	-0.19	-0.04	+0.09	+0.09
I.	female	+3.05	-4.21	+0.17	+0.28	+0.04	+0.02	-0.06	-0.06

fossils

J.	Pithecanthropus	-6.73	+3.63	+1.14	+2.11	-1.90	+0.24	+1.23	-0.55
K.	pekinensis	-5.90	+3.95	+0.89	+1.58	-1.56	+1.10	+1.53	+0.58
L.	Paranthropus robustus	-7.56	+6.34	+1.66	+0.10	-2.23	-1.01	+0.68	-0.23
M.	Paranthropus crassidens	-7.79	+4.33	+1.42	+0.01	-1.80	-0.25	+0.04	-0.87
N.	Meganthropus palaeojavanicus	-8.23	+5.03	+1.13	-0.02	-1.41	-0.13	-0.28	-0.13
O.	Proconsul africanus	+1.86	-4.28	-2.14	-1.73	+2.06	+1.80	+2.61	+2.48

Exhibit 33b. Representations of the fossil groups and the unknowns in the space of the first two CRIMCOORDS (Ashton et al., 1957; Andrews, 1972)

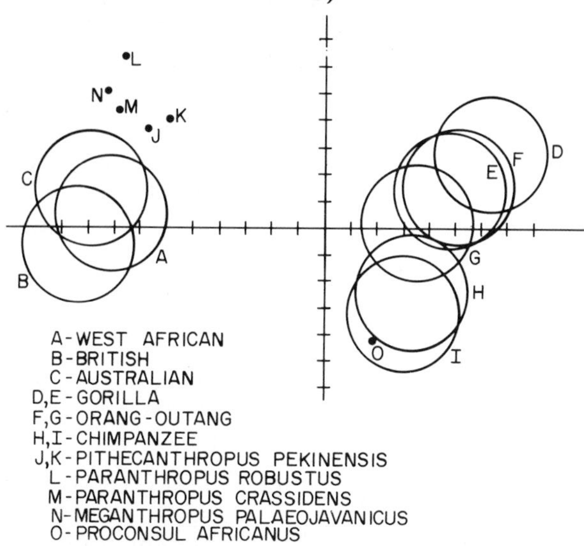

Exhibit 33c. High-dimensional plot of the centroids of fossil groups, using Eq. 83 for coefficients (Andrews, 1972)

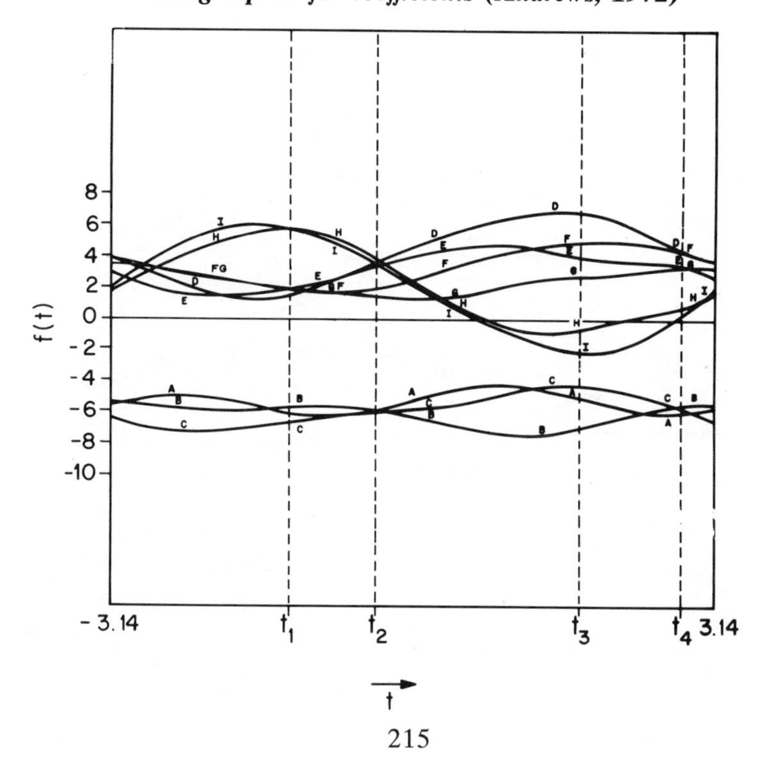

remaining two groups of apes. Thus Exhibit 33*c* has been useful for detecting directions of clusterings among the groups.

In Exhibit 33*d* the curves for the six fossils in the study have been superimposed on the curves of Exhibit 33*c*. Immediately, one fossil, *Proconsul africanus*, stands out as being different, although for specific (but different) values of t it comes close to each of the groups. The remaining fossils are quite similar to man, especially at t_2 and t_4. Andrews (1972) discusses more formal tests (based on certain confidence bands obtained by using the σ_f discussed earlier) to support the conclusions drawn from the function plot shown in Exhibit 33*d*. The feature that *Proconsul africanus* does not seem to belong to any of the groups illustrates the fact that in some applications of discriminant analysis it is wise to have the option of not classifying an unknown as necessarily belonging to any one of the prespecified groups. (See also Rao, 1960, 1962, regarding this issue.)

Exhibits 33*e* (See page 217) and *f* (See page 218) show function plots that correspond to Exhibits 33*c* and *d* when $\mathbf{a}'_t$ is specified according to

Exhibit 33d. High-dimensional plot of the centroids of the fossil groups and the unknowns, using Eq. 83 for coefficients (Andrews, 1972)

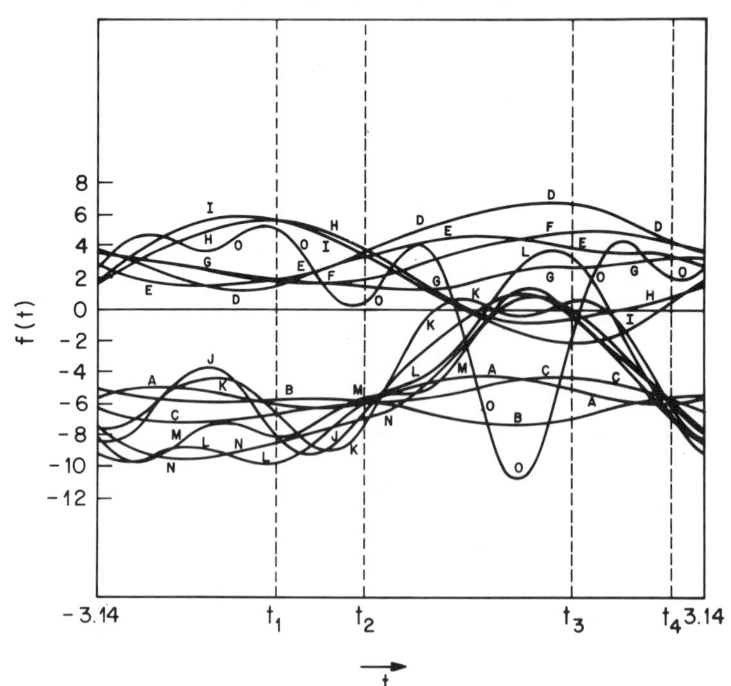

Exhibit 33e. High-dimensional plot of the centroids of fossil groups using Eq. 84 for coefficients

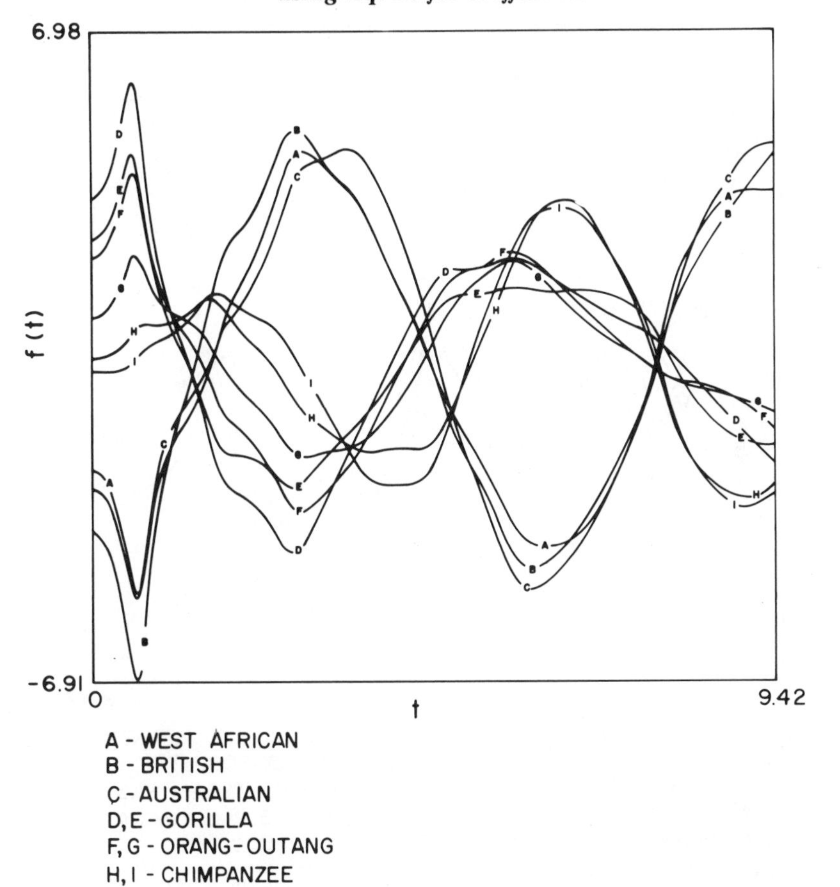

Eq. 84 rather than Eq. 83. Although the appearance of the plots in Exhibits 33*e* and *f* is more noisy and the separation of the human and ape groups is no longer as striking, the general indications and conclusions in this example are quite similar for the two choices of $\mathbf{a}_i'$.

Example 34. The data collected by Anderson (1935), and also used by Fisher (1936), shown in Exhibit 34*a* (see pages 219–220) consist of 50 quadrivariate observations (viz., logarithms of sepal length and width and of petal length and width) for *Iris setosa.* The original data on *Iris setosa,* as well as on two other species of iris (*Iris versicolor* and *Iris virginica*), are well known in the multivariate literature and have been utilized by many authors as the basis for testing different classification and

Exhibit 33f. High-dimensional plot of the centroids of the fossil groups and the unknowns, using Eq. 84 for coefficients

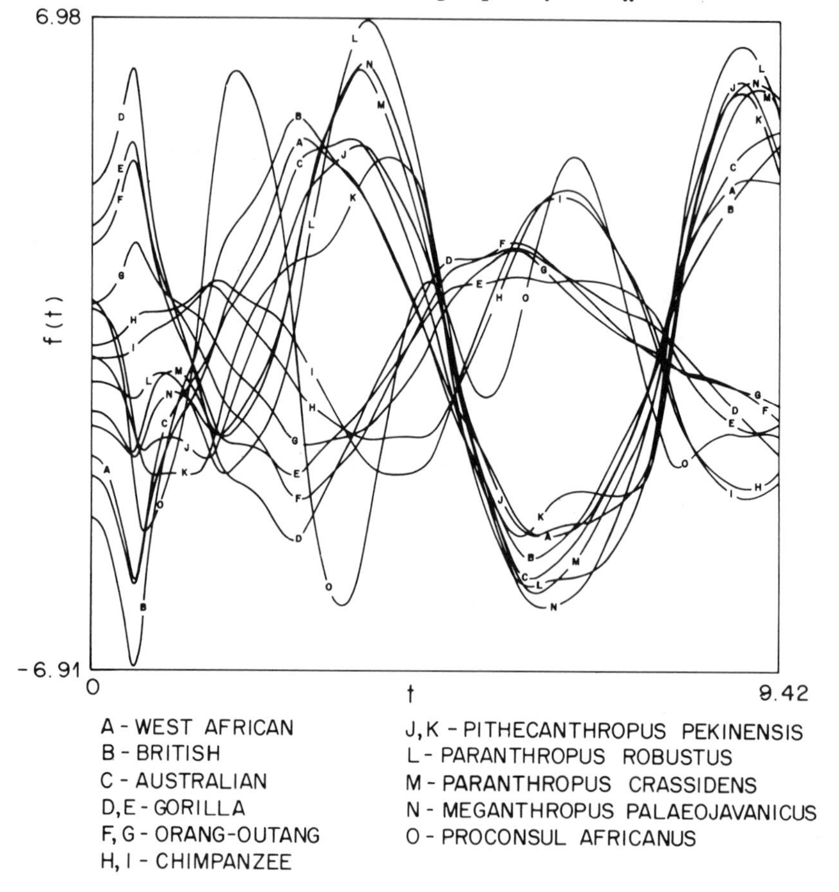

clustering algorithms (e.g., Friedman and Rubin, 1967). The data set is considered to be generally well behaved with no particular peculiarities, and it has been found that *Iris setosa* is easily distinguishable from the other two species (see, e.g., Fisher, 1936; Friedman and Rubin, 1967; and Exhibit 34*c*).

For present purposes the data of Exhibit 34*a* were initially "standardized" on each variable by substracting the median from each observation and then dividing by the interquartile range. With as many as the 50 observations in this example, a quantile contour plot rather than a function plot is the appropriate choice, and Exhibit 34*b* (see page 220) shows such a plot. The choice for $\mathbf{a}_t'$ in this case was {sin t, cos t, sin $2t$, cos $2t$}, and the quantiles chosen for display were the median, the lower and upper quartiles, and the lower and upper tenths. In Exhibit 34*b* the median is labeled *M*, the two quartiles are denoted as *Q*, and the tenths are shown as *T*'s. Exhibit 34*b* is actually a printer plot and is an example of graphical output that does not require any exotic, expensive, or specialized hardware.

Exhibit 34a. *Iris setosa* data (Anderson, 1935; Fisher, 1936)

Sepal Length (ln cm)	Sepal Width (ln cm)	Petal Length (ln cm)	Petal Width (ln cm)
1.629	1.253	0.336	−1.609
1.589	1.099	0.336	−1.609
1.548	1.163	0.262	−1.609
1.526	1.131	0.405	−1.609
1.609	1.281	0.336	−1.609
1.686	1.361	0.531	−0.916
1.526	1.224	0.336	−1.204
1.609	1.224	0.405	−1.609
1.482	1.065	0.336	−1.609
1.589	1.131	0.405	−2.303
1.686	1.308	0.405	−1.609
1.569	1.224	0.470	−1.609
1.569	1.099	0.336	−2.303
1.459	1.099	0.095	−2.303
1.758	1.386	0.182	−1.609
1.740	1.482	0.405	−0.916
1.686	1.361	0.262	−0.916
1.629	1.253	0.336	−1.204
1.740	1.335	0.531	−1.204
1.629	1.335	0.405	−1.204
1.686	1.224	0.531	−1.609
1.629	1.308	0.405	−0.916
1.526	1.281	0.	−1.609
1.629	1.194	0.531	−0.693
1.569	1.224	0.642	−1.609
1.609	1.099	0.470	−1.609
1.609	1.224	0.470	−0.916
1.649	1.253	0.405	−1.609
1.649	1.224	0.336	−1.609
1.548	1.163	0.470	−1.609
1.569	1.131	0.470	−1.609
1.686	1.224	0.405	−0.916
1.649	1.411	0.405	−2.303
1.705	1.435	0.336	−1.609
1.589	1.131	0.405	−1.609

Exhibit 34a. (Continued)

Sepal Length (ln cm)	Sepal Width (ln cm)	Petal Length (ln cm)	Petal Width (ln cm)
1.609	1.163	0.182	−1.609
1.705	1.253	0.262	−1.609
1.589	1.281	0.336	−2.303
1.482	1.099	0.262	−1.609
1.629	1.224	0.405	−1.609
1.609	1.253	0.262	−1.204
1.504	0.833	0.262	−1.204
1.482	1.163	0.262	−1.609
1.609	1.253	0.470	−0.511
1.629	1.335	0.642	−0.916
1.569	1.099	0.336	−1.204
1.629	1.335	0.470	−1.609
1.526	1.163	0.336	−1.609
1.668	1.308	0.405	−1.609
1.609	1.194	0.336	−1.609

Exhibit 34b. Quantile contour plot of Iris setosa data

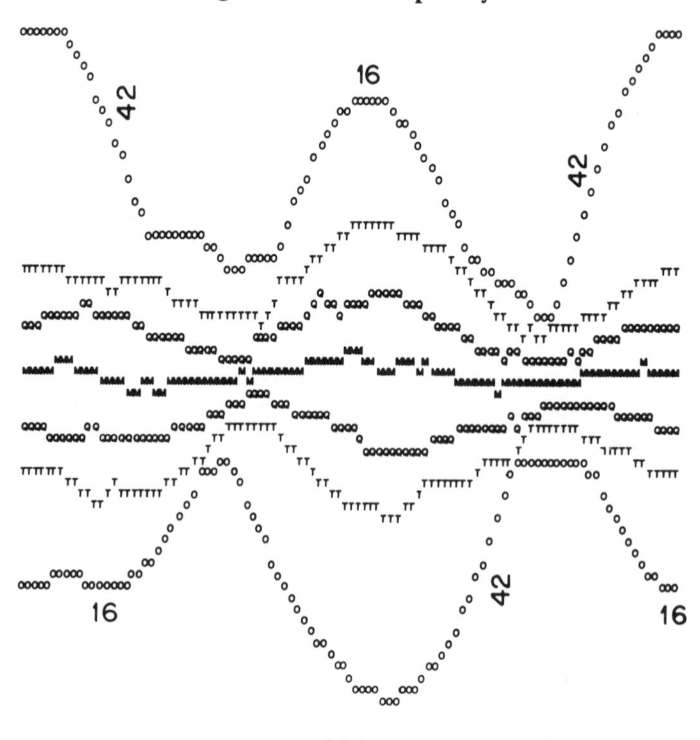

Also shown on Exhibit 34*b* are the two "outermost" points, plotted as 0's. An investigation of these points showed that one of them corresponded to the 16th observation in the original set, and the other to the 42nd observation. The process of identifying such consistently outlying observations, if they exist, can be automated by requiring the computer program to superimpose, with appropriate labels, the curves for all observations that are consistently (by some quantitive definition such as "for more than half the values of t") well separated from the majority. At any rate Exhibit 34*b* shows directly and simply that the 16th and 42nd observations are symmetrically and oppositely situated observations which seem to be quite clearly separated from the remaining observations as one views the data in several unidimensional directions.

Exhibit 34*c* shows a representation of the three groups of irises in the two-dimensional discriminant space for this problem. The fifty *Iris setosa*

Exhibit 34c. Representation of the three groups of irises in the space of the two CRIMCOORDS

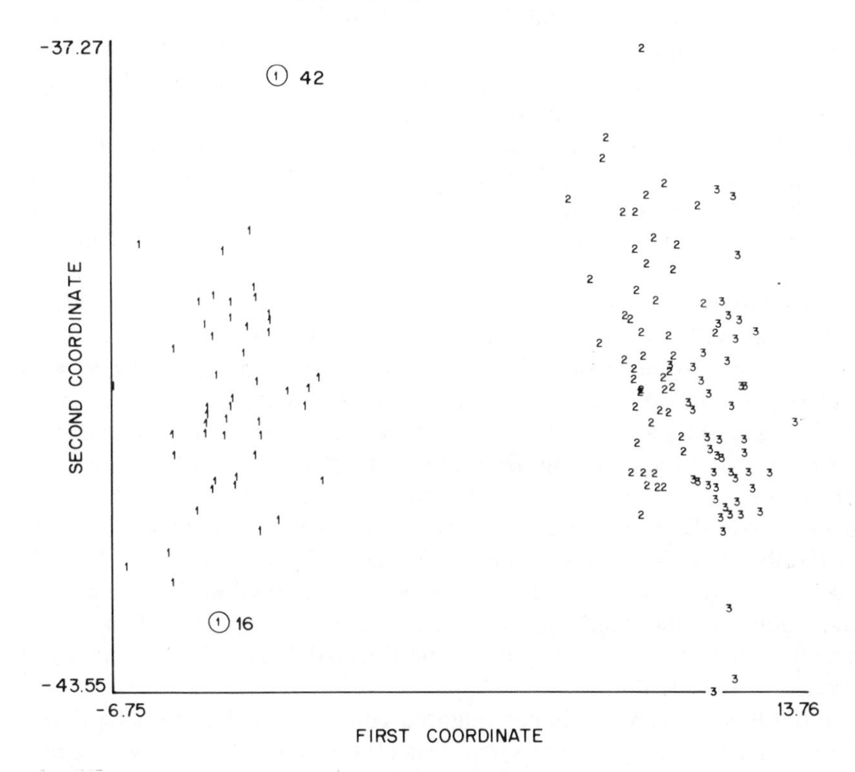

points are labeled 1 in this figure; those for *Iris versicolor*, 2; and those for *Iris virginica*, 3. In this picture the 16th and 42nd observations in the *Iris setosa* group are seen to lie at opposite ends of the data configuration with respect to the second discriminant coordinate, although the separation of the two points is by no means as striking as it is in the quantile contour plot.

Aside from the indications regarding the 16th and 42nd observations, by scanning the spacings among the quantiles across the whole picture in Exhibit 34*b* one can get a "feeling" for the shape of the quadrivariate distribution of the data in Exhibit 34*a*. The configurations of the *M*'s, *Q*'s, and *T*'s in Exhibit 34*b* indicate that the data in this example are quite symmetrically, although not spherically (perhaps because of the known intercorrelations among the variables here), distributed in 4-space.

Example 35. This example derives from a study of empirical groupings of corporations (see Chen et al., 1970, 1974, and also Examples 17, 18, and 22 in Chapter 4) on the basis of yearly observations on several variables chosen to represent the financial and economic characteristics of the firms. One question of interest to this study was what an appropriate classification would be for AT&T (American Telephone & Telegraph Company) vis-à-vis the dichotomy of corporations as either industrials or utilities. Standard and Poor's COMPUSTAT tape was used for deriving annual values for 13 variables (see Chen et al. 1970, 1974, for a list and definition of the variables), and the investigation was carried out by performing separate analyses of each of the years 1960–1969 in particular.

Quantile contour plots can be employed to summarize the findings regarding AT&T's classification. Exhibit 35*a* shows a quantile contour plot of the 13-dimensional data for 495 industrial firms for 1969; also included on the plot is the curve for AT&T, labeled *A*. [*Note*: As a preliminary standardization of each of the 13 variables, the median was subtracted and the resulting deviations were divided by the inter-quartile range; the 13-dimensional observation for AT&T was subjected to the same standardization as was performed for the industrial firms displayed in Exhibit 35*a*.] The choice of $\mathbf{a}_t'$ for Exhibit 35*a* was the one in Eq. 83, and the displayed quantiles (whose associated probabilities are defined in the legend for the figure) are in fact deviations from the median, which therefore appears as a steady level line (labeled *H* for "half") across the middle of the picture.

Exhibit 35*b* shows a similar quantile contour plot for the 94 utilities involved in the study for the same year (1969), and again AT&T's curve is shown as a series of *A*'s across the plot. [*Note*: The preliminary

Exhibit 35a. Quantile contour plot of 495 industrials and AT&T for 1969

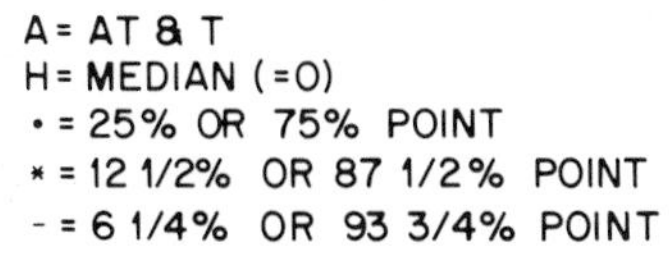

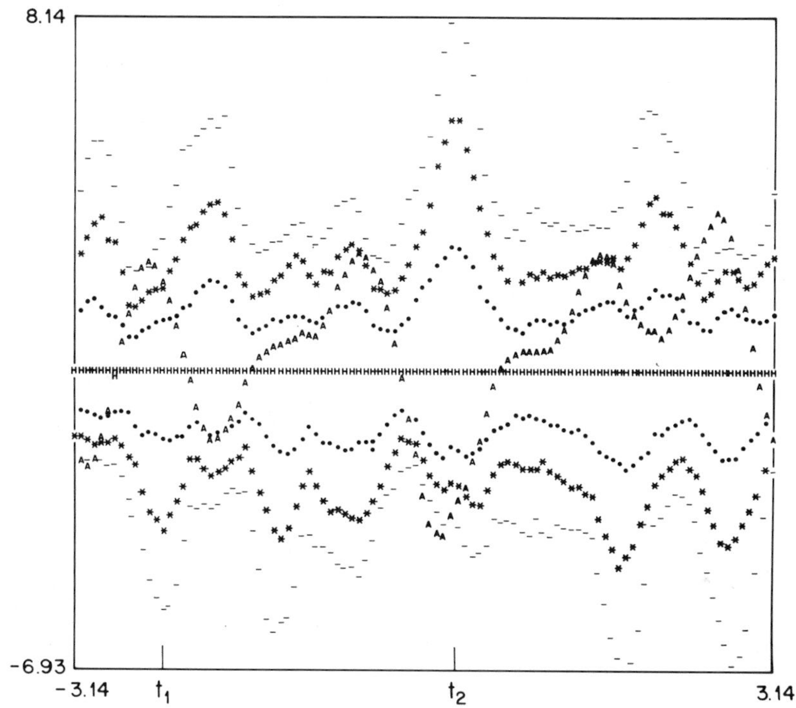

standardization for Exhibit 35*b* (see page 224) was, of course, based on the 94 utilities in this case.] A comparison of Exhibits 35*a* and *b* gives a clear visual impression that AT&T fits in better with the industrials than with the utilities. A more quantitative summary of this point is that for the industrials AT&T falls within the lower and upper quartiles for about 50% of the t values, within the lower and upper $12\frac{1}{2}\%$ points about 75% of the time, and within the band defined by the lower and upper $6\frac{1}{4}\%$ points at a frequency of about 88%. [*Note*: These frequencies are exactly those to be expected for a typical industrial firm.] The corresponding frequencies for the utilities shown in Exhibit 35*b* are much smaller, being, respectively, 10%, 20%, and 40%.

Also, the configurations of the quantiles exhibit strong asymmetries of

Exhibit 35b. Quantile contour plot of 94 utilities and AT&T for 1969

A = AT & T
H = MEDIAN (=0)
• = 25% OR 75% POINT
* = 12 1/2% OR 87 1/2% POINT
- = 6 1/4% OR 93 3/4% POINT

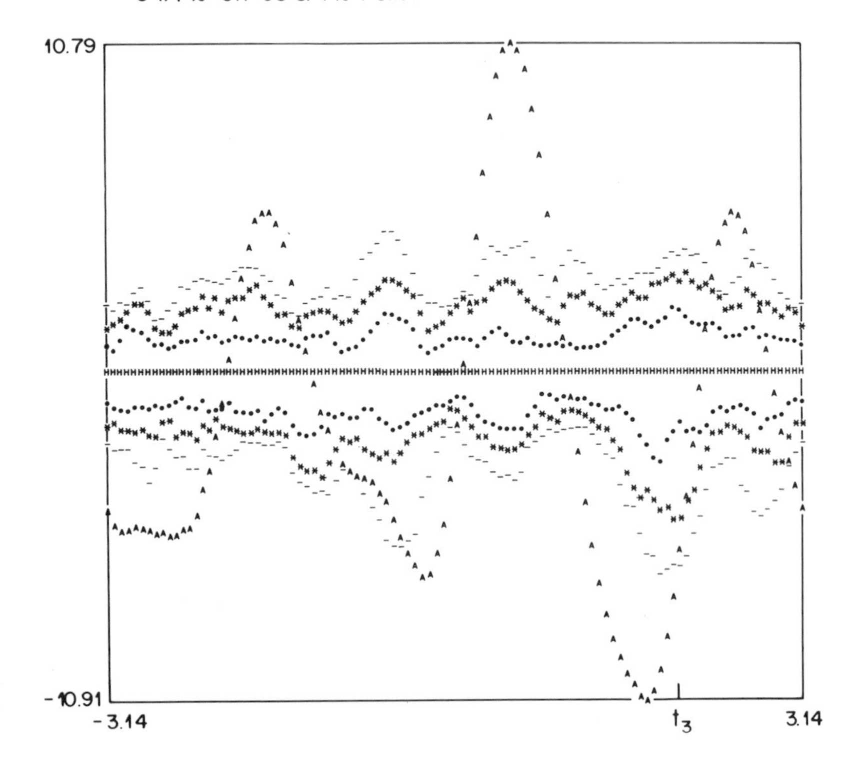

the data distributions along several directions (e.g., those that correspond to values of $t = t_1$ and t_2 in Exhibit 35a and $t = t_3$ in Exhibit 35b), and thus as an adjunct indication one has clear evidence of departures from joint normality of the distribution of the initial 13-dimensional data. The evidence for nonnormality of the distribution of these data is indeed plentiful, and Exhibits 35a and b are by no means the only indicators of this facet of the data. (See also Example 30 in Section 5.4.2.)

The above examples have demonstrated the utility of function plots and quantile contour plots. Despite their limitations the techniques have significant appeal because, at least in part, of the simplicity involved in their two-dimensional character, although the data being represented in them may be, and often are, quite high-dimensional. More work leading to other choices for $\mathbf{a}_t'$ (to provide, for instance, better coverage of "interesting" directions in p-space) and to methods for choosing $\mathbf{a}_t'$ in the light of the data would indeed be worthwhile.

6.3. COMPARISON OF SEVERAL MULTIRESPONSE SAMPLES

Many situations require the presentation of data from several identified groups for comparative purposes. The analysis of variance is a widely used technique for comparing two or more samples. In the uniresponse case Student's t and F statistics are familiar examples of summary statistics that are used for formal comparisons among two or more samples. It will often be useful in analyzing uniresponse data to supplement the computation of such summary statistics by probability plotting techniques, such as a Q-Q plot of one sample versus another, or perhaps superimposed normal probability plots of several samples on a single picture.

In the multiresponse situation Mahalanobis' D^2 or Hotelling's T^2 can be computed and utilized for formally comparing the locations of two groups. Also, for the two-group location problem, when the dimensionality (i.e., number of responses) exceeds the degrees of freedom available for estimation of the error covariance matrix, Dempster (1958) has proposed a test. Multivariate analysis of variance (MANOVA) is concerned with certain generalizations of the two-group procedures proposed by Mahalanobis and by Hotelling (see Sections 5.2.1 and 5.2.2). However, the formal analyses involved in MANOVA are often not sufficiently revealing. They need to be augmented by various graphical analyses, and the discussion of such graphical tools is the concern of this section.

It is perhaps typical of analysis of variance situations that one wishes to ask not one or two questions of the same body of data but several. One would also like to have a climate for the statistical analysis in such a situation that would allow unanticipated characteristics to be spotted. Examples of nonobvious but interesting indications are the presence of possibly real treatment effects, the existence of outliers in the data, and heteroscedasticity.

For these reasons it is reasonable to provide statistical procedures that use some sort of statistical model to aid in comparisons of various collections of comparable quantities, and yet enable one to make such comparisons without the need to commit oneself to any narrow specification of objectives. Examples of collections of comparable quantities are a collection of single-degree-of-freedom contrasts, a collection of mean squares in ANOVA, and a collection of sum-of-products matrices in MANOVA. Procedures for such comparisons have been called *internal comparisons methods* by Wilk and Gnanadesikan (1961, 1964). Specifically, some kinds of probability plotting techniques have been developed for internal comparisons of the relative magnitudes that are involved in ANOVA and MANOVA. These procedures provide a statistical measure for facilitating the assessment of relative magnitudes, which probably becomes rather nonintuitive when one is dealing with a large collection of comparable quantities. Moreover, the procedures can provide some insight into various possible inadequacies of the statistical model used to generate the analysis. The procedures are not excessively influenced by some data-independent aspects, such as the need to prechoose an error term.

In particular, Table 2 shows a categorization of orthogonal analysis of variance situations according to the multiplicity of response and the degrees-of-freedom decomposition of the experimental design or model for the data. Also given beneath the two-way categorization is a list of specific references that describe techniques of relevance to each of the cells in the table.

Table 2. Categorization of ANOVA and MANOVA cases

	Response Structure	
DF Decomposition	Uniresponse	Multiresponse
All 1 df	I	IV
All ν df	II	V
Mixed df	III	VI

I. A half-normal plot of absolute values of contrasts—C. Daniel, *Technometrics* **1** (1959), 311–41.

II. A ν-df chi-squared plot of sums of squares (or gamma plot of sums of squares with shape parameter $\eta = \nu/2$)—M. B. Wilk, R. Gnanadesikan, and M. J. Huyett, *Technometrics* **4** (1962), 1–20.

III. A generalized probability plot of mean squares—R. Gnanadesikan and M. B. Wilk, *J. R. Stat. Soc.* B**32** (1970), 88–101.

IV. A gamma plot of squared distances with an estimated shape parameter—M. B. Wilk and R. Gnanadesikan, *Ann. Math. Stat.* **35** (1964), 613–31; also, Chapter VII of Roy et al. (1971).

V. Gamma plots of certain functions of eigenvalues with an estimated shape parameter—R. Gnanadesikan and E. T. Lee, *Biometrika* **57** (1970), 229–37.

The two subsections that follow will be concerned with describing the methods for cells IV and V, respectively. Methods for cell VI are not yet available.

6.3.1. Graphical Internal Comparisons Among Single-Degree-of-Freedom Contrast Vectors

The techniques to be described in this subsection may be employed in any situation in which there is a meaningful decomposition of effects (in the sense of the analysis of variance) into orthogonal single-degree-of-freedom components. For instance, in multifactor experiments in which the factors are quantitative and are used at several levels for obtaining the treatment combinations, one has the familiar decomposition into linear, quadratic, etc., components for the treatment effects, and these form a natural set of orthogonal single-degree-of-freedom components that one may wish to intercompare. Two-level factorial (full and/or fractional) experiments, of course, yield a meaningful decomposition into main effects and interactions which together constitute an orthogonal single-degree-of-freedom set of effects whose interpretations are of prime interest in such experiments. For simplicity the prototype experimental situation for developing the methodology here will be taken to be that of a 2^N factorial experiment, i.e., N factors each of which has two levels, but it should be kept in mind that the methods have wider applicability, as indicated by the preceding discussion. More specifically, the setup will be one in which there are $n = 2^N$ treatment combinations in all, and corresponding to the ith treatment combination one has a p-dimensional observation, $\mathbf{y}_i'$ $(i = 1, \ldots, n)$, whose coordinates are the observed values of the p responses for the particular treatment combination.

In fact, for motivating some of the basic concepts underlying the methodology, consider a 2^3 experiment ($N = 3$, $n = 8$) with two responses ($p = 2$) measured on each experimental unit after "application" of one of the eight treatment combinations involved. Hence one has eight bivariate observations which can be represented as points in a two-dimensional space, as shown, for example, in Figure 9. The eight points in the plot are labeled by the respective treatment combinations associated with them, and the notation for the treatment combinations is the standard one for two-level factorial experiments. To illustrate how one may think about a treatment effect in this bivariate case, consider the problem of defining the main effect of factor A. The set of eight points can be divided into two groups of four observations each; in one group (shown in Figure 9 as unshaded circles) all the treatment combinations are ones in which the factor A is at its lower level, and in the other group (shown in Figure 9 as

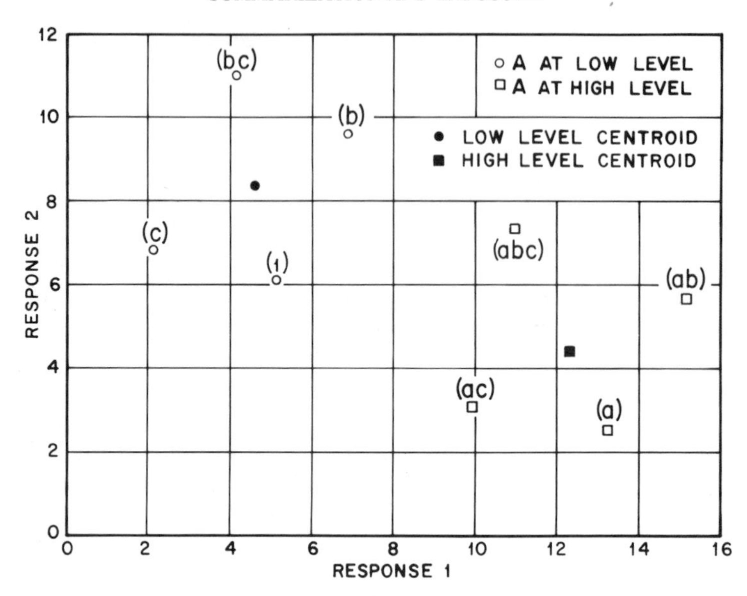

Fig. 9. Pictorial representation for two-dimensional main effect of A in a 2^3 experiment involving bivariate observations.

unshaded squares) the factor A is at its higher level. One can define a centroid of each of the sets of four observations, and these are shown in Figure 9 as a filled-in circle and a filled-in square. The univariate estimate of the main effect of A with respect to the first response, for instance, is just the distance between the projections of these two centroids on the horizontal axis. Similarly, the main effect of A with respect to the second response is the distance between the projections of these two centroids on the vertical axis. Next, a natural and not unreasonable conceptualization of the main effect of A in the two-dimensional situation would be as a distance between the two centroids in the two-dimensional space. Hence, in particular, one can think of a vector going from the filled-in circle to the filled-in square and consider that the "larger" (in some sense) this vector is, the greater is the two-dimensional main effect of factor A.

Clearly one can partition the eight observations in the example of the 2^3 experiment in other specific ways to get various pairs of groups of four observations each, and by analogous reasoning to that used above define the main effects of, as well as the various interactions among, all the factors. One will then have vectors going between the centroids for these different partitions corresponding to the seven effects involved, and these

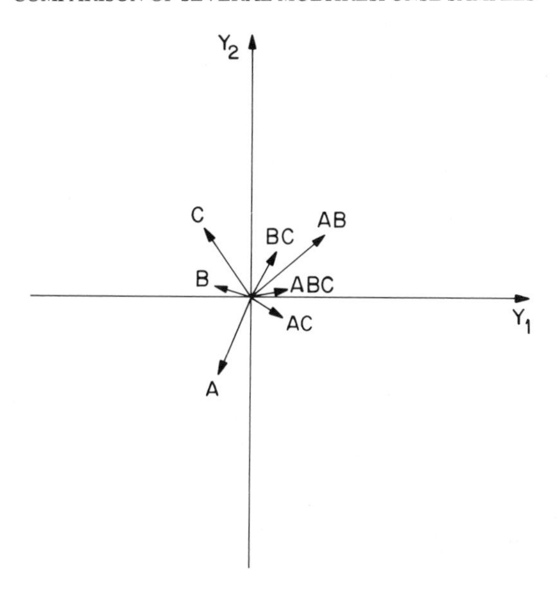

Fig. 10. Representation of contrast vectors in a 2^3 experiment involving bivariate observations.

vectors, labeled according to the effects to which they correspond, can be represented in a two-dimensional space. There will be seven such vectors emanating from the origin, corresponding to the three main effects, the three two-factor interactions, and the one three-factor interaction in this case of the 2^3 experiment (see Figure 10). The transformation involved in obtaining these vector effects from the initial observations is essentially an orthogonal transformation—in fact, the same one that is well known in the analysis of univariate two-level factorial experiments. Explicitly, if $\mathbf{y}_1'$, $\mathbf{y}_2', \ldots, \mathbf{y}_8'$ denote the bivariate observations in a 2^3 experiment, where the treatment combinations are taken in so-called standard order [i.e., (1), (*a*), (*b*), (*ab*), (*c*), (*ac*), (*bc*), (*abc*)], the seven bivariate treatment effect vectors, $\mathbf{x}_1', \ldots, \mathbf{x}_7'$ (such as the ones in Figure 10), are defined by the transformation

$$\begin{pmatrix} \mathbf{m}' \\ \mathbf{x}_1' \\ \cdot \\ \cdot \\ \cdot \\ \mathbf{x}_7' \end{pmatrix} = \mathbf{R} \begin{pmatrix} \mathbf{y}_1' \\ \cdot \\ \cdot \\ \cdot \\ \mathbf{y}_8' \end{pmatrix},$$

where

$$\mathbf{R} = \begin{bmatrix} + & + & + & + & + & + & + & + \\ - & + & - & + & - & + & - & + \\ - & - & + & + & - & - & + & + \\ + & - & - & + & + & - & - & + \\ - & - & - & - & + & + & + & + \\ + & - & + & - & - & + & - & + \\ + & + & - & - & - & - & + & + \\ - & + & + & - & + & - & - & + \end{bmatrix} \tag{85}$$

with + and − standing, respectively, for +1 and −1. Thus $\mathbf{m}'$, associated with an overall effect, is proportional to (i.e., 8 times) the mean vector for the data, while $\mathbf{x}_1', \mathbf{x}_2', \ldots, \mathbf{x}_7'$ are the vectors associated with the effects A, B, AB, C, AC, BC, and ABC, respectively. Since each of the rows of $\mathbf{R}$ that leads to one of the treatment effect vectors defines a *contrast* (i.e., the number of +1's is equal to the number of −1's, so that their sum is 0), the treatment effect vectors, $\mathbf{x}_1', \ldots, \mathbf{x}_7'$, may also be called *single-degree-of-freedom contrast vectors.* [*Note*: The usual definitions of the effects generally multiply the first row of $\mathbf{R}$ by $\frac{1}{8}$ so as to yield the mean vector itself, and also the remaining rows by $\frac{1}{4}$ so as to yield differences in the means of four observations as described in the discussion of Figure 9. Also, to make $\mathbf{R}$ an orthogonal matrix all that is required is to multiply it by the scalar $1/\sqrt{8}$.] In practice, the transformation involved in obtaining the $\mathbf{x}_i'$ from the initially observed $\mathbf{y}_i'$ is generally carried out by Yates's algorithm, which is not only simple but also computationally sound in the sense of numerical accuracy and stability.

Comparisons of the relative magnitudes of treatment effects are one important goal of the analysis of variance; and, returning to Figure 10, for this purpose one needs some way of measuring the "sizes" of the treatment effect vectors (or single-degree-of-freedom contrast vectors), $\mathbf{x}_1', \ldots, \mathbf{x}_7'$, displayed there. The problem here can be treated as being just the same as the one of choosing a distance measure for classification procedures (see Section 4.2.1), since the issues in choosing a measure of "size" for the vectors displayed in Figure 10, so that a "large" vector will correspond to a large effect, also arise in choosing a metric for measuring distances between centroids of various partitions of the observations in the two-dimensional space of the observations shown in Figure 9. At any rate, the methodology developed here depends on using a squared distance measure, $\mathbf{x}'\mathbf{A}\mathbf{x}$, wherein $\mathbf{A}$ is some positive semidefinite matrix, for measuring the size of the treatment effect, $\mathbf{x}$.

More generally, with $n = 2^N$ treatment combinations and p responses observed on each experimental unit, if $\mathbf{Y}'$ denotes the $n \times p$ matrix whose rows are the p-dimensional observations, let

$$\mathbf{Y}' = \begin{pmatrix} \mathbf{y}_1' \\ \cdot \\ \cdot \\ \cdot \\ \mathbf{y}_n' \end{pmatrix} = [\mathbf{Y}_1, \mathbf{Y}_2, \ldots, \mathbf{Y}_p],$$

so that $\mathbf{Y}_j$ consists of the n observations on the jth response ($j = 1, \ldots, p$). Then the univariate analysis of variance for the jth response will yield

$$\begin{pmatrix} m_j \\ \mathbf{X}_j \end{pmatrix} = \mathbf{R}\mathbf{Y}_j, \qquad j = 1, \ldots, p, \tag{86}$$

where $\mathbf{R}$ is an $n \times n$ matrix that can be built up by analogy with the one in Eq. 85, m_j corresponds to an overall (or mean) effect, and the $(n-1)$ elements of $\mathbf{X}_j$ correspond to the measures of the main effects and interactions for the jth response. Thus one obtains

$$\mathbf{R}\mathbf{Y}' = \begin{pmatrix} m_1 & m_2 & \cdots & m_p \\ \mathbf{X}_1 & \mathbf{X}_2 & \cdots & \mathbf{X}_p \end{pmatrix} = \begin{pmatrix} \mathbf{m}' \\ \mathbf{X}' \end{pmatrix}, \tag{87}$$

where the $(n-1) \times p$ matrix, $\mathbf{X}'$, has as its columns $\mathbf{X}_1, \mathbf{X}_2, \ldots, \mathbf{X}_p$, and as its rows the $(n-1)$ *single-degree-of-freedom contrast vectors* $\mathbf{x}_1', \mathbf{x}_2', \ldots, \mathbf{x}_{n-1}'$, i.e.,

$$\mathbf{X}' = \begin{pmatrix} \mathbf{x}_1' \\ \mathbf{x}_2' \\ \cdot \\ \cdot \\ \cdot \\ \mathbf{x}_{n-1}' \end{pmatrix} = [\mathbf{X}_1, \mathbf{X}_2, \ldots, \mathbf{X}_p]. \tag{88}$$

Equations 86, 87, and 88 suggest that one way of obtaining the vectors $\mathbf{x}_1, \ldots, \mathbf{x}_{n-1}$ is to perform univariate analyses of variance (perhaps via Yates's algorithm) of each response separately and then collect the p individual measures for each effect (main or interaction) together as a p-dimensional vector.

For assessing the relative magnitudes of the contrast vectors, the values

$$d_i = \mathbf{x}_i' \mathbf{A} \mathbf{x}_i, \qquad i = 1, \ldots, n-1, \tag{89}$$

for some choice of the positive semidefinite compounding matrix $\mathbf{A}$ (more will be said later regarding the choice of $\mathbf{A}$) are obtained. Exactly as in the simple 2^3 example discussed earlier, the d_i's, which are measures of the sizes of the $\mathbf{x}_i$'s, can be interpreted as squared distances between the centroids of certain partitions of the original observations for defining the different treatment effects.

To assess the contrast vectors by means of these squared distances, one needs an "evaluating distribution" or a "null distribution" of such squared distances. In other words, one needs a distribution that is reasonable under the usual kinds of null assumptions, such as multivariate normality (which may be a more reasonable assumption for the contrast vectors than for the original observations because of the "averaging" involved in obtaining the contrast vectors), homoscedasticity, and the absence of any real treatment effects.

More explicitly, under the usual linear model assumptions (see Section 5.2.1), the observations $\mathbf{y}_i'$ are p-variate normally distributed, with location parameters (or expected values) that reflect their factorial experimental structure and a common unknown covariance matrix, $\mathbf{\Sigma}$. Under these assumptions, taken in conjunction with the further null assumption that there are no real treatment effects, the contrast vectors $\mathbf{x}_1, \ldots, \mathbf{x}_{n-1}$ will be mutually independently distributed as $N[\mathbf{0}, \mathbf{\Sigma}]$. [*Note*: In order for the contrast vectors to have exactly the same covariance structure as the initial observations, the transformation matrix $\mathbf{R}$ in Eq. 86 must be specified to be orthogonal, i.e., with the multiplicative constant $1/\sqrt{n}$.] These null assumptions are to be used only as a basis for generating internal comparisons techniques, and an appealing characteristic of these techniques is that, in any specific application, they provide some indications of the appropriateness and adequacy of the assumptions themselves.

The question of an evaluating distribution thus can be formulated as follows: given that $\mathbf{x}_1, \ldots, \mathbf{x}_{n-1}$ are a random sample from $N[\mathbf{0}, \mathbf{\Sigma}]$, what is the distribution of $d_1, \ldots, d_{n-1}$, where $d_i = \mathbf{x}_i'\mathbf{A}\mathbf{x}_i$? In developing an answer to this question, one needs to consider the role of the compounding matrix $\mathbf{A}$, which itself may be, as mentioned earlier, and in practice often is, computed from the observations, $\mathbf{y}_i'$. For instance, one may decide to use $\mathbf{A} = \mathbf{I}$ (which amounts to measuring squared Euclidean distances between centroids of partitions of the data), or in order to reflect differences in the variances of the responses one may wish to use a diagonal matrix of reciprocals of either prespecified variances or estimates of these from the current data. More generally, one may wish to scale the contrast vectors to allow both for different variances of the responses and for intercorrelations among them, and then the choice for $\mathbf{A}$ will be the inverse of a prespecified or estimated covariance matrix of the responses. In general, the use of several choices of $\mathbf{A}$ in analyzing a single set of data may be productive since the different choices may lead to different findings about the data. Whatever the choice is for $\mathbf{A}$, however, since it is common to all the squared distances that are to be internally compared, it is treated in the approach taken here as being a fixed (i.e., nonrandom) quantity.

With this in mind, under the null assumptions outlined above, each of the squared distances d_i (for a selected compounding matrix $\mathbf{A}$) is distributed as the linear combination $c_1\chi_1^2+c_2\chi_2^2+\cdots+c_r\chi_r^2$, where $c_1, \ldots, c_r$ are the positive eigenvalues of $\mathbf{A\Sigma}$, r is the rank of $\mathbf{A}$, and the χ^2's are mutally independent central chi-squared variates, each with 1 degree of freedom. This well-known distributional result is not very useful as it stands; rather, its value lies in suggesting an equally well-known approximate result (see Satterthwaite, 1941; Patnaik, 1949; Box, 1954). The approximate result in question is that the distribution is represented reasonably adequately by a gamma distribution. Thus, specifically, under the null assumptions one can consider $d_1, d_2, \ldots, d_{n-1}$ approximately as a random sample from the gamma distribution with scale parameter λ and shape parameter η, i.e., with density

$$f(d; \lambda, \eta)\begin{cases} =\dfrac{\lambda^\eta}{\Gamma(\eta)}\, d^{\eta-1}\exp(-\lambda d) & \text{for } d\geq 0, \\ =0 & \text{for } d<0, \end{cases} \tag{90}$$

where both λ and $\eta>0$.

To be able to use this evaluating distribution, one needs estimates of λ and η since these are in general unknown. In particular, if a "proper" estimate, $\hat{\eta}$, of η is available, one can obtain a gamma probability plot (i.e., a *Q-Q* plot whose abscissa corresponds to a gamma distribution; see the discussion of *Q-Q* plotting in Section 6.2) of the ordered squared distances. The ordinate on such a plot will correspond to values of the ordered square distances, and the abscissa will represent the corresponding quantiles of a standard gamma distribution with a shape parameter equal to this "properly estimated" value, $\hat{\eta}$. Under null conditions the resulting configuration will be linear, oriented toward the origin with a slope that is an estimate of $1/\lambda$. [*Note*: For gamma probability plotting one does not need a knowledge of the scale parameter since it affects, not the linearity, but only the slope of the configuration; however, the estimation of λ and of η will be carried out simultaneously, although only the estimate of η is needed.]

If the null conditions are not in accord with the data—say, for instance, that there are some real treatment effects—the largest squared distances will be too "large" and will exhibit themselves as departures from a linear "error" configuration defined by the smaller squared distances. Departures from other null assumptions (e.g., homoscedasticity, normality) may also be expected to show up as systematic departures from the "null" linear configuration of the "null" d_i's (i.e., those that conform adequately to the null assumptions).

The next question is what a "proper" estimate of η (and λ) might be. It is desirable that the estimate be based on a null subset of the squared distances (i.e., only those d_i's that satisfy the null assumptions), so that the d_i's which do not conform to such assumptions will stand out against a background defined by d_i's that do. In particular, when a d_i reflects a real treatment effect, its distribution will also be a linear combination of independent χ^2's, but now involving a noncentral χ^2. It is, however, known (Patnaik, 1949) that such a combination involving a noncentral χ^2 can also be approximated by a suitably chosen gamma distribution. Hence, to minimize the influence of possibly real treatment effects on the estimation of the parameters required for the evaluating gamma distribution, it is wise to base the estimation on an order statistics formulation. In other words, one orders the squared distances to obtain $0 \leq d_{(1)} \leq d_{(2)} \cdots \leq d_{(M)} \cdots \leq d_{(K)} \cdots \leq d_{(n-1)}$. Then, on the basis of judgment, one chooses a number, $K[\leq (n-1)]$, as the number of squared distances that are likely to conform to the null assumptions. As additional insurance one bases the actual estimation on the M smallest squared distances considered as the M smallest observations in a random sample of size K. The actual method of estimation to be used with this formulation will be maximum likelihood. The maximum likelihood estimates, $\hat{\lambda}$ and $\hat{\eta}$, obtained from $d_{(1)}, \ldots, d_{(M)}$ considered as the M smallest order statistics in a random sample of size K [where $M \leq K \leq (n-1)$] from the gamma distribution with density as specified in Eq. 90 are functions only of $d_{(M)}$ and the ratios of the geometric and arithmetic means of $d_{(1)}, \ldots, d_{(M)}$ to $d_{(M)}$, i.e.,

$$P = \frac{\prod_{i=1}^{M} [d_{(i)}]^{1/M}}{d_{(M)}} \quad \text{and} \quad S = \frac{\sum_{i=1}^{M} d_{(i)}}{M d_{(M)}}.$$

It is necessarily true that $0 \leq P \leq S \leq 1$. Wilk et al. (1962b) provide tables that enable one to obtain the maximum likelihood estimates, $\hat{\lambda}$ and $\hat{\eta}$.

The above discussion has been cast in terms of an interest in internal comparisons among all $(n-1)$ single-degree-of-freedom contrast vectors. This is, however, not a requisite in any application, and in some situations either one may wish to consider all n single degrees of freedom (although in most analysis of variance situations the overall mean effect would be real a priori and hence set aside in making the other assessments), or, more realistically, one may wish to compare internally only a subset of the $(n-1)$ single-degree-of-freedom contrast vectors, $\mathbf{x}_1, \ldots, \mathbf{x}_{n-1}$. The steps involved in the graphical internal comparisons procedure for assessing the relative magnitudes of $L[\leq(n-1)]$ contrast vectors can therefore be summarized now:

1. Calculate the $(n-1)$ single-degree-of-freedom contrasts for each response separately.
2. Form the p-dimensional contrast vectors, $\mathbf{x}_1, \ldots, \mathbf{x}_{n-1}$.
3. Choose the subset of $L[\leq(n-1)]$ contrast vectors to be internally compared—$\mathbf{x}_i$, $i=1, \ldots, L$.
4. Select the positive semidefinite compounding matrix, $\mathbf{A}$.
5. Compute the measures of size (or the squared distances), $d_i = \mathbf{x}_i'\mathbf{A}\mathbf{x}_i$, $i=1, 2, \ldots, L$, and order them to obtain $d_{(1)} \leq d_{(2)} \leq \cdots \leq d_{(L)}$.
6. Select the number K $(\leq L)$ on the basis of judgment.
7. Select the number M $(\leq K)$, and, using $d_{(1)} \leq \cdots \leq d_{(M)}$, calculate

$$P = \frac{\prod_{i=1}^{M} [d_{(i)}]^{1/M}}{d_{(M)}} \quad \text{and} \quad S = \frac{\sum_{i=1}^{M} d_{(i)}}{M d_{(M)}}.$$

8. Using K/M, P, and S, determine the maximum likelihood estimates, $\hat{\lambda}$ and $\hat{\eta}$.
9. Plot $d_{(1)}, d_{(2)}, \ldots, d_{(L)}$ against the corresponding quantiles of the gamma distribution with parameters $\lambda = 1$, $\eta = \hat{\eta}$; i.e., plot the points $\{\tilde{x}_i, d_{(i)}\}$ for $i=1, \ldots, L$, where $\tilde{x}_i$ is defined by

$$\int_0^{\tilde{x}_i} \frac{1}{\Gamma(\hat{\eta})} u^{\hat{\eta}-1} \exp(-u)\, du = p_i,$$

for a specified cumulative probability p_i [e.g., $(i-\frac{1}{2})/L$, $(i-\frac{1}{3})/L+\frac{1}{3}$ or $i/(L+1)$].

Before presenting examples of application of this graphical internal comparisons method, a few comments on certain features of the method may be appropriate. First, with regard to the choice of the compounding matrix $\mathbf{A}$, it has already been stated that data-analytic wisdom suggests the use of several $\mathbf{A}$'s in analyzing any given set of data. Once again, the point is that any truly multivariate situation cannot usually be fully described by any single unidimensional representation, and the implication of this here is that different choices for $\mathbf{A}$ may lead to quite different and possibly interesting insights into the multivariate nature of the data. A flexible collection to use has, however, been developed (see Section II of Appendix C in Roy et al., 1971, and also Wilk et al., 1962), and the following is a list of the set:

1. $\mathbf{A}_1 = \mathbf{I}$, the identity matrix.
2. $\mathbf{A}_2 = \mathbf{S}_L^{-1}$, the inverse of a covariance (or sum-of-products) matrix obtained from all L contrast vectors to be internally compared.

3. $\mathbf{A}_3 = \mathbf{S}_R^{-1}(\mathbf{A}_1)$, the inverse of a sum-of-products matrix obtained from the R ($<L$) contrast vectors whose associated distances based on the compounding matrix $\mathbf{A}_1$ are the R smallest distances; (see also the discussion of robust estimators of dispersion in Section 5.2.3).

4. $\mathbf{A}_4 = \mathbf{S}_R^{-1}(\mathbf{A}_2)$, the inverse of a sum-of-products matrix obtained from the R ($<L$) contrast vectors whose associated distances based on the compounding matrix $\mathbf{A}_2$ are the R smallest distances.

5. $\mathbf{A}_5 = \mathbf{S}^{-1}$, the inverse of a sum-of-products matrix based on a user-specified subset of contrast vectors.

6. $\mathbf{A}_6 = \mathbf{D}(1/s_{ii}(\mathbf{A}_2))$, a diagonal matrix of the reciprocals of the diagonal elements of $\mathbf{A}_2^{-1}$.

7–9. For $j = 7, 8, 9$, $\mathbf{A}_j = \mathbf{D}(1/s_{ii}(\mathbf{A}_l))$, a diagonal matrix of the reciprocals of the diagonal elements of $\mathbf{A}_l^{-1}$, $l = 3, 4, 5$.

10. $\mathbf{A}_{10} = \mathbf{a}_l\mathbf{a}_l'$, where $\mathbf{a}_l$ is the eigenvector (or principal component) corresponding to the lth largest eigenvalue of either the correlation matrix or the covariance matrix computed from all L contrast vectors (see Section 2.2.1 for a discussion of principal components).

A second issue in using the gamma probability plotting procedure outlined earlier involves the choice of values for K and M, which is deliberately left to the user's discretion. Since this is an informal statistical tool, which is not concerned with such things as the precise significance levels to be associated with formal tests of hypotheses, it would be expected that both prior and posterior (i.e., after seeing the data) considerations would influence the choice of values for K and M. The choice of these values would, of course, affect the estimates of λ and η, and one concern may be the sensitivity of these estimates to the choices of K and M. Figure 11 shows a plot of the maximum likelihood estimate of η as a function of K/M for various values of P and S (defined earlier), and it is seen that the estimate of η is quite insensitive to the value of K provided that M is not too close to K. For many two-level factorial experiments a choice of K and M such that $K/M > 3/2$ seems to be recommendable as a relatively safe rule. In most situations the loss of efficiency in estimating η due to choosing a small value of M relative to K appears to have little or no effect on the interpretations of the configurations observed on the gamma probability plots (see Wilk and Gnanadesikan, 1964; and Chapter VII of Roy et al., 1971).

A third facet of the method is that as a multiresponse technique it is intended to augment rather than to replace analyses of various subsets of the responses, including the responses considered individually. The analyses of subsets by comparable gamma probability plots are accomplished quite easily by suitably modifying the $p \times p$ compounding matrix $\mathbf{A}$. For

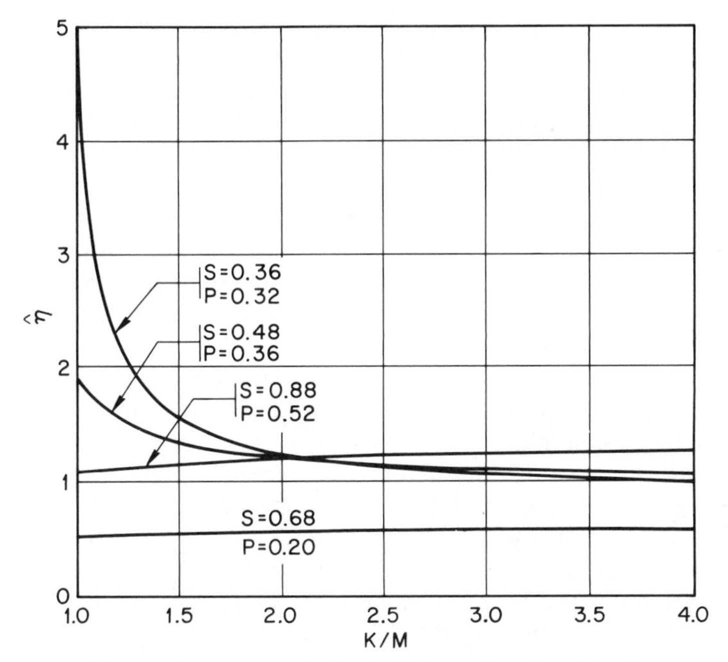

Fig. 11. Dependence of $\hat{\eta}$ on K/M for various P and S values.

example, if one is interested in studying a subset of q ($<p$) of the initial variables, a choice of zero, for all the elements in rows and columns of **A** that correspond to the complementary subset of $(p-q)$ of the variables, will yield squared distances based only on the chosen set of q variables. [*Note*: If **A** is to be an inverse of a covariance or sum-of-products matrix, an appropriate procedure may be to invert the matrix for the q responses of interest rather than extracting a $q \times q$ matrix from the inverse of the full $p \times p$ matrix.] When $q=1$ (i.e., a single response is to be analyzed by itself), **A** may be specified as a matrix all of whose elements are 0 except for a single element (which can be taken to be equal to 1) in the diagonal position corresponding to the particular response. For this choice of **A**, to treat the case when $q=1$, the squared distances are just the squared contrasts for the particular response, and it is customary to treat these as having a chi-squared distribution with 1 degree of freedom, which is equivalent to specifying $\eta=\frac{1}{2}$ instead of estimating a value for it. Example 36 contains some discussion of the issues pertaining to the productive interplay between an overall multiresponse analysis and separate univariate analyses of the individual responses.

Example 36. The data are from a study of nine factors thought to affect Picturephone® quality, and the experiment was organized as a one-half replicate of a 2^9 factorial in a split-plot design (see Wilk and Gnanadesikan, 1964, and also Chapter VII of Roy et al., 1971). There were $p=8$ responses per experimental unit, and each was a subjective assessment of picture quality on a 10-point scale. Exhibit 36*a* shows a gamma probability plot of squared distances obtained by the above method for the 129 main effects and two- and three-factor interactions involved. The squared distances are labeled by the treatment effects to which they correspond. The choice for **A** in this example was the identity matrix, and the values of K/M, M, and the estimated shape parameter are all shown on the figure.

One interpretation of this figure is that the "large" points (viz., those that correspond to the larger squared distances and appear in the plot toward the right-hand top) all correspond to real treatment effects. On the other hand, one might also think that the configuration is suggestive of two intersecting straight lines. Since the experiment was of a split-plot type, there are whole-plot factors and subplot factors. Typically, in a split-plot experiment, the whole-plot factors will have a variance or, in this multiresponse case, a covariance structure that is more dispersed than the covariance structure for the subplot factors. It turns out here that 14 of the 17 points on the suggested line of steeper slope correspond to

Exhibit 36a. Gamma probability plot for PICTUREPHONE® data; $L=129$, $\mathbf{A}=\mathbf{I}$, $M=64$, $K/M=2$, $\hat{\eta}=2.33$

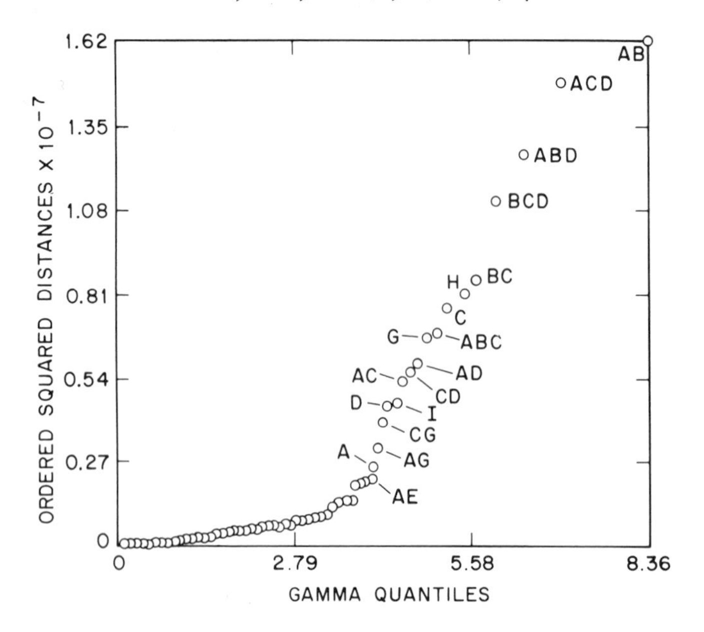

Exhibit 36b. Gamma probability plot for PICTUREPHONE® data; $L=115$, $\mathbf{A}=\mathbf{I}$, $M=57$, $K/M=2$, $\hat{\eta}=2.40$

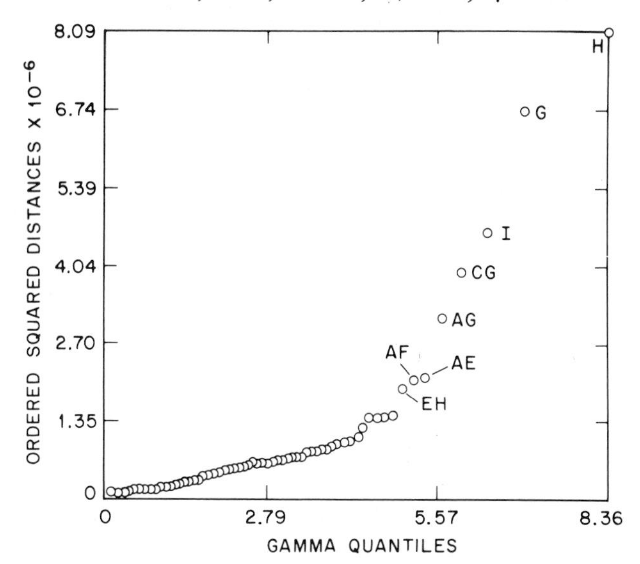

effects involving the whole-plot factors, A, B, C, and D. Thus the two intersecting lines correspond to groups of treatment effects with different covariance structures, and one ought to split up the collection of treatment effects into those that have a whole-plot covariance structure and another set that has a subplot covariance structure. A gamma probability plot of squared distances for the 115 main effects and two- and three-factor interactions not solely confined to the whole-plot factors is shown in Exhibit 36*b*. On this plot one can identify the "top" 7 or 8 points as being indicative of real treatment effects.

As with all multiresponse methods, it would be legitimate in this example to inquire what might have happened if one had carried out the analysis of this experiment by doing separate analyses of the eight responses involved, perhaps using the uniresponse probability plotting technique of cell I in Table 2. Exhibits 36*c*, *d*, and *e* (see pages 240–241) show typical $\chi^2_{(1)}$ (actually, gamma with shape parameter $\frac{1}{2}$) probability plots of the 115 squared contrasts for three of the responses. [*Note*: A chi-squared-with-one-degree-of-freedom probability plot of the squared contrasts should yield a configuration "equivalent" to one obtained by making a half-normal plot of the absolute contrasts.] These three plots are typical of the eight plots that one can get by analyzing the responses separately. Comparing Exhibit 36*b* with these three figures, one sees that one is able to identify many more possibly real effects in the multivariate

Exhibit 36c. Gamma probability plot ($\eta=\frac{1}{2}$) of 115 squared contrasts for second variable

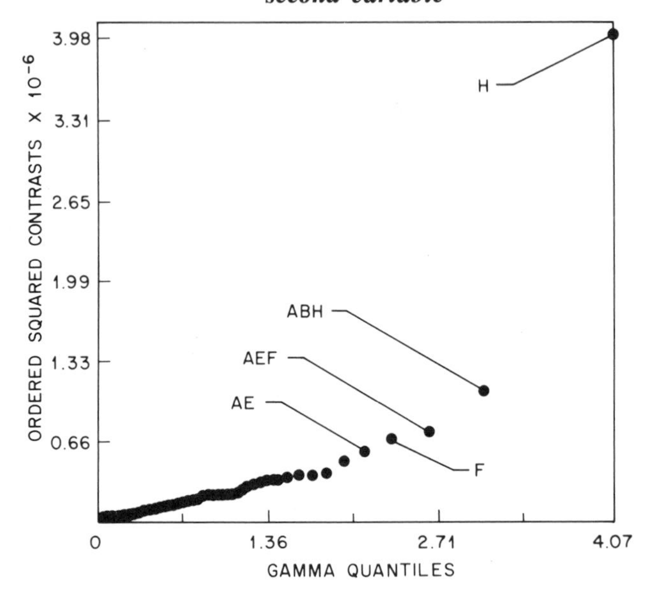

Exhibit 36d. Gamma probability plot ($\eta=\frac{1}{2}$) of 115 squared contrasts for third variable

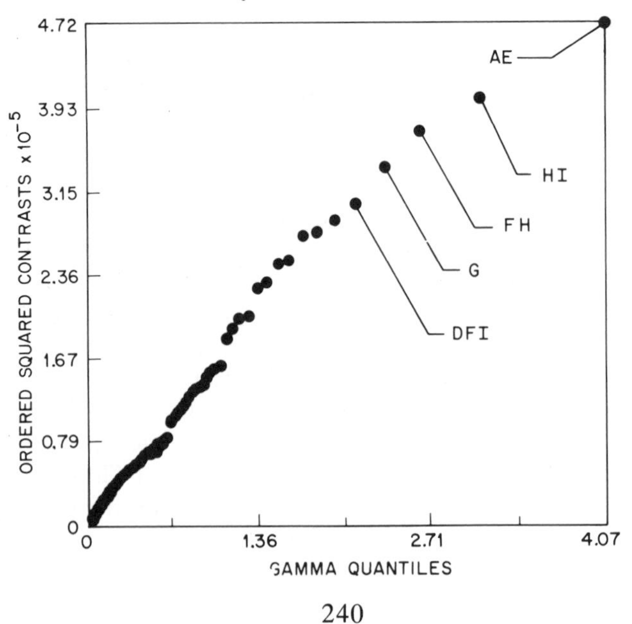

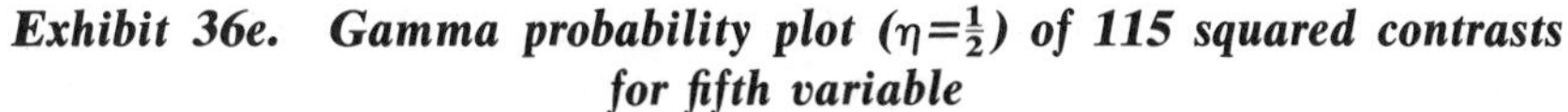

Exhibit 36e. Gamma probability plot ($\eta = \frac{1}{2}$) of 115 squared contrasts for fifth variable

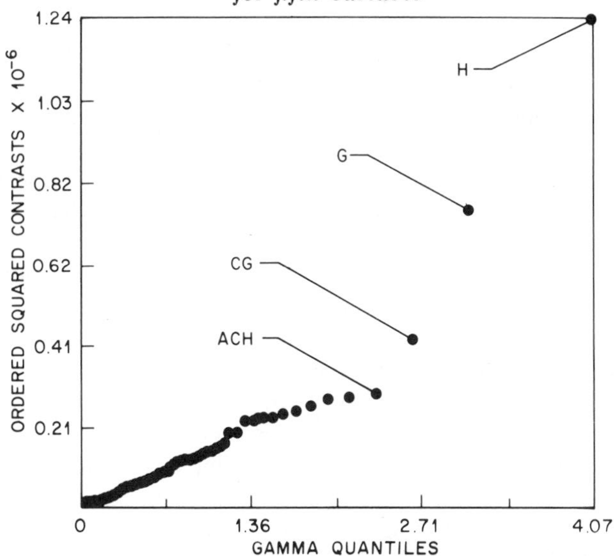

analysis than in the separate univariate analyses. In this example the combined evidence from the eight separate uniresponse analyses is that three or four of the treatment effects are possibly real, whereas the multiresponse analysis provides evidence that seven or eight of the effects may be real.

One possible explanation for the greater sensitivity of the multiresponse analysis in this example is provided by the estimated value of the shape parameter, viz., $\hat{\eta} = 2.4$. If the responses were indeed statistically independent, one might expect the squared distances to be distributed as a chi-squared variate with 8 ($= p$) degrees of freedom or, equivalently, as a gamma variate with shape parameter 4 ($= p/2$). Thus the lower value for $\hat{\eta}$ suggests that there is probably an accumulation of several fairly small real effects on the separate response scales into a smaller-dimensional space, which is then revealed better by the multiresponse analysis. [*Note*: An interesting modification of the separate uniresponse analyses suggested by this is to estimate a shape parameter for the probability plot of the squared contrasts, rather than using the prespecified $\chi^2_{(1)}$ distribution for them.] Also, there is perhaps a stabilizing effect on the error configuration (i.e., the linear part of the plot) due to the intercorrelations among the responses.

Example 37. This example is based on data (cf. Chapters IV and VII of Roy et al., 1971) from a one-quarter replicate of a 2^7 experiment concerned with seven factors which might affect the operation of a detergent manufacturing process. The original study involved measurements on seven responses, but for present purposes only a bivariate subset of the original seven is considered. The two responses are called *rate* (bins/hour) and *stickiness*, and Exhibit 37*a* shows the 32 bivariate observations involved, together with the treatment combinations that label them. The seven experimental factors involved were as follows: *A*—air injection, *B*—nozzle temperature, *C*—crutcher amperes, *D*—inlet temperature, *E*—tower air flow, *F*—number of baffles, and *G*—nozzle pressure.

Andrews et al. (1971) used these data as an example for applying their methods (see Section 5.3) for developing data-based transformations, and the discussion here is drawn largely from their paper. An indirect way of assessing the data-based transformation methods described earlier in Section 5.3 is to study the effects of the transformations on the outputs of statistical analyses, such as analyses of variance, performed before and after the transformations. In the present example, for instance, one can obtain 31 estimated treatment effects (or single-degree-of-freedom contrasts) of interest for each response. One can use the graphical internal comparisons technique of cell I in Table 2 for simultaneously assessing the 31 effects, viz., via either a half-normal probability plot of the absolute values of the estimated effects, or, equivalently, a $\chi^2_{(1)}$ probability plot of the squared values. One can do this for effects estimated on both the untransformed and the transformed scales of the responses and compare the resulting configurations. It is perhaps reasonable to expect that, because of the averaging involved in obtaining the estimated treatment effects, except for bad nonnormality of the original observations the estimated effects would be adequately normal in distribution. For instance, in the last example, despite the initial 10-point scale for the eight responses, all the probability plots associated with the contrasts (viz., Exhibits 36*c*, *d*, and *e*) are quite linear for the major part, and the lack of systematic curvilinearity in these plots lends credence to the "expected" normality of the contrasts. However, in the present example, Exhibits 37*b* and *c*, which are $\chi^2_{(1)}$ (or corresponding gamma) probability plots of the squared contrasts on each of the two untransformed response scales, appear to indicate the presence of considerable distributional peculiarities. The extremely "choppy" appearance of the lower left-hand end of Exhibit 37*b* (see page 244) can perhaps be attributed to the essentially discrete nature of the response termed *rate*, as evident in Exhibit 37*a*, whereas the distributional departure indicated in Exhibit 37*c* (see page 244) and associated with the other response seems to be more subtle.

Exhibit 37a. Data matrix for 2^{7-2} experiment on detergent manufacturing process (cf. Roy et al., 1971, p. 54)

Run No.	Treatment Combination	Rate (bins/hour)	Stickiness
1	(1)	38.0	5.40
2	*afg*	38.0	5.90
3	*bfg*	35.0	2.95
4	*ab*	36.0	5.38
5	*cg*	38.0	5.22
6	*acf*	37.0	5.33
7	*bcf*	37.0	4.90
8	*abcg*	36.0	4.50
9	*df*	34.5	3.15
10	*adg*	38.0	3.06
11	*bdg*	36.0	5.70
12	*abdf*	37.0	4.20
13	*cdfg*	38.5	4.70
14	*acd*	38.0	4.20
15	*bcd*	38.0	5.17
16	*abcdfg*	39.0	5.66
17	*eg*	37.0	4.60
18	*aef*	38.0	5.20
19	*bef*	32.0	2.49
20	*abeg*	39.0	6.10
21	*ce*	39.0	3.84
22	*acefg*	37.0	4.90
23	*bcefg*	35.0	4.30
24	*abce*	34.0	3.50
25	*defg*	37.0	3.24
26	*ade*	37.0	3.79
27	*bde*	39.0	5.80
28	*abdefg*	39.0	5.30
29	*cdef*	39.0	5.60
30	*acdeg*	40.0	6.20
31	*bcdeg*	40.0	5.47
32	*abcdef*	40.0	4.77

The transformation method of Box and Cox (1964) to improve marginal normality and the one proposed by Andrews et al. (1971) for enhancing joint normality (see Section 5.3) were employed, and the estimated values of the power transformation parameters are shown in Exhibit 37*d* (see page 244). In estimating these transformations, in addition to enhancing normality an attempt was made to reduce nonadditivities at the same time by specifying a fit (or linear model) solely in terms of the seven main effects on the transformed scales of the responses.

Exhibit 37b. Gamma probability plot ($\eta=\frac{1}{2}$) of 31 squared contrasts for untransformed rate data

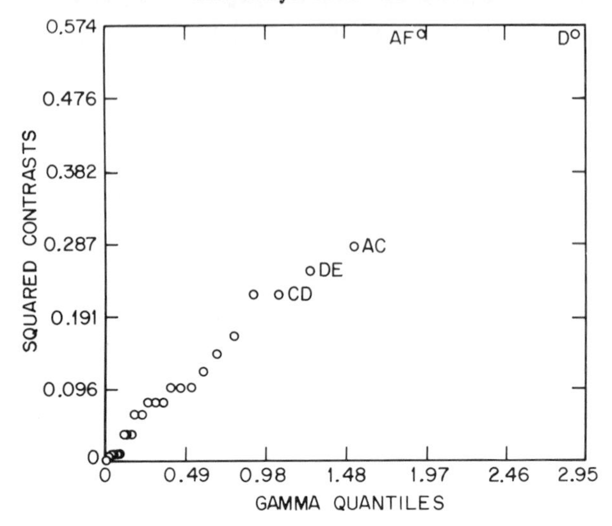

Exhibit 37c. Gamma probability plot ($\eta=\frac{1}{2}$) of 31 squared contrasts for untransformed stickiness data

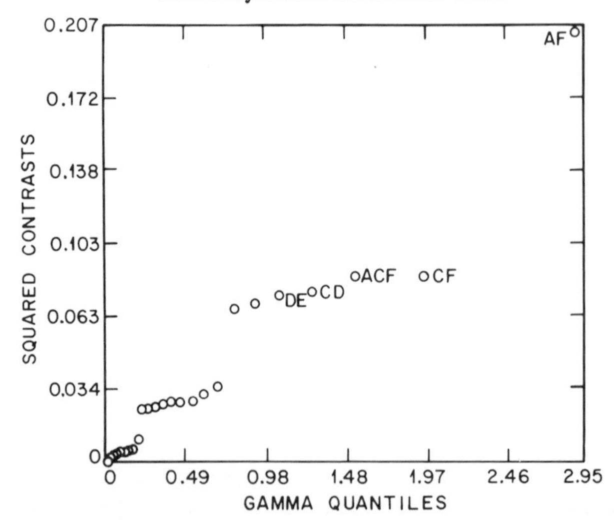

Exhibit 37d. Estimates of transformation parameters for detergent manufacture data (cf. Andrews et al., 1971)

Box-Cox Method Estimates		Andrews et al. Method Estimates	
$\hat{\lambda}_1$	$\hat{\lambda}_2$	$\hat{\lambda}_1$	$\hat{\lambda}_2$
8.88	2.06	7.22	1.88

Exhibit 37e. Gamma probability plot ($\eta=\frac{1}{2}$) of 31 squared contrasts for rate data transformed by Method I

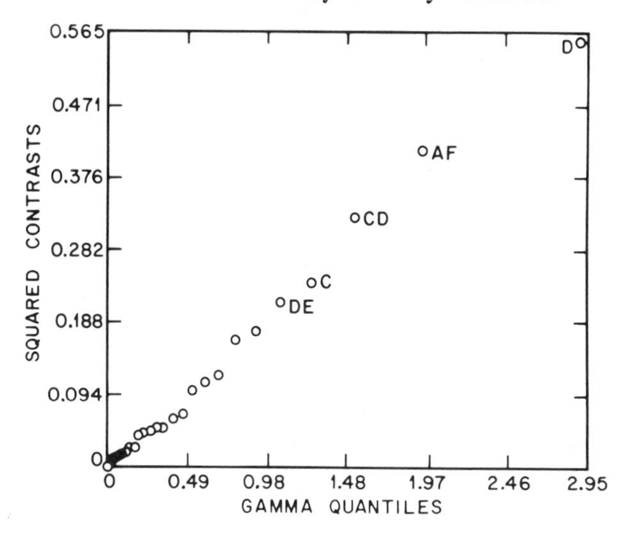

The improvements achieved by using the transformations determined by the Box and Cox method are evident in Exhibits 37*e* and *f*, which show the $\chi^2_{(1)}$ probability plots for squared effects on the transformed scales of the two variables involved. The smoother configurations of these two plots, especially at the lower end, suggest not only possible improvement of underlying normality but also the delineation of a more homogeneous grouping of fairly small effects, from which one can hopefully derive a "cleaner" estimate of error variance.

A similar evaluation of the method proposed by Andrews et al. (1971) for enhancing joint normality can be made by comparing the two gamma probability plots of the squared distances obtained from both the untransformed bivariate data and the transformed bivariate observations obtained by using the powers in the second set of columns of Exhibit 37*d* for the two variables. Exhibits 37*g* and *h* (see pages 246–247) show the gamma probability plots, the former derived from the untransformed observations and the latter from observations transformed by using $\hat{\lambda}_1 = 7.22$ and $\hat{\lambda}_2 = 1.88$. The choice for the compounding matrix in both these plots was of the $\mathbf{A}_3$ type, involving a subselection of R contrast vectors with smallest Euclidean lengths (see the list of choices for $\mathbf{A}$ given earlier); in particular, R was taken to be 15. Also, for both plots, K was considered to be 31 and the value of M was taken as 15, so that $K/M = 2.07$. The estimated values of the shape parameter are indicated in the captions for the figures.

Not only is the null configuration of the "smaller" points (i.e., the ones

Exhibit 37f. Gamma probability plot ($\eta=\frac{1}{2}$) of 31 squared contrasts for stickiness data transformed by Method I

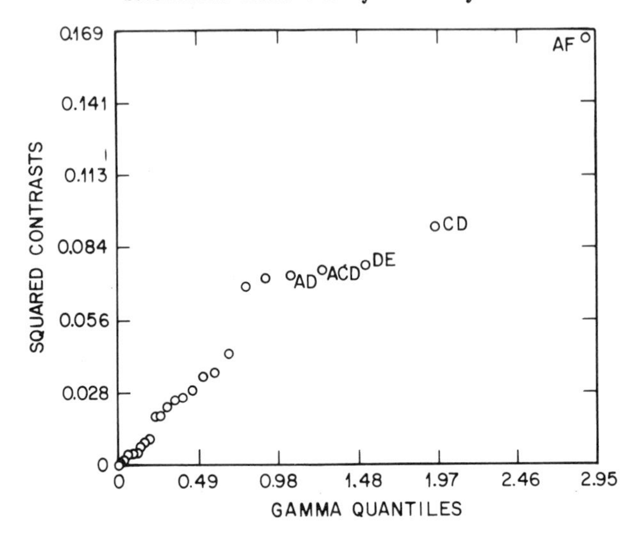

Exhibit 37g. Bivariate gamma probability plot ($\hat{\eta}=0.75$) for detergent data untransformed

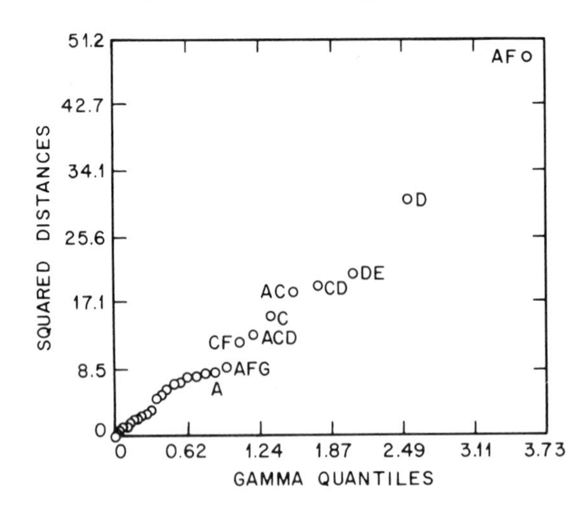

Exhibit 37h.* *Bivariate gamma probability plot ($\hat{\eta}$=1.01) for detergent data transformed by Method II

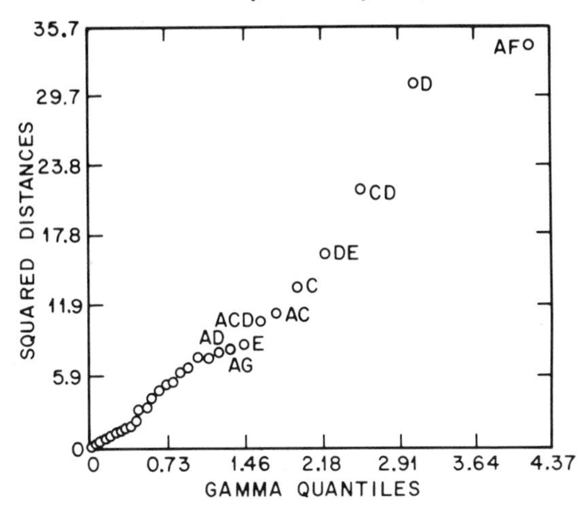

in the lower left-hand corner) in Exhibit 37h smoother, but also the delineation of the departures at the "large" end (viz., the upper right-hand corner) is clearer than in Exhibit 37g. An interesting feature, however, is that the improvement in the null configuration (i.e., the lower end) in going from Exhibit 37g to h is nowhere near as striking as the improvements from Exhibit 37b to e or from Exhibit 37c to f. This suggests that the approach used in the internal comparisons method for estimating the shape parameter may be introducing a very valuable robustness into the process, inasmuch as the suspected marked nonnormality of the responses does not seem to unduly distort the configurations on the gamma probability plot.

Example 38. The main purpose of this example, taken from Wilk and Gnanadesikan (1964), is to provide a rather dramatic illustration of the importance of using more than one choice for the compounding matrix **A**. Sixty random deviates were generated from a five-dimensional normal distribution with zero mean vector and a distinctly nonspherical covariance matrix,

$$\mathbf{\Sigma} = \begin{pmatrix} 1 & & & & \\ 2 & 5 & & & \\ -3 & -7 & 11 & & \\ 4 & 10 & -16 & 25 & \\ 2 & 5 & -8 & 9 & 16 \end{pmatrix},$$

where the elements above the diagonal are of course obtainable by symmetry. To a random selection of 10 of these 60 observations were added certain constants to shift their means and thus simulate "real" effects. The shifts chosen were as follows: three vectors equal to (3, 7, 10, 12, 11), three others equal to (5, 5, 5, 5, 5), three more equal to (7, 2, 0, 5, 4), and a last one equal to (5, 8, 15, 20, 18).

Exhibits 38*a* and *b* are gamma probability plots of the squared distances derived from these observations for two choices of $\mathbf{A}$, viz., $\mathbf{A}=\mathbf{I}$ and $\mathbf{A}=\mathbf{S}^{-1}$, the inverse of a sum-of-products matrix based on a random selection of 30 out of the 50 "central" (i.e., zero mean) observations. Each figure is actually a plot of only the 55 smallest squared distances instead of all 60 of them, and the values of M and K/M, as well as the resulting estimate of η, are all indicated in the captions. The ranges of the values of the squared distances are quite different, as are the two values of $\hat{\eta}$, but the most striking thing about the two figures is that Exhibit 38*a* contains no indication of the five known nonnull observations, whereas Exhibit 38*b* clearly delineates them. In light of the prespecified nonsphericity of the covariance matrix used in generating the observations, it is perhaps not surprising that the identity matrix is not a very appropriate choice. The main implication in practice, however, is that without a considerable amount of knowledge about the data a safe rule

Exhibit 38a. Gamma probability plot for artificial five-dimensional data; $L=55$, $\mathbf{A}=\mathbf{I}$, $M=30$, $K/M=1.4$, $\hat{\eta}=1.4$

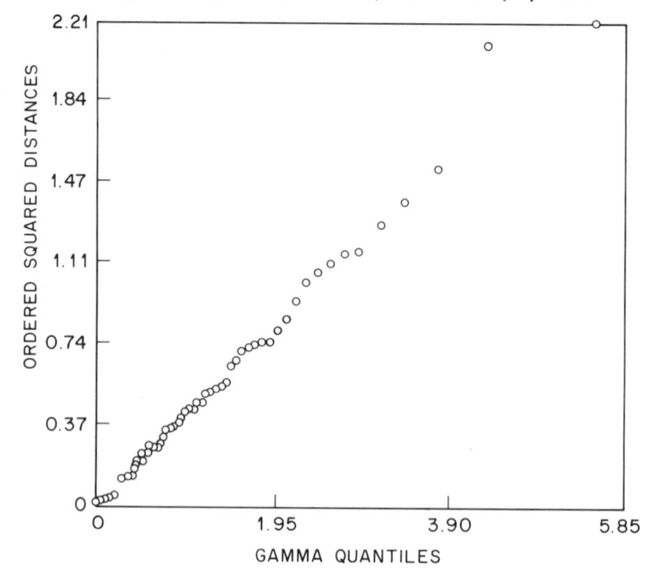

Exhibit 38b. Gamma probability plot for artificial five-dimensional data; $L=55$, $\mathbf{A}=\mathbf{S}_{30}^{-1}$, $M=30$, $K/M=1.4$, $\hat{\eta}=3.83$

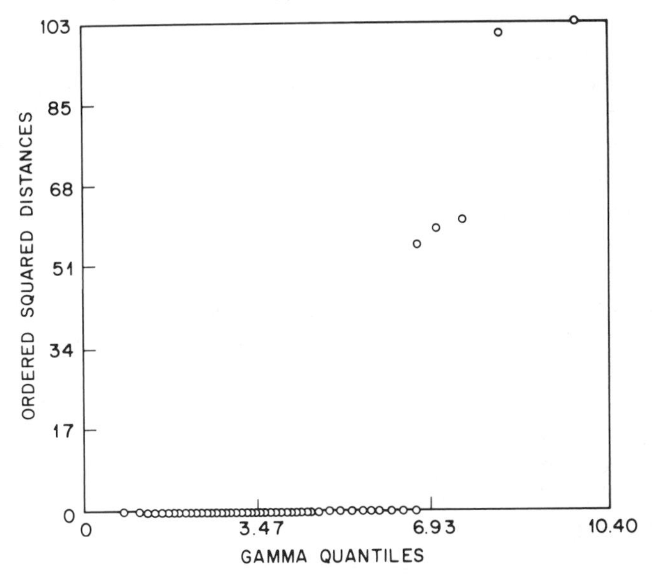

would be to use different choices of **A** and then to compare the results to gain further insights into the structure of the data.

6.3.2. Graphical Internal Comparisons Among Equal-Degree-of-Freedom Groupings

This subsection deals with probability plotting techniques for Cell V of Table 2. In the geometrical terms used in Section 5.2.1 for describing the concepts and processes of orthogonal multivariate analysis of variance, the situation represented by this cell arises when the decomposition of n-space into orthogonal subspaces contains a set of r mutually orthogonal linear subspaces, each of dimensionality ν (>1), and the observations are p-dimensional. Thus the prototype here is a situation in which there are r $p\times p$ sum-of-products matrices, $\mathbf{S}_1, \ldots, \mathbf{S}_r$, each based on ν degrees of freedom, and one wishes to compare simultaneously the "sizes" of the dispersions summarized by these matrices, or equivalently, by the mean sum-of-products matrices, $\mathbf{S}_i/\nu$'s, using probability plotting techniques. One example of this occurs when internal comparisons are desired among all the main effects, or interactions of the same order, in an m-level factorial experiment. In this case all the main effects will have $\nu=(m-1)$ degrees of freedom, and each qth order interaction will have

$\nu=(m-1)^{q+1}$. Another example occurs when one has $(\nu+1)$ replications within cells and wishes to assess the validity of the assumption of a common within-cell covariance structure.

An intrinsic difficulty of the present problem is to define measures of "size" of a dispersion matrix. One should not expect that any single measure will provide an adequate summary of the dispersion information contained in the matrix. Certain functions of the eigenvalues of a dispersion matrix may be used as unidimensional summaries of the size of the dispersion; see, for example, Roy et al. (1971, Chapter II, Section 3). Two such functions are the arithmetic mean, or sum, and the geometric mean of the eigenvalues. Since the arithmetic mean is sensitive to very large and very small eigenvalues, whereas the geometric mean tends to be particularly sensitive only to very small eigenvalues, the two functions may lead to different insights concerning the dispersion structure. The use of both functions is recommended for data-analytic purposes, and the two methods described below are based, respectively, on the two functions.

In analysis of variance applications, such as the factorial experiment mentioned earlier, the dimensionality of response p may often exceed the value ν. In this case the matrices $\mathbf{S}_1, \ldots, \mathbf{S}_r$ will have ν positive eigenvalues and $(p-\nu)$ zero eigenvalues. A natural modification of the second function, therefore, is to consider the geometric mean of the nonzero eigenvalues. Specifically, then, the two functions to be considered as measures of size of a sum-of-products matrix, $\mathbf{S}=((s_{ij}))$, with eigenvalues $c_1\geq\cdots\geq c_t>0$, are

$$\mathcal{A}=\sum_{i=1}^{t} c_i=\text{tr}(\mathbf{S})=\sum_{j=1}^{p} s_{jj},$$

$$\mathcal{G}=\left(\prod_{i=1}^{t} c_i\right)^{1/t}.$$

When the different responses in a multiresponse analysis of variance are measured on very different scales, it may be desirable to weight the responses accordingly, so that deviations from null conditions on the different response scales are not given the same weight. For incorporating this feature in the present mode of analysis, one can use as starting points in the analysis not just the sum of products matrices, $\mathbf{S}_i$'s, but also scaled versions of them, viz., $\mathbf{S}_1\mathbf{A}, \ldots, \mathbf{S}_r\mathbf{A}$, where $\mathbf{A}$ is a positive semidefinite matrix. [*Note*: Computationally, the eigenvalues required for $\mathcal{G}$ may be obtained either from a singular-value decomposition appropriate to problems, such as discriminant analysis, that involve two covariance matrices, or from eigenanalyses of the symmetric matrices, $\mathbf{Z}_i'\mathbf{A}\mathbf{Z}_i$, where $\mathbf{S}_i=\mathbf{Z}_i\mathbf{Z}_i'$, rather than eigenanalyses of the asymmetric forms, $\mathbf{S}_i\mathbf{A}$, using the mathematical property that the nonzero eigenvalues of $\mathbf{Z}_i\mathbf{Z}_i'\mathbf{A}$ are also the nonzero eigenvalues of $\mathbf{Z}_i'\mathbf{A}\mathbf{Z}_i$.] However, when the $\mathbf{S}_i$'s and $\mathbf{A}$ are all positive definite, then, since the product of the eigenvalues of $\mathbf{S}_i\mathbf{A}$ is

$$\prod_{j=1}^{p} c_j = |\mathbf{S}_i\mathbf{A}| = |\mathbf{S}_i|\,|\mathbf{A}|,$$

the scaling by $\mathbf{A}$ is immaterial for purposes of internal comparisons among the $\mathbf{S}_i$'s in terms of the statistic $\mathscr{G}$.

The matrix $\mathbf{A}$ plays the same role here as the compounding matrix $\mathbf{A}$ did in the method discussed in Section 6.3.1. Hence, as before, possible choices for the $p \times p$ matrix $\mathbf{A}$ include (i) the identity matrix, which may be appropriate when $\mathbf{S}_1, \ldots, \mathbf{S}_r$ pertain to a decomposition of the error covariance structure; (ii) a diagonal matrix of reciprocals of variances of the responses; and (iii) the inverse of a covariance matrix of the responses. Once again, under choices (ii) and (iii), the matrix $\mathbf{A}$ may be either prespecified or estimated from the data on hand, and in either case, since it is used as a common factor to scale all the $\mathbf{S}_i$'s, it is considered a fixed matrix for the subsequent internal comparisons analyses, just as in Section 6.3.1. For the rest of this subsection, it is to be understood that the internal comparisons of the "magnitudes" of $\mathbf{S}_1, \ldots, \mathbf{S}_r$ are to be made via the r associated values of either $\mathscr{A}$ or $\mathscr{G}$,

$$a_i = \text{tr}(\mathbf{S}_i\mathbf{A}), \qquad g_i = \left\{\prod_{j=1}^{t} c_j(\mathbf{S}_i\mathbf{A})\right\}^{1/t}.$$

Next an evaluating distribution is needed for each of these collections. If such a distribution were available for the statistic $\mathscr{A}$, for instance, one could obtain a probability plot of the ordered values, $0 < a_{(1)} \leq \cdots \leq a_{(r)}$, against the corresponding quantiles of the distribution. A similar use may be made of the distribution of $\mathscr{G}$. Under null conditions the distributions of both $\mathscr{A}$ and $\mathscr{G}$ turn out to be well approximated by gamma distributions.

That this is so for $\mathscr{A}$ can be seen by recognizing that $\mathscr{A}$ is the sum of ν mutually independent positive semidefinite quadratic forms, each of whose distributions may itself be adequately approximated by a gamma distribution, as stated in Section 6.3.1. Specifically, each matrix $\mathbf{S}_i$ can be represented as $\mathbf{Z}_i\mathbf{Z}_i'$, where $\mathbf{Z}_i' = \mathbf{R}_i\mathbf{Y}'$. Furthermore, $\mathbf{Y}'$ is the $n \times p$ matrix of original observations, and the $\nu \times n$ matrix, $\mathbf{R}_i$, is such that $\mathbf{R}_i\mathbf{R}_i' = \mathbf{I}(\nu)$ and $\mathbf{R}_i\mathbf{R}_j' = \mathbf{O}$ $(i \neq j)$. Then

$$a_i = \text{tr}(\mathbf{Z}_i\mathbf{Z}_i'\mathbf{A}) = \sum_{j=1}^{\nu} \mathbf{z}_{ij}'\mathbf{A}\mathbf{z}_{ij},$$

where $\mathbf{z}_{ij}$ is the jth column of $\mathbf{Z}_i$. [*Note*: This way of looking at the measure of size $\mathscr{A}$ is discussed also in Section 5 of Chapter VII of Roy et al., 1971.] Null assumptions, which may be employed to develop methodology for studying specific departures from them, are that the original observations are mutually independent with identical, but unknown, covariance matrices $\mathbf{\Sigma}$ and that there are no real effects associated with the r groups. Under such null conditions, $\mathbf{z}_{ij}$ $(i=1, \ldots, r; j=1, \ldots, \nu)$ may be considered as a random sample from $N(\mathbf{0}, \mathbf{\Sigma})$, so that a_i is the sum of ν mutually independent positive semidefinite quadratic forms, and, furthermore, $a_1, \ldots, a_r$ are mutually independent. The normality assumption concerning the $\mathbf{z}_{ij}$'s is not unreasonable since they are linear combinations of the original observations. As stated in Section 6.3.1, under null conditions the distribution of each quadratic form can be adequately approximated by a gamma distribution with scale parameter λ_a and shape parameter η_a/ν, and hence $a_1, \ldots, a_r$ may be considered as approximately a random sample from a gamma distribution with scale parameter λ_a and shape parameter η_a.

The use of $\mathscr{A}$ is thus seen to be a direct extension of the method discussed in Section 6.3.1 for the single-degree-of-freedom case, and one may wonder why an analysis of the νr quadratic forms, $\mathbf{z}'_{ij}\mathbf{A}\mathbf{z}_{ij}$, by that method is not adequate for the present problem. The issue, of course, is that the orthogonal decomposition yielding the individual p-dimensional vectors, $\mathbf{z}_{ij}$, is arbitrary and may not have any meaningful interpretation, whereas the a_i's are defined uniquely and meaningfully. The problem is the same here as in uniresponse analysis of variance, where a sum of squares with ν degrees of freedom does not necessarily have a unique meaningful decomposition into ν orthogonal single degrees of freedom.

In connection with the null distribution of $\mathscr{G}$, Hoel (1937) suggests approximating the distribution of the geometric mean of the eigenvalues of a $p\times p$ sample covariance matrix based on a sample of size n with $p\leq(n-1)$ by a gamma distribution whose shape parameter is a function only of p and n but whose scale parameter is $|\mathbf{\Sigma}|^{1/p}$ times a quantity involving p and n, where $\mathbf{\Sigma}$ is the unknown underlying covariance matrix. The unknown scale and shape parameters are then obtained by equating the first two moments. For present purposes the null distribution is approximated by a gamma distribution with unknown scale and shape parameters, λ_g and η_g, respectively, and the parameters are then estimated by maximum likelihood instead of the method of moments. A Monte Carlo investigation was carried out to check on the adequacy of the approach. The dots in Figure 12 constitute a typical gamma probability plot of geometric means from the Monte Carlo study, and the reasonably good linear configuration indicates that the present approach is adequate.

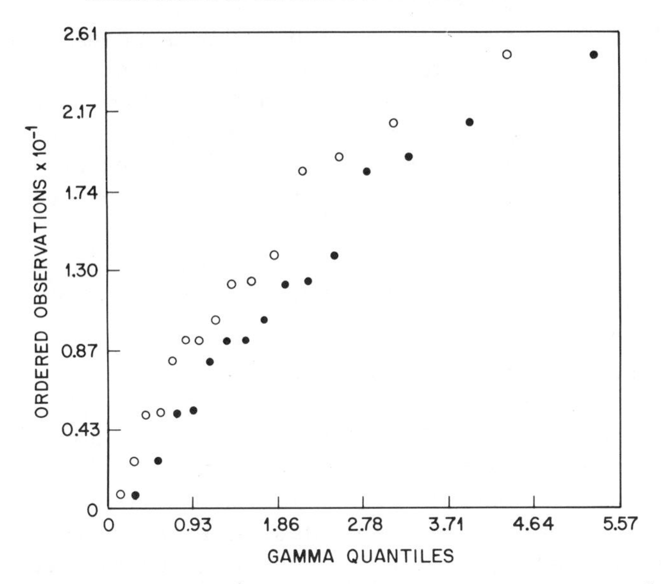

Fig. 12. Gamma probability plots for two estimates of shape parameter; O—maximum likelihood estimate; ●—method of moments estimate.

The o's in Figure 12 provide the probability plot for the same set of geometric means, employing Hoel's estimate of the shape parameter instead of the maximum likelihood estimate. A comparison of the two configurations suggests that the maximum likelihood method of fitting the approximating distribution is to be preferred to the method of moments.

In the analysis of variance application, in order to minimize the effects of possibly real sources of variation on the estimation of the scale and shape parameters, an order statistics formulation along the lines of Section 6.3.1 may be employed once again. Specifically, if

$$a_{(1)} \leq \cdots \leq a_{(r)} \quad \text{or} \quad g_{(1)} \leq \cdots \leq g_{(r)}$$

denote the ordered values of the trace or geometric mean, then, considering the $M(\leq K)$ smallest of these as the M smallest order statistics in a random sample of size $K(\leq r)$ from a gamma distribution, one can find the maximum likelihood estimate of the scale and shape parameters using only these M values (see Wilk et al., 1962*b*).

Next, using the estimate of the shape parameter, one can obtain a gamma probability plot of the r ordered values, $a_{(1)} \leq \cdots \leq a_{(r)}$ or $g_{(1)} \leq \cdots \leq g_{(r)}$. Under null conditions the resulting configuration would be expected to be linear with zero intercept and slope $1/\lambda_a$ on the plot of the a_i's (and $1/\lambda_g$ on the plot of the g_i's). Departures from linearity may then be studied for pinpointing violations of the null assumptions, such as the presence of possibly real sources of variation, the existence of more than one underlying error covariance structure, and other distributional peculiarities. The interpretation of these probability plots is similar to that of other probability plotting techniques that have been proposed for augmenting analyses of variance; in particular, it is quite analogous to the technique discussed in Section 6.3.1.

Two examples, taken from Gnanadesikan and Lee (1970), are given next to demonstrate the use of the techniques described above.

Example 39. This example (see also Example 4 in Chapter VII of Roy et al., 1971) consists of computer-generated trivariate normal data that simulate the results of a 30-cell experiment with four replications per cell. The data for 15 of the 30 cells had an underlying covariance matrix **I**, while in the remaining cells the covariance matrix was 9**I**. Exhibit 39a shows a gamma probability plot of the 30 ordered values of the trace of the within-cell covariance matrices for these data. The shape parameter required for this plot was based on the 15($=M$) smallest observed trace values, and K was taken as 30. Each point on the plot is labeled 1 or 2 according as it derives from a cell with one or the other of the two

Exhibit 39a. Gamma probability plot for $\mathcal{A}$ in the artificial data example; $\mathbf{A}=\mathbf{I}$, $r=K=30$, $M=15$, $\hat{\eta}=1.996$

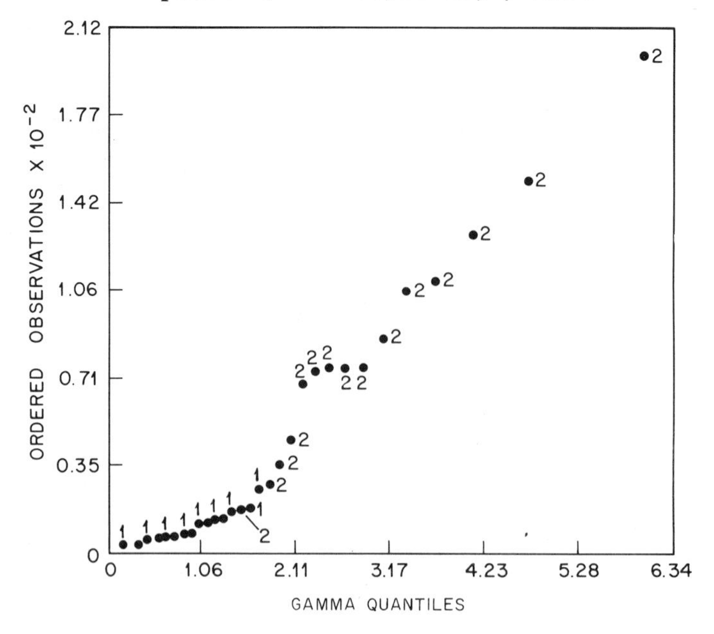

Exhibit 39b. ***Gamma probability for*** $\mathscr{G}$ ***in the artificial data example;*** $\mathbf{A}=\mathbf{I}$, $r=K=30$, $M=15$, $\hat{\eta}=1.514$

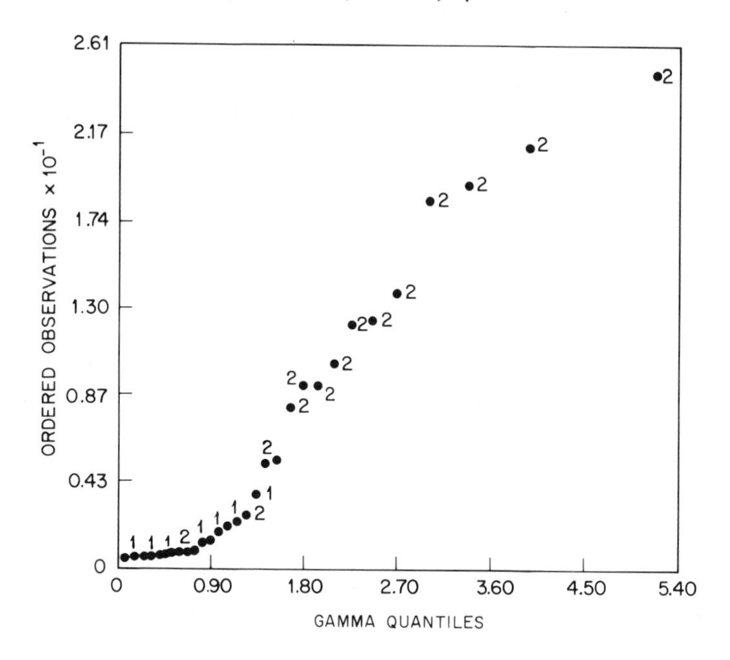

covariance structures employed in generating the data. The configuration is suggestive of two intersecting straight lines, each of which consists of points that correspond to the cells with a common covariance structure. Exhibit 39*b* shows the analogous gamma probability plot for the 30 ordered values of the geometric mean statistic, and the same phenomenon of two intersecting straight lines is seen again, although the general configuration in Exhibit 39*b* is smoother than the one in Exhibit 39*a*.

Example 40. The set of data derives from the talker-identification problem used also as the basis for earlier examples (e.g., Examples 16 and 19 in Chapter 4). As part of the analysis for obtaining a discriminant space for representing the utterances (see Example 16) and for assigning an unknown to one of the contending speakers, it is usual to pool the within-speaker covariance matrices of the utterances to obtain an overall within-speakers covariance matrix. It is legitimate in such multivariate classification problems to inquire about the validity of such a pooling procedure, and it would be useful to have an informal statistical procedure for a preliminary, simultaneous intercomparison of the covariance matrices from the different speakers. Specifically, for a set of 10 speakers, one input to a classification analysis consisted of a six-dimensional representation of each utterance of a given word, and there were seven utterances available per speaker. As a preliminary to pooling the 10 within-speaker covariance matrices, one can assess their similarity in "size" by using the methods described above in this subsection. Exhibits 40*a* and *b* show the gamma probability plots for the $10(=r)$ values of each of the functions $\mathscr{A}$ and $\mathscr{G}$, respectively, in this example. No scaling was performed on the covariance matrices, so that $\mathbf{A}=\mathbf{I}$, and the estimation of the parameters of the evaluating gamma distribution was based, in each case, on the $M=5$ smallest observed values with $K=r=10$. In this example the configurations obtained by using the estimates from the complete sample, i.e., $M=K=10$, were quite similar to the ones in Exhibits 40*a* and *b*.

The points in Exhibits 40*a* and *b* are labeled 1 through 10 to correspond with the speaker from whose covariance matrix a particular point derives. Although the point corresponding to speaker 9 appears to depart from the linear configuration suggested by the other points in Exhibit

Exhibit 40a. Gamma probability plot for $\mathscr{A}$ in talker-identification example; $\mathbf{A}=\mathbf{I}$, $r=K=10$, $M=5$, $\hat{\eta}=4.236$

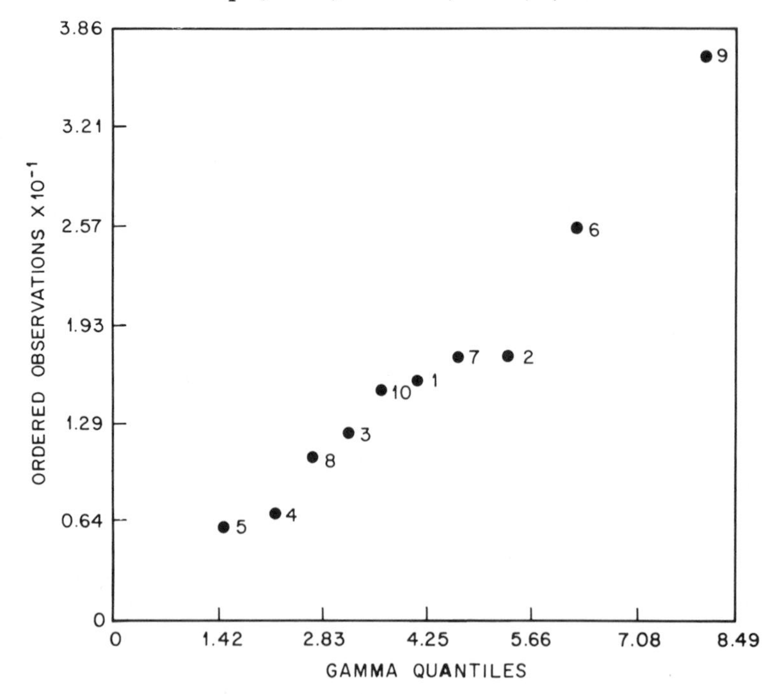

Exhibit 40b. Gamma probability plot for $\mathscr{G}$ in talker-identification example; $\hat{\eta}=4.245$

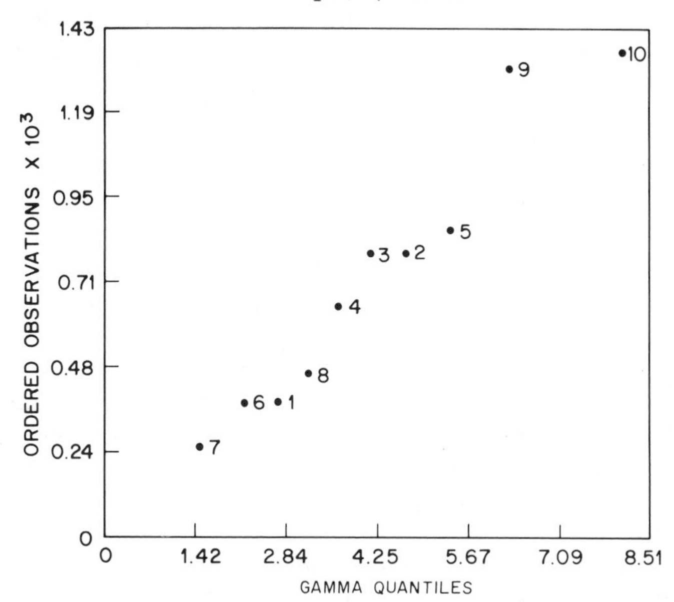

40a, the departure is not marked. The general indication of both plots is that the 10 covariance structures form a reasonably homogeneous group in terms of their "sizes." The two plots exhibit different internal orderings of the covariance matrices for the 10 speakers, in accordance with the sensitivities of $\mathscr{A}$ and $\mathscr{G}$ to different aspects of the covariance structures. Thus, for example, speaker 6 who is second from the top in Exhibit 40a, turns out to be second from the bottom in Exhibit 40b, suggesting that the covariance matrix for that speaker may have a noticeably small eigenvalue, as was indeed the case. Also, the covariance matrix of speaker 9, the top point in Exhibit 40a and the next-to-top point in Exhibit 40b, is indicated as possibly having a markedly large eigenvalue and no significantly small eigenvalue, and this again was found to be true.

The essential concepts in the probability plotting approaches described in the last two subsections (and indeed also the others mentioned in Table 2) are first to obtain meaningful summary statistics to serve as the medium for making the simultaneous assessments and, second, to display the internal comparisons against a null background by means of a probability plot of the ordered observed values of a statistic versus the corresponding quantiles of an appropriate null statistical distribution. Thus, with the method of Section 6.3.1, the d_i's constitute the summary statistics, whereas the a_i's and g_i's are the corresponding entities for the method of Section 6.3.2, and the appropriate evaluating distribution in each case turns out to be a gamma distribution whose parameters are fitted by maximum likelihood, using an order statistics approach.

The probability plotting methods discussed here have been concerned with internal comparisons of relative magnitudes and not with orientational aspects of multiresponse data. Orientational information is contained in eigenvectors associated with covariance matrices, whereas summary statistics such as $\mathcal{A}$ and $\mathcal{G}$ are based on eigenvalues. Data-analytic techniques for assessing orientational similarities among covariance matrices need to be developed. Probability plotting methods based on comparing the angles between corresponding eigenvectors seem to be feasible and particularly worth considering.

The technique described in Section 6.3.2 can be used, as demonstrated in Example 40, for assessing equality (in "size") of covariance matrices, provided that all the different estimates are based on the same numbers of observations. If all the samples are large, the effect of unequal numbers of observations may not be serious. The gamma probability plots of $\mathcal{A}$ and $\mathcal{G}$ in this context may turn out to be more robust indicators than the classical tests for equality of covariance matrices, which are known to be generally quite nonrobust in the sense that they tend to be more sensitive to nonnormality of the data than to heteroscedasticity.

6.4. MULTIDIMENSIONAL RESIDUALS AND METHODS FOR DETECTING MULTIVARIATE OUTLIERS

With large bodies of data, although models are appealing as parsimonious representations that may lead to simple interpretations of the data, it is very important to have means of gauging the appropriateness and sensitivities of the models under consideration. The useful role of residuals in exposing any inadequacies of a fitted model in the analysis of uniresponse problems has come to be widely recognized (see, e.g., Terry, 1955; Anscombe, 1960, 1961; Draper and Smith, 1966).

One use of univariate residuals is to detect so-called outliers or extremely deviant observations, which are not uncommon in large data sets. Robust fitting of models or robust estimation (see Section 5.2.3) is one approach for handling outliers, namely, by minimizing the influence of such outliers on the fitted model. Often, however, pinpointing an outlier for further investigation and pursuit can be a valuable outcome of the statistical analysis of the data, and procedures directed specifically at detecting outliers can be useful (see, e.g., Grubbs, 1950, 1969; Dixon, 1953).

The purpose of this section is to discuss multiresponse residuals and describe some techniques for identifying multivariate outliers. The discussion here draws on the work of Gnanadesikan and Kettenring (1972) and Devlin et al. (1975).

6.4.1. Analysis of Multidimensional Residuals

Given some summarizing fit to a body of multiresponse data, there exists, in principle, a vector of multivariate residuals between the data and the fit; but, more than in the univariate case, the important issue arises of how to express these multivariate residuals. Although experience is still rudimentary on these matters, some things can be done, and the discussion in this section will be concerned with some statistical methods for analyzing multivariate residuals.

For the discussion here and in Section 6.4.2, it is convenient to distinguish two broad categories of statistical analyses of multiresponse problems: (i) the analysis of internal structure, and (ii) the analysis of superimposed or extraneous structure (see also Section 3.1). The first category includes techniques, such as principal components, factor analysis, and multidimensional scaling (see Chapter 2), that are useful for studying internal dependencies and for reduction of the dimensionality of response. Multivariate multiple regression and multivariate analysis of variance (see Section 5.2.1), which are the classical techniques for investigating and specifying the dependence of multiresponse observations on design characteristics or extraneous independent variables, are examples of the second category.

Each category of analysis gives rise to multivariate residuals. For instance, as discussed in Chapter 2 (see pp. 58–59), linear principal components analysis may be viewed as fitting a set of mutually orthogonal hyperplanes by minimizing the sum of squares of orthogonal deviations of the observations from each plane in turn. At any stage, therefore, one has residuals that are perpendicular deviations of data from the fitted hyperplane. On the other hand, in analyzing superimposed structure (i.e., the second category above) by multivariate multiple regression, one has the well-known least squares residuals, viz., (observations) − (predictions from a least squares fit). For purposes of data analysis it is often desirable to use the least squares residuals as input to a principal components analysis, which, in turn, will lead to the orthogonal residuals mentioned earlier. Augmenting multivariate multiple regression fitting by a principal components transformation of the residuals from fit may help in describing statistical correlations in the errors of the combined original variables, or in indicating inadequacies in the fit of the response variables by the design variables. For present purposes principal components residuals and least squares residuals are considered separately.

Principal Components Residuals. Equation 6 (in Chapter 2) defines the linear principal components transformation of the data in terms of the eigenvectors of the sample covariance matrix, **S**. Each row, $\mathbf{a}'_j$ $(j = 1, \ldots, p)$, of $\mathbf{A}'$ provides a principal component coordinate, and each row of **Z** gives the deviations of the projections of the original sample from the projection of the sample centroid, $\bar{\mathbf{y}}$, onto a specific principal component coordinate. Using standardized variables as the starting point would lead to corresponding interpretations of the principal components analysis of **R**, the sample correlation matrix.

When principal components analysis is viewed as a method of fitting linear subspaces, or as a statistical technique for detecting and describing possible linear singularities in the data, interest lies especially in the projections of the data onto the principal component coordinates corresponding to the small eigenvalues (i.e., the last few rows of **Z**). Thus, for instance, with $p = 2$ the essential concepts are illustrated in Figure 13, where y_1 and y_2 denote the original coordinates and z_1 and z_2 denote the two principal components derived from the covariance matrix of the bivariate data. The straight line of closest fit to the data (where closeness

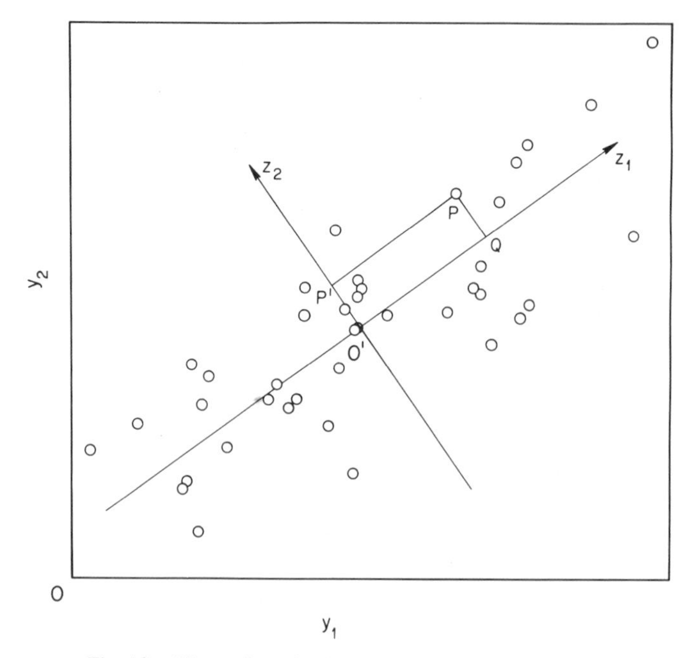

Fig. 13. Illustration of principal components residuals.

is measured by the sum of squares of perpendicular deviations) is the z_1-axis. The orthogonal residual of a typical data point, P, as shown in the figure, is the vector $\overrightarrow{QP}$, which is seen to be equivalent to the vector $\overrightarrow{O'P'}$, where P' is the projection of P onto the z_2-axis, the second principal component. More generally, with p-dimensional data, the projection onto the "smallest" principal component (i.e., the one with least variance) will be relevant for studying the deviation of an observation from a hyperplane of closest fit, while projections on the "smallest" q principal component coordinates will be relevant for studying the deviation of an observation from a fitted linear subspace of dimensionality $(p-q)$.

For detecting lack of fit of individual observations, one method suggested by Rao (1964) is to study the sum of squared lengths of the projections of the observations on the last few, say q, principal component coordinates. For each initial observation, $\mathbf{y}_i$ $(i=1, \ldots, n)$, the procedure consists of computing

$$d_i^2 = \sum_{j=p-q+1}^{p} [\mathbf{a}_j'(\mathbf{y}_i - \bar{\mathbf{y}})]^2$$

$$= (\mathbf{y}_i - \bar{\mathbf{y}})'(\mathbf{y}_i - \bar{\mathbf{y}}) - \sum_{j=1}^{p-q} [\mathbf{a}_j'(\mathbf{y}_i - \bar{\mathbf{y}})]^2,$$

and considering inappropriately large values of d_i^2 as indicative of a poor $(p-q)$-dimensional fit to the observation (or, equivalently, that the observation is possibly an aberrant one). An informal graphical technique, which might have value as a tool for exposing other peculiarities of the data in addition to assessing the fit, would be to make a gamma probability plot of the d_i^2's, using an appropriately chosen or estimated shape parameter. One method of obtaining a suitable estimate of the shape parameter would be to base it on a collection of the smallest observed d_i^2's.

In addition to looking at a single summary statistic, such as d_i^2 above, it may often be useful to study the projections of the data on the last few principal component coordinates (i.e., the last few rows of $\mathbf{Z}$ in Eq. 6) in other ways. These might include the following:

1. Two- and three-dimensional scatter plots of bivariate and trivariate subsets of the last few rows of $\mathbf{Z}$ with points labeled in various ways, such as by time if it is a factor.

2. Probability plots of the values within each of the last few rows of $\mathbf{Z}$. Because of the linearity of the transformation involved, it may not be unreasonable to expect these values to be distributed more nearly normally than the original data, and normal probability plotting will provide a reasonable starting point for the analysis. This analysis may help in

pinpointing specific "smallest" principal component coordinates, if any, on which the projection of an observation may look abnormal, and thus may augment the earlier-mentioned gamma probability plotting analysis of the d_i^2's.

3. Plots of the values in each of the last few rows of **Z** against certain distances in the space of the first few principal components. If, for example, most of the variability of a set of five-dimensional data is associated with the first two principal components, it may be informative to plot the projections on each of the three remaining principal component axes against the distance from the centroid of each of the projected points in the two-dimensional plane associated with the two largest eigenvalues. This may show a certain kind of multidimensional inadequacy of fit—namely, if the magnitude of the residuals in the coordinates associated with the smaller eigenvalues is related to the clustering of the points in the two-dimensional space of the two eigenvectors corresponding to the largest two eigenvalues.

An important issue concerning the analyses suggested above is their robustness. Clearly, if an aberrant observation is detected, one may want to exclude it from the initial estimate of **S** (or **R**) and then repeat the process of obtaining and analyzing the principal components residuals. In some circumstances one may also decide to use a robust estimate of the covariance (or correlation) matrix, such as the ones considered in Section 5.2.3, even for the initial analysis, in the hope that the aberrant observations will become even more conspicuous in the subsequent analysis of residuals.

Example 41. To illustrate the use of some of the methods for analyzing principal components residuals, two sets of data are taken from a study by Chen et al. (1970, 1974) concerned with grouping corporations. (See also Examples 17, 18, 22, and 35.) As part of the study, the appropriateness of prespecified groupings (e.g., chemicals, oils, drugs) was examined initially, and, as mentioned in the discussion of Example 17, a preliminary attempt was made to develop core groups of companies from an internal analysis of each prespecified category. One approach for forming core groups was to identify and eliminate outliers by studying the principal components residuals.

There were 14 variables per company per year in the study. Specifically, for 1963 data were available for 20 drug companies, and Exhibit 41*a* shows a scatter plot of all the drug companies in the space of the last two principal components of the 14×14 correlation matrix derived from these data. Companies 8 and 9 are indicated as possible outliers with respect to the configuration of the remaining companies in this plot. Company 9

Exhibit 41a. Plot of 20 drug companies in the space of the last two principal components

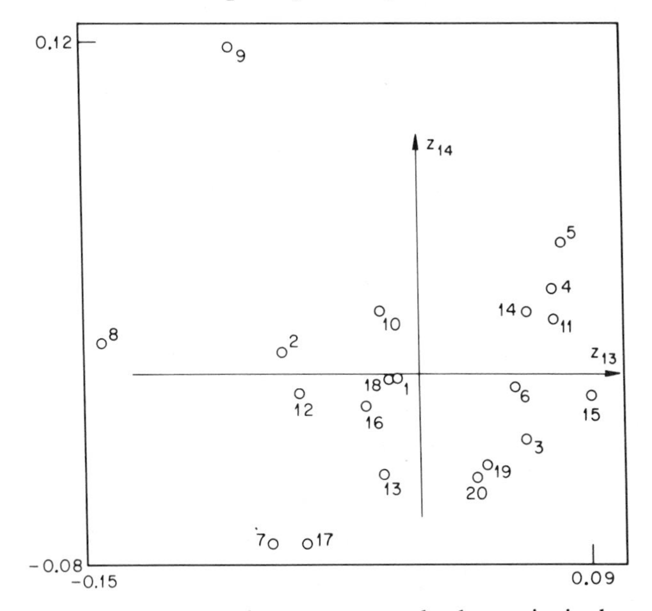

appears to be an outlier with respect to the last principal component in particular, while company 8 seems to be a moderate outlier on the penultimate principal component.

The second set of data is from 23 drug companies for the year 1967 and illustrates the use of probability plotting of the elements in each of the last few rows of $\mathbf{Z}$. Exhibit 41*b* (see page 264) shows a normal probability plot of the tenth principal component of the sample correlation matrix. The points corresponding to companies 11 and 19 are seen to deviate at the top right-hand end of the plot from the reasonably good linear configuration of the remaining points. The original data in this example exhibited considerable nonnormality (see Examples 30 and 35), and the earlier-mentioned aspect of improved normality induced by the principal components transformation is evident in Exhibit 41*b* by the linearity of the configuration of most of the points, with just a mild indication of a distribution with shorter tails that the normal.

Least Squares Residuals. In the notation employed earlier (see Eqs. 49, 50, and 63) in discussing the multivariate multiple regression model, the n multivariate least squares residuals (called residuals hereafter) are the p-dimensional rows (denoted as $\mathbf{e}_1'$, $\mathbf{e}_2'$, ..., $\mathbf{e}_n'$) of

$$\hat{\boldsymbol{\varepsilon}} = \mathbf{Y}' - \mathbf{X}\hat{\boldsymbol{\Theta}}. \tag{91}$$

Exhibit 41b. Normal probability plot of the tenth principal component of the 14×14 correlation matrix for 23 drug companies

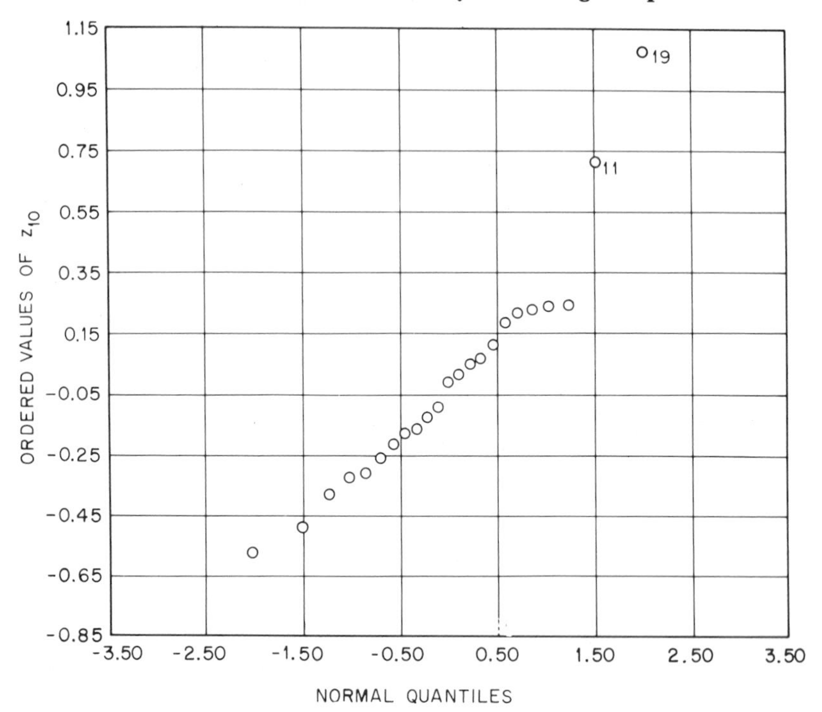

Depending on the structure of $\mathbf{X}$, there will be certain singularities among the residuals in that certain linear combinations of the rows of $\hat{\boldsymbol{\varepsilon}}$ will be $\mathbf{0}'$. Depending on the correlational structure and functional dependencies among the p responses, there could be singularities in the other direction (viz., the columns of $\hat{\boldsymbol{\varepsilon}}$), and the existence and nature of such singularities may be investigated by principal components transformations of the p-dimensional residuals.

In some applications there may be a natural ordering among the responses, which may lead one to consider the use of a step-down analysis (see Section 4.c of Chapter IV in Roy et al., 1971). The analysis at each stage is a univariate analysis of a single response, utilizing all the responses that have been analyzed at the preceding stages as covariates. At each stage, therefore, step-down residuals may be obtained from this approach and studied by any of the available techniques for analyzing univariate least squares residuals.

Larsen and McCleary (1972) have proposed the concept of partial residuals and ways of using them. Entirely analogous definitions of multivariate partial residuals and methods of analyzing them may be suggested.

As a first approach to analyzing the residuals defined in Eq. 91, one may wish to consider the entire collection of them as an unstructured multivariate sample. Sometimes such a view may be more appropriate for subsets of the residuals than for the totality of them. For instance, in a two-way table the residuals within a particular row (or column) may be considered as an unstructured sample. At any rate, with such a view one can then employ methods applicable to the study of unstructured multivariate samples (see Section 6.2), including the following:

1. Separate plotting of uniresponse residuals, perhaps against values of certain independent or extraneous variables (e.g., time) or against the predicted values. Augmenting such scatter plots with curves of locally smoothed quantiles (e.g., moving median and quartiles) can be very useful (see Cleveland & Kleiner, 1975).
2. One-dimensional probability plotting of the uniresponse residuals. Full-normal plots of the uniresponse residuals or half-normal plots of their absolute values (or, equivalently, $\chi^2_{(1)}$ plots of squared residuals) provide natural starting points. Residuals generally seem to tend to be "supernormal" or at least more normally distributed than original data, and such probability plots may be useful in delineating outliers or other peculiarities in the data.
3. The use of one, or preferably several, distance functions to convert the multiresponse residuals to single numbers, followed by the probability plotting of these. The idea here is simply to treat the residuals (the $\mathbf{e}$'s defined in Eq. 91) as single-degree-of-freedom vectors and to use the gamma probability plotting methodology described in Section 6.3.1 for analyzing them. The presence of outliers and of heteroscedasticity will be revealed by departures from linearity of the configuration on an appropriately chosen gamma probability plot of the values of a quadratic form in the $\mathbf{e}$'s. For instance, an aberrant observation may be expected to yield a residual for which the associated quadratic form value will be unduly large, thus leading to a departure of the corresponding point from the linearity of the other points on the gamma probability plot (see Example 42). Heteroscedasticity will be indicated by a configuration that is piecewise linear, with the points corresponding to the residuals derived from observations with the same covariance structure belonging to the same linear piece.

The approach of gamma plotting quadratic forms of the residuals assumes a particularly simple, and already encountered, form for an unstructured sample. The residuals in this case are just deviations of the individual multiresponse observations from the sample mean vector, $\mathbf{e}_i = \mathbf{y}_i - \bar{\mathbf{y}}$ $(i = 1, \ldots, n)$. The study of the generalized squared distance of the observations from the sample mean (see Example 7 in Chapter 2 and the procedure for plotting radii described on pp. 172–173) is thus a special case. (See also Cox, 1968; Healy, 1968.)

Example 42. The data derive from an experiment on long-term aging of a transistor device used in submarine cable repeaters (see Abrahamson et al., 1969). Sets of 100 devices, in a configuration of 10 rows by 10 columns, were aged, and a characteristic called the *gain* of each device was obtained at each of several test periods. An initial transformation to logarithms was made, and the aging phenomenon of interest was then the behavior of the log gain as a function of time. One approach to studying the aging behavior for purposes of identifying devices with peculiar aging characteristics was to fit a polynomial (specifically, a cubic was used) to the data on log gain versus time for each device, and to study the fitted coefficients by analysis of variance techniques. A separate univariate analysis of variance of each coefficient, as well as a multivariate analysis of variance of the four coefficients simultaneously, was performed. The multivariate approach was employed partly because of the high intercorrelations observed among the fitted coefficients. It was not used as a substitute for the separate univariate analyses of the individual coefficients. For present purposes attention is confined to the multivariate approach.

A simple one-way (i.e., rows and columns-within-rows were the sources of variation) multivariate analysis of variance (MANOVA), when used as a means for obtaining formal tests of hypotheses, revealed very little. None of the usual MANOVA tests of the null hypothesis of no row effects (see Section 5.2.1 and also Chapter IV of Roy et al., 1971) had an associated p-value smaller than 0.3. The danger in basing an analysis solely on such tests, which are based on single summary statistics, is revealed by the use of the informal gamma plotting technique described above.

Exhibit 42*a* shows a gamma probability plot of the 100 values of a quadratic form, $\mathbf{e}_i'\mathbf{S}^{*-1}\mathbf{e}_i$ $(i = 1, \ldots, 100)$, in the four-dimensional residuals. The covariance matrix, $\mathbf{S}^*$, of the residuals is a robust estimate (of the type discussed in Section 5.2.3) obtained from the residuals themselves. [*Note*: Since $\mathbf{S}^{*-1}$ is common to all 100 values of the quadratic form being analyzed, it is not necessary to multiply $\mathbf{S}^*$ by the "unbiasing" constant for the present application.] The shape parameter required for the plot was estimated by maximum likelihood based on the 50 smallest values of the quadratic form, considered as the 50 smallest order statistics in a random sample of size 100.

The point that stands out clearly from the configuration of the others in Exhibit 42*a* corresponds to the seventh device in the first row, and the implication is that the four-dimensional residual for this device is inordinately "large," i.e., a possibly aberrant observation has been pinpointed! This residual (and other such if they exist) has, of course, contributed to the estimate of the columns-within-rows dispersion matrix that was employed as the error dispersion matrix in the formal tests of significance mentioned earlier. The effect would be to inflate the error dispersion inappropriately, and it is not surprising, therefore, that the tests revealed no significant departures from the null hypothesis. Upon verification the aging configuration of device 7 in row 1 was found to be indeed abnormal in relation to the behavior of the majority of devices.

To facilitate further study of the residuals, a replot, shown in Exhibit 42*b*, (see page 268) may be made of the 99 points left after omitting the point corresponding to the aberrant device. The configuration on this plot may lead one to conclude that device 9 (the one from which the top right-hand corner point derives) and also the other devices (1–6 and 8) in row 1 are suspect, i.e., all 10 devices in row 1 are associated with peculiar residuals. Such a conclusion, however, may not be warranted, and the

Exhibit 42a. Gamma probability plot derived from three-dimensional residuals scaled by a robust covariance matrix

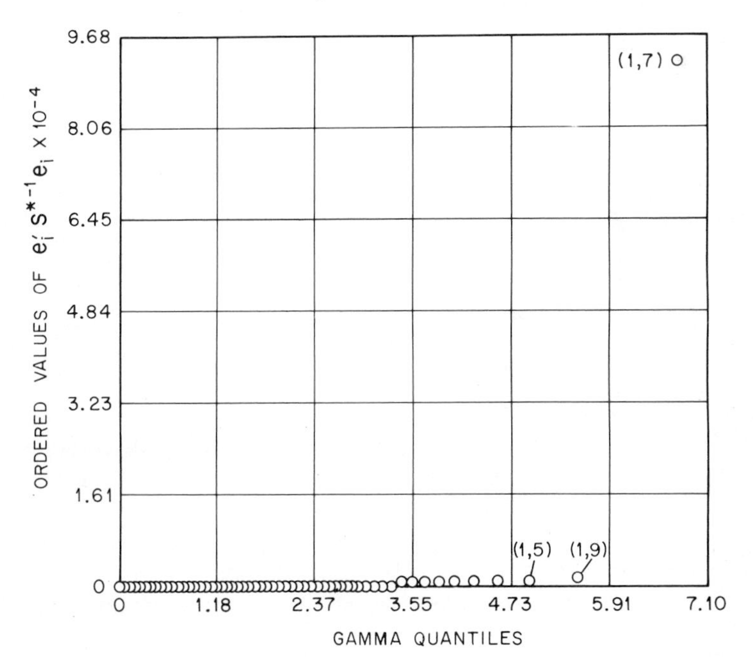

Exhibit 42b. Replot obtained from Exhibit 42a after omitting point (1,7)

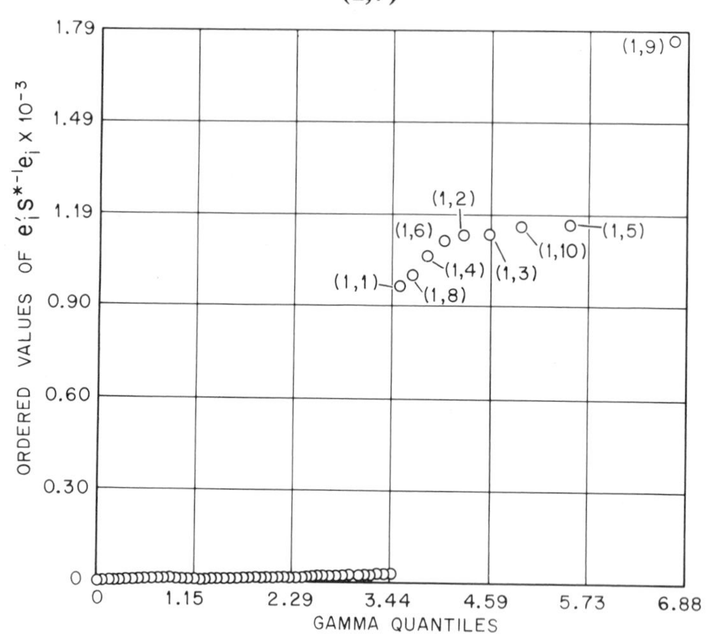

discussion that follows will clarify the issue involved. The analysis of the data is then continued in Example 43.

When the matrix **X** in Eq. 49, the so-called design matrix or matrix of values of the regressors, corresponds to more structured situations (e.g., a multiway classification), there are at least two sources of statistical difficulty in analyzing the residuals. First, there are constraints on subsets of the residuals (e.g., the sum of the residuals in a row of a two-way table is the null vector), which imply correlations among the residuals. Second, the presence of outliers may seriously bias the usual effects which are subtracted from an observation (e.g., row, column, and overall mean vectors in a two-way classification) so as to mask the local effect of an outlier on the corresponding residual. The first source of difficulty (viz., the singularities among residuals) is especially critical when the numbers of levels of the factors involved (e.g., the number of rows or columns in a two-way table) are small, but the second source can be important even when each of the factors has a moderate number of levels.

Thus in Example 42 the extreme outlier (viz., the observation for device 7 in row 1) may have so badly biased the mean vector for the first row that all the residuals [=(observation vector)−(row mean vector)] in that row have been unduly biased. If the outlier is extreme enough, this can indeed happen, and a method is needed for insuring against such masking effects of the outliers on the residuals.

One way of accomplishing this is to combine the ideas and methods of robust estimation discussed in Section 5.2.3 with the desirability of analyzing the residuals. Specifically, instead of using the usual least squares estimates of the elements of $\mathbf{\Theta}$ in the linear model (Eq. 49), one could use robust estimates of them, thus obtaining $\hat{\mathbf{\Theta}}^*$, and then define a set of *robustified residuals* (see also the discussion at the end of Section 5.2.3) as the rows of

$$\hat{\boldsymbol{\varepsilon}}^* = \mathbf{Y}' - \mathbf{X}\hat{\mathbf{\Theta}}^*. \tag{92}$$

If one were to utilize the simplest direct approach to developing $\hat{\mathbf{\Theta}}^*$, which was described toward the end of Section 5.2.3, $\hat{\mathbf{\Theta}}^*$ would just be a matrix each of whose elements, $\hat{\theta}^*_{ij}$, is a uniresponse robust estimator of a univariate location-type parameter.

Example 43. To illustrate the use of robustified residuals, the data used in Example 42 are employed again. Instead of using the row mean vectors for defining the residuals, the vector of midmeans, $\mathbf{y}^*_{T(.25)}$, discussed in Section 5.2.3, for each row is used, and the robustified four-dimensional residuals are obtained as the difference between the four-dimensional observation (viz., the four coefficients of the aging curve for a device) and the vector of midmeans for the row in which the observation appears.

The 100 four-dimensional robustified residuals thus obtained in this example can then be analyzed by the gamma probability plotting technique described and illustrated earlier in the context of analyzing the regular residuals. Exhibit 43*a* (see page 270) shows a gamma probability plot of the 100 values of a quadratic form in the modified residuals, $\mathbf{e}^{*\prime}_i\mathbf{S}^{*-1}\mathbf{e}^*_i$ $(i = 1, \ldots, 100)$, where $\mathbf{S}^*$ as before is a robust estimate of the covariance matrix, and the shape parameter required for the plot is estimated once again using the smallest 50 observed values of the quadratic form. In Exhibit 43*a* the point corresponding to device 7 in row 1 again stands out, and Exhibit 43*b* (see page 271) shows a replot obtained after omitting this point. Comparing Exhibits 43*b* and 42*b*, it is seen that the biasing effect on all the residuals in the first row caused by the extremely deviant observation for device 7 in that row is no longer evident. The configuration in Exhibit 43*b* may be used to delineate additional outliers, such as device 1 in row 7, by looking for points in the top right-hand corner that deviate noticeably from the linear configuration of the points in the lower left-hand portion of the picture.

Exhibit 43a. Gamma probability plot derived from three-dimensional robust residuals scaled by a robust covariance matrix

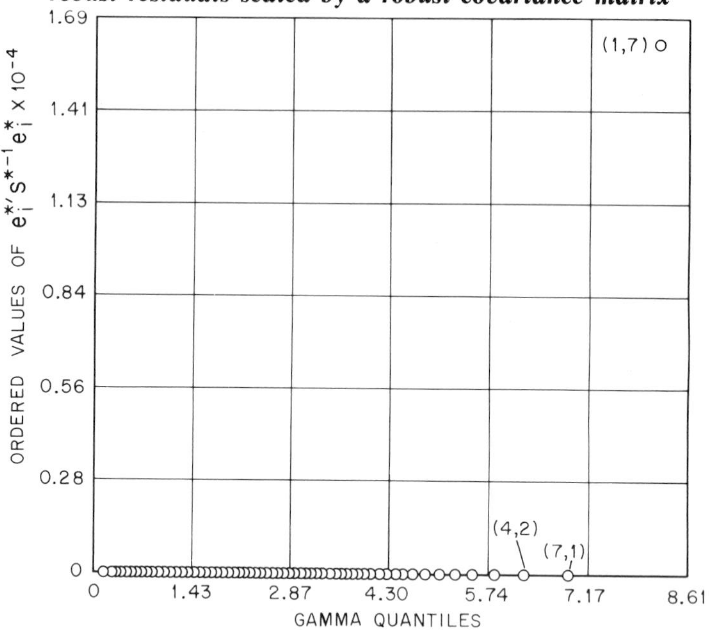

Whether $\hat{\boldsymbol{\varepsilon}}^*$ as defined in Eq. 92 is the most appropriate set of robust residuals for purposes of analysis, or whether one needs to modify them (e.g., by weighting them), is a question for further investigation. They do at least constitute a simple starting point. The robustified residuals defined by Eq. 92 will not necessarily satisfy the constraints satisfied by the usual residuals. For example, in a two-way classification they will not necessarily add up to the null vector, either by rows or by columns, or even across all cells. The robustified residuals do not form a cohesive group unless there are no outliers in the data, and in the latter case the usual least squares estimator, $\hat{\boldsymbol{\Theta}}$, and the robust estimator, $\hat{\boldsymbol{\Theta}}^*$, will not be very different, so that the usual residuals, $\hat{\boldsymbol{\varepsilon}}$, and the robustified residuals, $\hat{\boldsymbol{\varepsilon}}^*$, will also be expected to be very similar when there are no outliers. The main use of the robustified residuals is, in fact, to accentuate the presence of outliers, and hence the fact that they do not satisfy the same constraints as the usual residuals is perhaps unimportant. If, however, one desires to have modified residuals satisfy these constraints as nearly as possible, then iterating the analysis in certain ways may help. Tukey (1970) has suggested such a scheme for using midmeans in analyzing multiway tables with uniresponse data, and an extension of this approach to the multiresponse case may be feasible.

Exhibit 43b. Replot obtained from Exhibit 43a after omitting point (1,7)

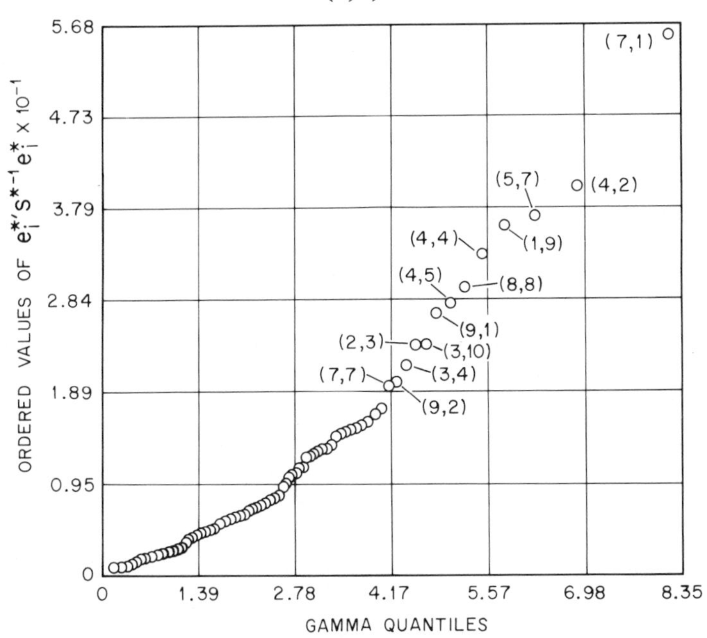

6.4.2. Other Methods for Detecting Multivariate Outliers

In the preceding section ways of pinpointing maverick observations through an analysis of multivariate residuals were discussed. In this section some additional techniques are suggested for detecting multivariate outliers.

The consequences of having defective responses are intrinsically more complex in a multivariate sample than in the much-discussed univariate case. One reason is that a multivariate outlier can distort not only measures of location and scale but also those of orientation (i.e., correlation). A second reason is that it is much more difficult to characterize a multivariate outlier. A single univariate outlier may typically be thought of as "the one that sticks out on the end," but no such simple concept suffices in higher dimensions. A third reason is the variety of types of multivariate outliers that may arise: a vector response may be faulty because of a gross error in one of its components or because of systematic mild errors in all of its components.

The complexity of the multivariate case suggests that it would be fruitless to search for a truly omnibus outlier detection procedure. A more reasonable approach seems to be to tailor detection procedures to protect against specific types of situations, e.g., correlation distortion, thus building up an arsenal of techniques with different sensitivities. This approach recognizes that an outlier for one purpose may not necessarily be one for another purpose! However, if several analyses are to be performed on the same sample, the result of selective segregation of outliers can be a more efficient and effective use of the available data.

It is essential that the procedures be computationally inexpensive enough to allow for routine screening of large data sets. Those that can simultaneously expose other features of the data, such as distributional peculiarities, have added economic appeal.

Following the dichotomy of multivariate methods mentioned in Section 6.4.1, the proposed procedures will be presented under the general headings of internal and external analysis techniques. In the former category are the techniques, such as principal components analysis, that are appropriate for examining an unstructured sample of data; in the latter category are techniques, such as canonical correlation analysis, that are applicable in the presence of some superimposed structure.

An approach that can lead to outlier-detection methods for both categories of problems is one which exploits the feature that outliers tend to have an unduly large or distorting influence on summary statistics. Gnanadesikan and Kettenring (1972) propose a variety of statistics, addressed to different multivariate problems, for assessing the influence of each observation on several standard multiresponse analyses. A few of these will be described later in this section.

The influence function, advocated by Hampel (1968, 1973, 1974), is a useful device for considering the effect of observations on a statistic. As such it can be useful not only as a tool for motivating and designing specific types of robust estimators but also as a means for developing methods for outlier detection (see Devlin et al., 1975).

For a general parameter $\theta = T(F)$, expressed as a functional of the distribution function, F, the influence function $I(\mathbf{y};\, \theta)$ at $\mathbf{y}$ is defined (see Hampel, 1974) as

$$I(\mathbf{y};\, \theta) = \lim_{\varepsilon \to 0} \left(\frac{\tilde{\theta} - \theta}{\varepsilon} \right),$$

where $\tilde{\theta} = T(\tilde{F})$ and $\tilde{F} = (1-\varepsilon)F + \varepsilon\delta_{\mathbf{y}}$ is a "perturbation" of F by $\delta_{\mathbf{y}}$, the distribution function for a point mass of 1 at $\mathbf{y}$. The essential concept in this "theoretical" influence function is that one can use it to assess the influence of the point $\mathbf{y}$ on the parameter θ.

Three finite sample versions of the influence function may be distinguished. The first, termed the *empiric* influence function by Mallows

(1973), is obtained by replacing F in the above definition by the empirical cumulative distribution function, F_n, which is a step-function with a step of height $1/n$ at each of the observations $\mathbf{y}_1, \ldots, \mathbf{y}_n$.

In the second finite sample version, the desire is to study the difference between $\hat{\theta}$ (an estimator of θ obtained from the n observations in the sample) and $\hat{\theta}_+$, an estimator of the same form as $\hat{\theta}$ obtained from the n original observations plus a conceptualized additional observation, $\mathbf{y}$. Specifically, this version of the influence function is defined as

$$I_+(\mathbf{y}; \theta) = (n+1)(\hat{\theta}_+ - \hat{\theta}),$$

and it is essentially the so-called sensitivity curve used by Andrews et al. (1972) for studying the properties of various robust estimates of location.

The third version focuses on the individual effects of the actual observations in the sample and is particularly suited to assessing the influence of individual observations on the estimator $\hat{\theta}$. This version, called the *sample* influence function by Devlin et al. (1975), is defined as

$$I_-(\mathbf{y}_i; \hat{\theta}) = (n-1)(\hat{\theta} - \hat{\theta}_{-i}), \qquad i = 1, \ldots, n, \tag{93}$$

where $\hat{\theta}_{-i}$ is an estimator of the same form as $\hat{\theta}$ but is calculated by omitting the ith observation, $\mathbf{y}_i$. The quantity $(\hat{\theta} + I_-)$ is the ith pseudo-value in Tukey's (1958) jackknife technique (see also Miller, 1974, and references therein).

[*Note*: Both I_+ and I_- can be considered as approximations to the empiric influence function by taking ε in the latter to be $1/n+1$ and $-1/n-1$, respectively (see Mallows, 1973).]

Hampel (1968) discusses the use of the influence function in the contexts of estimating univariate location and scale. Devlin et al. (1975) describe its application in the context of bivariate correlation. Specifically, it can be established that the influence function of ρ, the population product moment correlation coefficient, for any bivariate distribution for which ρ is defined (viz., second moments are finite) is

$$I(y_1, y_2; \rho) = -\tfrac{1}{2}\rho(\tilde{y}_1^2 + \tilde{y}_2^2) + \tilde{y}_1\tilde{y}_2,$$

where $\tilde{y}_j$ is the standardized form of y_j [i.e., $\tilde{y}_j = (y_j - \mu_j)/\sqrt{\sigma_{jj}}$, $j = 1, 2$]. Furthermore, if z_1 and z_2 denote, respectively, the standardized sum of and difference between $\tilde{y}_1$ and $\tilde{y}_2$, and if $u_1 = (z_1 + z_2)/\sqrt{2}$, the above equation may be rewritten as

$$I(y_1, y_2; \rho) = (1 - \rho^2)u_1u_2.$$

Also, the influence function of $z(\rho) = \tanh^{-1} \rho$, Fisher's z-transform of ρ, may be shown to be free of ρ:

$$I(y_1, y_2; z(\rho)) = u_1u_2,$$

where u_1 and u_2 are as above. With the additional assumption that (y_1, y_2) has a bivariate normal distribution, it follows that the influence function of $z(\rho)$ has a psn (product of two independent standard normal variables) distribution.

The analogous sample influence function of r, the sample product moment correlation coefficient, is

$$I_{-}(y_{i1}, y_{i2}; r) = (n-1)(r - r_{-i}) \approx (1 - r^2) u_{i1} u_{i2}, \qquad i = 1, \ldots, n, \quad (94)$$

wherein the first equality follows from the definition of I_{-} given in Eq. 93, and the expression on the right is the value of the empiric influence function at the ith observation, $\mathbf{y}'_i = (y_{i1}, y_{i2})$. The quantity r_{-i} in Eq. 94 denotes the correlation coefficient based on all but the ith observation, and u_{i1}, u_{i2} are sample analogues of u_1 and u_2:

$$u_{i1} = \frac{\sqrt{n}}{2}\left(\frac{d_{i1} + d_{i2}}{\sqrt{1+r}} + \frac{d_{i1} - d_{i2}}{\sqrt{1-r}}\right),$$

$$u_{i2} = \frac{\sqrt{n}}{2}\left(\frac{d_{i1} + d_{i2}}{\sqrt{1+r}} - \frac{d_{i1} - d_{i2}}{\sqrt{1-r}}\right),$$

where

$$d_{ij} = \frac{y_{ij} - \bar{y}_j}{\sqrt{a_{jj}}}, \qquad \bar{y}_j = \frac{\sum_{i=1}^{n} y_{ij}}{n}, \qquad a_{jj} = \sum_{i=1}^{n} (y_{ij} - \bar{y}_j)^2.$$

For the Fisher transform, $z(r)$, the analogous approximate result is

$$I_{-}(y_{i1}, y_{i2}; z(r)) = (n-1)[z(r) - z(r_{-i})] \approx u_{i1} u_{i2}, \qquad (95)$$

which is free of r. From the forms of Eqs. 94 and 95 it follows that the contours of the sample influence functions of both r and $z(r)$ may be approximated by hyperbolas with axes oriented along the principal axes of the sample correlation matrix. Also, from Eq. 95 it follows that for a reasonably large sample of bivariate normal data one can approximate the distribution of $I_{-}(y_{i1}, y_{i2}; z(r))$ by a psn distribution. These properties of the sample influence function are used below to develop informal graphical tools for detecting bivariate observations that may unduly distort r.

Internal Analysis Techniques for Outlier Detection. A basic and widely used approach to displaying mutiresponse data is through two- and three-dimensional scatter plots of the original and the principal component variables. Of the principal components the first and last few are usually of greatest interest to study. The first few principal components are especially sensitive to outliers which are inappropriately inflating variances and covariances (if one is working with **S**) or correlations (if one is working with **R**). Motivation, in terms of residuals, for looking at the last few principal components was discussed in Section 6.4.1. The kind of outlier which can be detected along these axes is one that is adding unimportant dimensions to, or obscuring singularities in, the data.

Probability plots (e.g., normal plots) and standard univariate outlier tests (such as those due to Grubbs, 1950, 1969, and to Dixon, 1953) may be carried out on each row of the observation matrix, **Y**, or of the derived principal components, **Z** (see Eq. 6 in Chapter 2). Outliers that distort location, scale, and correlation estimates may be uncovered in this manner.

Two graphical methods, specifically addressed to detecting observations that may have a distorting influence on the correlation coefficient, r, have been proposed by Devlin et al. (1975). The first of these is to augment a simple x-y scatter plot of the data with contours of the sample influence function of r (see Eq. 94) so as to facilitate the assessment of the effect of individual bivariate observations on the value of r. Thus, treating the approximation to $I_{-}(y_{i1}, y_{i2}; r)$ in Eq. 94 as a function of two variables, y_1 and y_2, one superimposes selected contours (which will be hyperbolas) of the function directly onto the scatter plot. The contour levels chosen for display purposes will depend on the sample size and other considerations relevant to the particular application.

The second proposal is to make a suitable Q-Q probability plot of the n values, $I_{-}(y_{i1}, y_{i2}; z(r))$, for $i=1, \ldots, n$, utilizing the approximation in Eq. 95. From a data-analytic viewpoint, it is appropriate to use distributional assumptions (such as normality) for developing methodology as long as the effects of departures from such assumptions are themselves assessable in specific applications. With this in mind, it is proposed that bivariate normality of the data be assumed as a null background and a Q-Q plot be made of the n ordered values of the sample influence function of $z(r)$ against the corresponding quantiles of the distribution of the product of two independent standard normal deviates (see Eq. 95 and the discussion following it). Despite the facts that these sample influence function values, by definition, are just $(n-1)$ times the differences between $z(r)$ and $z(r_{-i})$, and that one would intuitively expect the appropriate null distribution to be normal [since $z(r)$ and $z(r_{-i})$ are themselves approximately normal], algebraic manipulations of the sample influence function reveal that the relevant distribution is psn rather than normal. In fact, initially Gnanadesikan and Kettenring (1972) proposed a normal probability plot for the $z(r_{-i})$ values, only to discover later that the more appropriate procedure would be to utilize a psn distribution, which is also symmetric but has much thicker tails than the normal.

However, there is an "equivalent" normal probability plot that can be made in place of the psn plot. This may be preferred for purposes of interpretation because of the greater familiarity of normal plots for many people. The idea follows from recognizing that the psn distribution is parameter free, so that one can transform the n sample influence function values involved to their equivalent standard normal deviates and then make a normal plot of these transformed quantities. Thus, if $i_{(1)} \leq i_{(2)} \leq \cdots \leq i_{(n)}$ denote the n ordered sample influence function values of $z(r)$ and if G denotes the distribution function of the psn distribution, the transformed values needed for the normal plot are the v's defined by

$$\Phi(v_{(l)}) = G(i_{(l)}), \qquad l = 1, \ldots, n,$$

where Φ is the distribution function of the standard normal distribution.

The configuration of the psn probability plot of the $i_{(l)}$, or the normal plot of the $v_{(l)}$, may be used for checking on possible departures from the assumed null conditions, such as the presence of outliers that distort r, or smoother departures of the data distribution from bivariate normality. For instance, if most of the data are reasonably well behaved, with the exception of a few outliers that have disproportionate effects on r, one would expect most of the points on the Q-Q plot to conform to a linear configuration, while the points that derive from omission of the outlying observations will depart from such a linear configuration by being either "too big" or "too small." On the other hand, if the entire data distribution is distinctly nonnormal, one will expect to see departures from linearity in most regions of the plot (see Examples 43 and 44).

The differences, $(r - r_{-i})$ and $[z(r) - z(r_{-i})]$, are two examples of unidimensional statistics that can aid in pinpointing the effects of individual observations on familiar summary statistics. A variety of others, including ones for detecting observations that distort eigenanalyses such as principal components analysis, are described by Gnanadesikan and Kettenring (1972). Also, Wilks (1963) proposed that a test for a single outlier in an unstructured sample be based on the statistic

$$w = \max_i \left\{ \frac{|\mathbf{A}_{-i}|}{|\mathbf{A}|} \right\},$$

where $\mathbf{A} = (n-1)\mathbf{S}$ denotes the sum-of-products matrix based on all n observations in the sample, and $\mathbf{A}_{-i}$ again denotes a matrix computed just like $\mathbf{A}$ but without the ith observation. The statistic w turns out to be equivalent to the maximum observed generalized squared distance in the sample, i.e., $\max_i \, (\mathbf{y}_i - \bar{\mathbf{y}})' \, \mathbf{S}^{-1}(\mathbf{y}_i - \bar{\mathbf{y}})$, thus establishing a connection with the procedure of studying these generalized squared distances as described in Section 6.4.1 in the context of least squares residuals. Focusing on the largest observed generalized squared distance would be natural for developing a formal single-statistic test, but studying a gamma probability plot of the collection of n generalized squared distances may be more revealing for data-analytic purposes. For carrying out the formal test, the work of Siotani (1959) on the asymptotic distribution of the maximum generalized squared distance in multivariate normal samples provides some useful results and tables of percentage points.

The cluster analysis techniques discussed in Chapter 4 provide a different type of tool for identifying outliers. If the outliers constitute a distinct group of observations that are far removed from the majority of the data, one would expect them to be delineated as a cluster of observations. For instance, in a hierarchical clustering scheme, if one uses interpoint distances [i.e., $(\mathbf{y}_i - \mathbf{y}_k)'\mathbf{S}^{-1}(\mathbf{y}_i - \mathbf{y}_k)$] as the input, the expectation is that outlier clusters, if any exist, will join the main body of points near or at the final level of clustering (see Example 47).

Example 44. The computer-generated data shown in Exhibit 44a are a sample of 60 observations, 58 of which are from a bivariate normal

Exhibit 44a. Scatter plot with influence function contours for sample of bivariate normal data with two outliers added; $n=60$, $\rho=0$, $r=0.026$

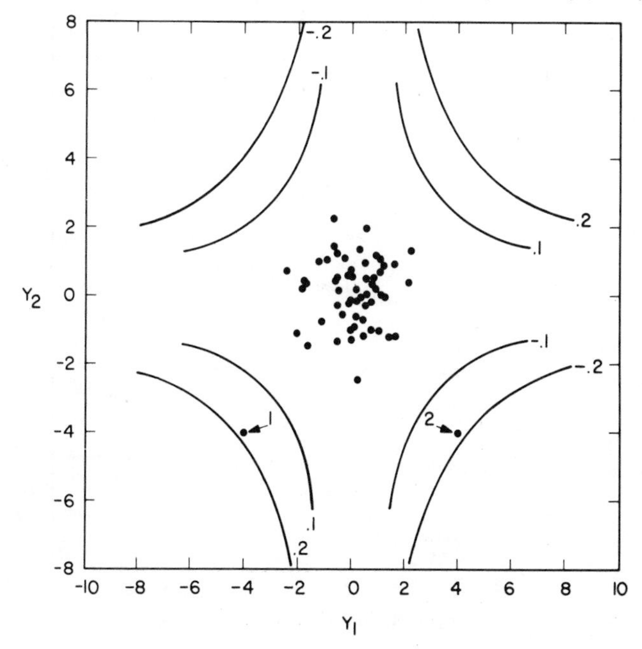

distribution with $\rho = 0.9$ and the remaining 2 observations simulate moderate outliers with opposite (i.e., inflation vs. deflation) effects on r. Also shown on the scatter plot are selected contours of the sample influence function of r.

The interpretation of the contours would be that the point labeled 1 would increase the value of r by about 0.2, while observation 2 would decrease r by about the same amount. These separate inferences concerning the two outliers are reasonably accurate in these data since the actual computed values are $r = 0.026$, $r_{-1} = -0.201$, and $r_{-2} = 0.253$. The value of the correlation coefficient when both outliers are omitted is, however, 0.029.

Exhibit 44*b* shows a psn probability plot of the sample influence function values of $z(r)$ for the same data, and Exhibit 44*c* shows the equivalent normal plot of the associated transformed quantities. Although the outliers stand out clearly on both these plots, the middle of the configuration (i.e., the linear part) in Exhibit 44*c* is stretched out more than the corresponding part of Exhibit 44*b*.

Exhibit 44b. PSN probability plot of values of z_{-i} for data of Exhibit 44a

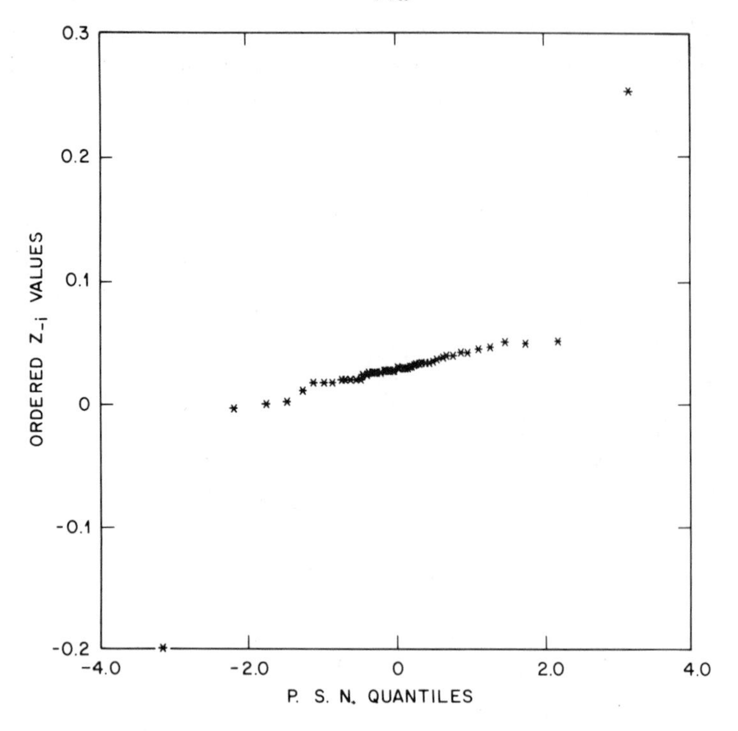

Exhibit 44c. Normal probability plot of transformed influence function values for data of Exhibit 44a

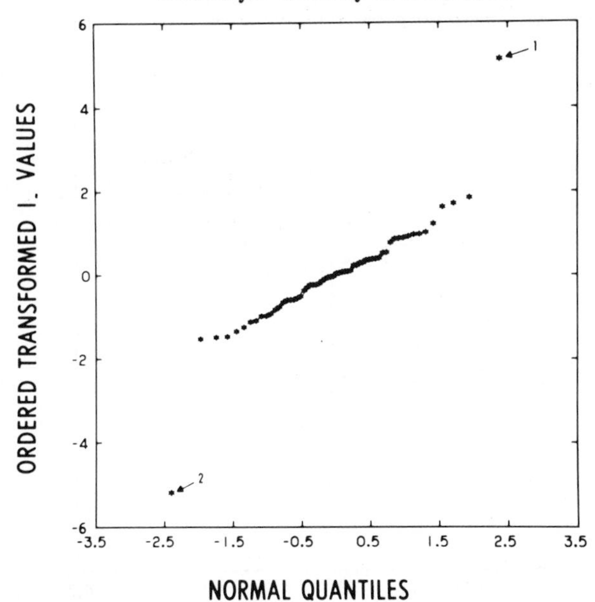

Example 45. The iris data (Anderson, 1935; Fisher, 1936) employed in Example 34 are used here to illustrate the issues and methods of outlier detection. Specifically, Exhibit 45*a* (see page 280) shows a scatter plot of values of the natural logarithms of 10 times the sepal lengths and widths for the 50 specimens of *Iris setosa.* Also shown in Exhibit 45*a* are contours of the sample influence function of *r*. Although the 42nd observation stands out clearly from the rest of the data, its location with respect to the contours suggests that it does not have a distorting influence on the value of *r*. Indeed, $r_{-42} = 0.723$, whereas $r = 0.730$.

Exhibit 45*b* (see page 280) which shows a normal plot of the transformed sample influence function values of $z(r)$, provides further confirmation of this. There are 11 points with more extreme influence on r than the 42nd observation. The most striking feature of Exhibit 45*b*, however, is the nonlinearity of the configuration even in the middle region, indicating that the logarithmically transformed data may be quite nonnormal, at least with respect to the two variables considered here.

Exhibit 45*c* (see page 281) shows a $\chi^2_{(2)}$ probability plot of the 50 generalized squared distances in these data. As expected from the scatter plot (Exhibit 45*a*), in terms of the elliptical distance measured by the generalized squared distance, the 42nd observation is indeed an outlier (see also

Exhibit 45a. Scatter plot with influence function contours for natural logarithms of sepal length and width of 50 Iris setosa

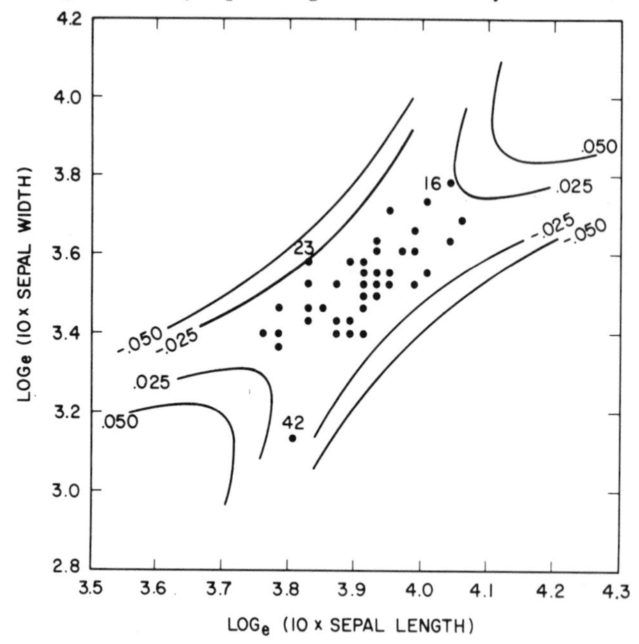

Exhibit 45b. Normal probability plot of transformed influence function values for the Iris setosa data of Exhibit 45a

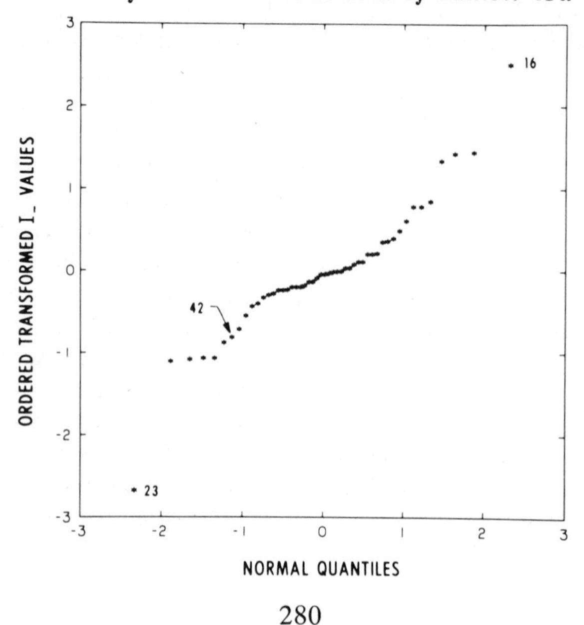

Exhibit 45c. Chi-squared (df=2) probability plot of the generalized squared distances for the Iris setosa data of Exhibit 45a

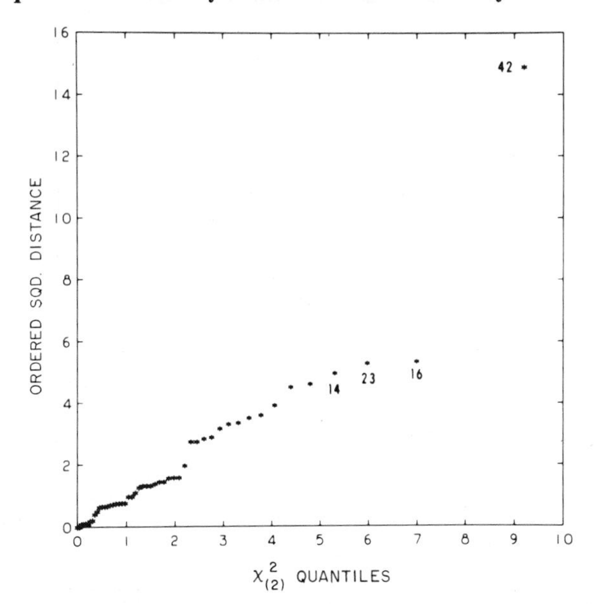

Example 34). This example thus illustrates the point that a multiresponse observation that is judged to be an outlier for one purpose may be quite a reasonable observation for other purposes.

Example 46. This example, taken from Devlin et al. (1976), illustrates the use of the psn probability plot (or the equivalent normal plot) of the sample influence function values of $z(r)$ to detect relatively smooth departures of the data distribution from normality. Five random samples, each of size 200, were generated from a bivariate t distribution with 5 degrees of freedom. Each sample yielded 200 sample influence function values of $z(r)$. By averaging the five corresponding ordered sample influence function values (i.e., average of the smallest in each set of 200, average of the second smallest, etc.), a smoothed set of ordered values was obtained. Exhibit 46 shows a psn probability plot of these, and the smoothly nonlinear configuration obtained indicates that the effect of the data having a bivariate t distribution rather than a bivariate normal distribution is to induce a longer-tailed (although still symmetric) distribution for the influence function values. Such an implication is accurate since it can also be established theoretically (Devlin et al., 1976).

Example 47. For illustrating the use of hierarchical clustering in identifying outliers, 14-dimensional data for 32 chemical companies for

Exhibit 46. ***PSN probability plot of averaged z_{-i} values from samples of bivariate t distribution (df=5)***

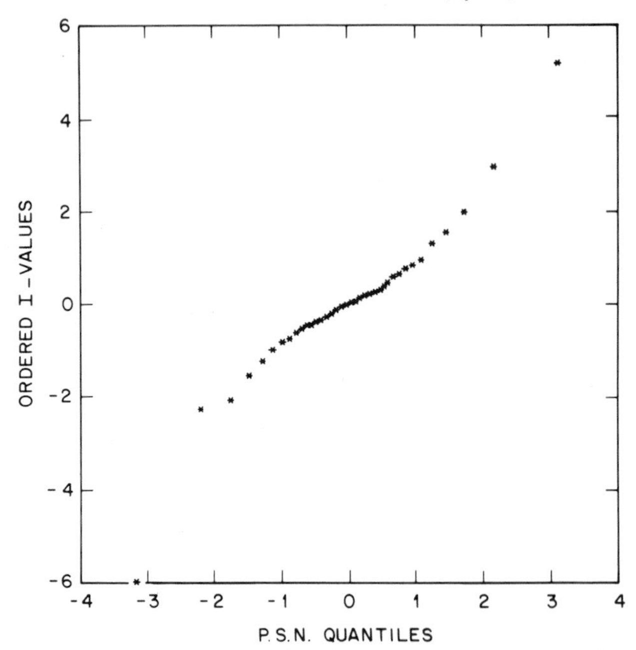

the year 1965 are taken from the study on grouping of corporations by Chen et al. (1970, 1974). Data from this study have been used repeatedly in earlier examples, and the present illustration is taken from Gnanadesikan and Kettenring (1972). The generalized squared intercompany distances in the 14-dimensional space were used as input to the minimum method of hierarchical clustering described in Section 4.3.1. The results, along with clustering strength values, are displayed in Exhibit 47. Company 14, which joins the cluster at the very end at a substantially higher clustering strength than the preceding value, appears to be an outlier, a finding that was corroborated by a variety of other analyses.

External Analysis Techniques for Outlier Detection. Discriminant analysis of two or more groups of multiresponse observations (see Section 4.2) and canonical analysis of two or more sets of variables (see Section 3.3) are among the basic multivariate external analysis techniques.

Valuable insight can be gleaned from two- and three-dimensional displays of the discriminant and canonical variables. Such views of the discriminant space, as illustrated in Examples 16 and 17, show the relative sizes, shapes, and locations of the groups, as well as possible

Exhibit 47. Hierarchical clustering tree for 32 chemical companies

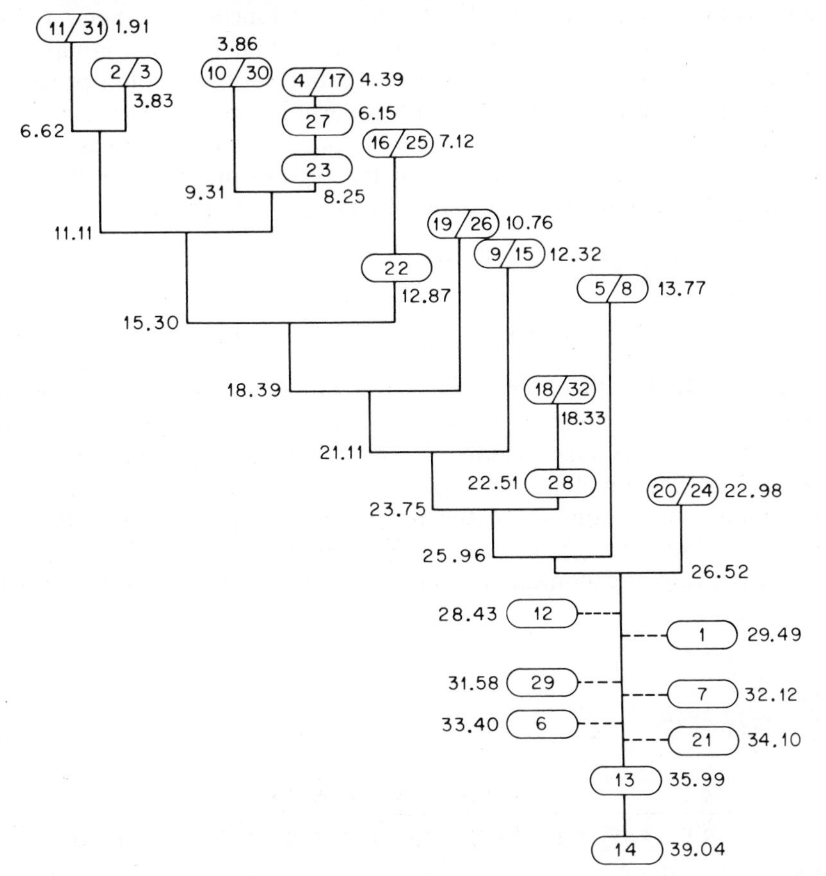

peculiarities in the positions of individual points. The discriminant analysis may be preceded by internal analyses of the individual groups for outliers, with the hope of making the dispersions within the individual groups similar, as is required for the validity of the standard multigroup discriminant analysis procedure (see the discussion in Example 17). The remaining observations can then be used to derive the discriminant coordinates, but the visual displays may profitably include the positions of all of the data in the transformed space. The canonical variable plot, another mechanisim for data exposure, can reveal outliers that are inducing an artificial linear relationship among the sets. Plots of the principal components (or other appropriate linear functions) of the canonical variables, as discussed in Kettenring (1971), are alternative summaries that have special appeal when the number of sets is large.

Normal probability plots and univariate outlier procedures can be applied to the canonical variables or to linear functions of them, and to the discriminant variables, making a separate plot for each group. The slopes of the configurations on the last-mentioned of these plots provide a partial check on the homogeneity of dispersion among the groups.

Gnanadesikan and Kettenring (1972) propose two examples of univariate statistics that are sensitive to the type of multivariate effects of interest in discriminant and canonical analyses. The first is

$$w_{ki}^2 = \sum_t c_t\{\mathbf{a}_t'(\mathbf{y}_{ki} - \bar{\mathbf{y}}_k)\}^2$$
$$= (\mathbf{y}_{ki} - \bar{\mathbf{y}}_k)'\mathbf{W}^{-1}(\mathbf{y}_{ki} - \bar{\mathbf{y}}_k), \qquad k = 1, \ldots, g;\ i = 1, \ldots, n_k,$$

where $\mathbf{y}_{ki}$ is the ith observation in the kth group, $\bar{\mathbf{y}}$ is the kth group mean, n_k is the number of observations in the kth group, and the eigenvalue c_t, the eigenvectors $\mathbf{a}_t$, and the matrices $\mathbf{B}$ and $\mathbf{W}$ are as defined in Section 4.2 (see Eqs. 51 and 52). The statistic, w_{ki}^2, is a weighted sum of squares of the projections of $(\mathbf{y}_{ki} - \bar{\mathbf{y}}_k)$ onto the discriminant axes, and $\sum\sum w_{ki}^2 = (n-g)\sum c_t$, where $n = \sum n_k$.

For the case of canonical analysis of two sets of variables, the proposed statistic is

$$x_i^2 = \frac{\prod_t (1 - r^{(t)^2})}{\prod_t (1 - r_{-i}^{(t)^2})}$$
$$= \frac{1 - (n/n-1)(\mathbf{y}_i - \bar{\mathbf{y}})'\mathbf{A}^{-1}(\mathbf{y}_i - \bar{\mathbf{y}})}{\{1 - (n/n-1)(\mathbf{y}_{1i} - \bar{\mathbf{y}}_1)'\mathbf{A}_{11}^{-1}(\mathbf{y}_{1i} - \bar{\mathbf{y}}_1)\}\{1 - (n/n-1)(\mathbf{y}_{2i} - \bar{\mathbf{y}}_2)'\mathbf{A}_{22}^{-1}(\mathbf{y}_{2i} - \bar{\mathbf{y}}_2)\}}$$
$$i = 1, \ldots, n,$$

where $r^{(t)}$ is the tth canonical correlation computed from all n observations, while $r_{-i}^{(t)}$ is based on all but the ith observation, and where

$$\mathbf{y}_i' = (\mathbf{y}_{1i}' \mid \mathbf{y}_{2i}'), \qquad \bar{\mathbf{y}}' = (\bar{\mathbf{y}}_1' \mid \bar{\mathbf{y}}_2'), \qquad \text{and} \qquad \mathbf{A} = (n-1)\mathbf{S} = \begin{pmatrix} \mathbf{A}_{11} & \mathbf{A}_{12} \\ \mathbf{A}_{21} & \mathbf{A}_{22} \end{pmatrix}$$

are partitioned in accordance with the dimensions of the two sets. (The subscript k, which designated the group in the definition of w_{ki}^2, now refers to the set.)

As aids for examining the collection of w_{ki}^2 and of x_i^2, it would seem reasonable to use gamma probability plots of the w_{ki}^2 and normal probability plots of the log x_i^2. These choices for the null distributions, however, need to be investigated more carefully for their appropriateness.

REFERENCES

Section 6.1	Cox (1973), Cox & Hinkley (1974).
Section 6.2	Anderson (1935), Andrews (1972), Ashton et al. (1957), Chen et al. (1970, 1974), Cleveland et al. (1975), Devlin et al. (1975), Fisher (1936), Fisherkeller et al. (1974), Friedman & Rubin (1967), Gnanadesikan (1968, 1973), Gnanadesikan et al. (1967), Gnanadesikan & Wilk (1969), Hartigan (1973), Hastings (1970), Mallows & Wachter (1970), Rao (1960, 1962), Stein (1969), Tukey (1970), Wilk et al. (1962a), Wilk & Gnanadesikan (1968), Zimmer & Larsen (1965).
Section 6.3	Daniel (1959), Dempster (1958), Gnanadesikan & Lee (1970), Gnanadesikan & Wilk (1970), Roy et al. (1971), Wilk et al. (1962a), Wilk & Gnanadesikan (1961, 1964).
Section 6.3.1	Andrews et al. (1971), Box (1954), Box & Cox (1964), Patnaik (1949), Roy et al. (1971), Satterthwaite (1941), Wilk et al. (1962b), Wilk et al. (1962), Wilk & Gnanadesikan (1964).
Section 6.3.2	Gnanadesikan & Lee (1970), Hoel (1937), Roy et al. (1971), Wilk et al. (1962b).
Section 6.4	Anscombe (1960, 1961), Devlin et al. (1975), Dixon (1953), Draper & Smith (1966), Gnanadesikan & Kettenring (1972), Grubbs (1950, 1969), Terry (1955).
Section 6.4.1	Abrahamson et al. (1969), Chen et al. (1970, 1974), Cleveland & Kleiner (1975), Cox (1968), Healy (1968), Larsen & McCleary (1972), Rao (1964), Roy et al. (1971), Tukey (1970).
Section 6.4.2	Anderson (1935), Andrews et al. (1972), Chen et al. (1970, 1974), Devlin et al. (1975, 1976), Dixon (1953), Fisher (1936), Gnanadesikan & Kettenring (1972), Grubbs (1950, 1969), Hampel (1968, 1973, 1974), Kettenring (1971), Mallows (1973), Miller (1974), Siotani (1959), Tukey (1958), Wilks (1963).

References

Abelson, R. P. & Tukey, J. W. (1959). Efficient conversion of non-metric information into metric information. *Proc. Soc. Stat. Sect. Am. Stat. Assoc.*, 226–30.

Abrahamson, I. G., Gentleman, J. F., Gnanadesikan, R., Walcheski, A. F., & Williams, D. E. (1969). Statistical methods for studying aging and for selecting semiconductor devices. *ASQC Tech. Conf. Trans.*, 533–40.

Aitchison, J. & Brown, J. A. C. (1957). *The Lognormal Distribution.* Cambridge University Press.

Aitkin, M. A. (1972). A class of tests for multivariate normality based on linear functions of order statistics. Unpublished manuscript.

Anderson, E. (1935). The irises of the Gaspe Peninsula. *Bull. Am. Iris Soc.* **59,** 2–5.

Anderson, E. (1954). Efficient and inefficient methods of measuring specific differences. In *Statistics and Mathematics in Biology* (O. Kempthorne, ed.), Iowa State College Press, Ames, pp. 98–107.

Anderson, E. (1957). A semi-graphical method for the analysis of complex problems. *Proc. Nat. Acad. Sci. USA* **43,** 923–7. [Reprinted in *Technometrics* **2** (1960), 387–92.]

Anderson, T. W. (1958). *An Introduction to Multivariate Statistical Analysis.* Wiley, New York.

Anderson, T. W. (1966). Some nonparametric multivariate procedures based on statistically equivalent blocks. In *Multivariate Analysis* (P. R. Krishnaiah, ed.), Academic Press, New York, pp. 5–27.

Anderson, T. W. (1969). Statistical inference for covariance matrices with linear structure. In *Multivariate Analysis* II (P. R. Krishnaiah, ed.), Academic Press, New York, pp. 55–66.

Anderson, T. W. & Rubin, H. (1956). Statistical inference in factor analysis. *Proc. 3rd Berkeley Symp. Math. Stat. Probab.* **5,** 11–50.

Andrews, D. F. (1971). A note on the selection of data transformations. *Biometrika* **58,** 249–54.

Andrews, D. F. (1972). Plots of high-dimensional data. *Biometrics* **28,** 125–36.

Andrews, D. F. (1973). Robust estimation for multiple linear regression models. *Bull. Int. Stat. Inst., Proc. 39th Sess. ISI at Vienna* **45,** Book 1, 105–11.

Andrews. D. F. (1974). A robust method for multiple linear regression. *Technometrics* **16,** 523–31.

Andrews, D. F., Bickel, P. J., Hampel, F. R., Huber, P. J., Rogers, W. H., & Tukey, J. W. (1972). *Robust Estimates of Location–Survey and Advances.* Princeton University Press.

Andrews, D. F., Gnanadesikan, R., & Warner, J. L. (1971). Transformations of multivariate data. *Biometrics* **27,** 825–40.

Andrews, D. F., Gnanadesikan, R., & Warner, J. L. (1972). Methods for assessing multivariate normality. Bell Laboratories Memorandum.

Andrews, D. F., Gnanadesikan, R., & Warner, J. L. (1973). Methods for assessing multivariate normality. In *Multivariate Analysis* III (P. R. Krishnaiah, ed.), Academic Press, New York, pp. 95–116.

Anscombe, F. J. (1960). Rejection of outliers. *Technometrics* **2,** 123–47.

Anscombe, F. J. (1961). Examination of residuals. *Proc. 4th Berkeley Symp. Math. Stat. Probab.* **1,** 1–36.

Ashton, E. H., Healy, M. J. R., & Lipton, S. (1957). The descriptive use of discriminant functions in physical anthropology. *Proc. R. Soc.* B **146,** 552–72.

Ball, G. H. (1965). Data analysis in the social sciences—what about details? *AFIPS Conf. Proc., Fall Joint Comput. Conf.* **27,** 533–60.

Ball, G. H. & Hall, D. J. (1965). ISODATA, a novel method of data analysis and pattern classification. Stanford Research Institute Report.

Bartholomew, D. J. (1959). A test of homogeneity for ordered alternatives. *Biometrika* **46,** 36–48.

Bartlett, M. S. (1951). The effect of standardization on an approximation in factor analysis. *Biometrika* **38,** 337–44.

Barton, D. E. & Mallows, C. L. (1961). The randomization bases of the amalgamation of weighted means. *J. R. Stat. Soc.* B**23,** 423–33.

Becker, M. H., Gnanadesikan, R., Mathews, M. V., Pinkham, R. S., Pruzansky, S., & Wilk, M. B. (1965). Comparison of some statistical distance measures for talker identification. Bell Laboratories Memorandum.

Bennett, R. S. (1965). The intrinsic dimensionality of signal collections. Ph.D. thesis, Johns Hopkins University.

Bickel, P. J. (1964). On some alternative estimates for shift in the *p*-variate one-sample problem. *Ann. Math. Stat.* **35,** 1079–90.

Blackith, R. E. (1960). A synthesis of multivariate techniques to distinguish patterns of growth in grasshoppers. *Biometrics* **16,** 28–40.

Blackith, R. E. & Roberts, M. I. (1958). Farbenpolymorphismus bei einigen Feldheuschrecken. *Z. Vererbungsl.* **89,** 328–37.

Box, G. E. P. (1954). Some theorems on quadratic forms applied in the study of analysis of variance problems—I. *Ann. Math. Stat.* **25,** 290–302.

Box, G. E. P. & Cox, D. R. (1964). An analysis of transformations. *J. R. Stat. Soc.* B**26,** 211–52.

Boynton, R. M. & Gordon, J. (1965). Bezold-Brüke hue shift measured by color-naming technique. *J. Opt. Soc. Am.* **55,** 78–86.

Bricker, P. D., Gnanadesikan, R., Mathews, M. V., Pruzansky, S., Tukey, P. A., Wachter, K. W., & Warner, J. L. (1971). Statistical techniques for talker identification. *Bell Syst. Tech. J.* **50,** 1427–54.

Bruntz, S. M., Cleveland, W. S., Kleiner, B., & Warner, J. L. (1974). The dependence of ambient ozone on solar radiation, wind, temperature, and mixing height. *Proc. Symp. Atmos. Diffus. Air Pollution, Am. Meteorol. Soc.* 125–8.

Burnaby, T. P. (1966). Growth invariant discriminant functions and generalized distances. *Biometrics* **22,** 96–110.

Businger, P. A. (1965). Algorithm 254. Eigenvalues and eigenvectors of a real symmetric matrix by the QR method. *Commun. ACM* **8,** 218–9.

Businger, P. A. & Golub, G. H. (1969). Algorithm 358. Singular value decomposition of a complex matrix. *Commun. ACM* **12,** 564–5.

Carroll, J. D. (1969). Polynomial factor analysis. *Proc. 77th Ann. Conv. Am. Psych. Assoc.*, 103–4.

Carroll, J. D. & Chang, J. J. (1970). Analysis of individual differences in multidimensional scaling via an N-way generalization of "Eckart-Young" decomposition. *Psychometrika* **35,** 283–319.

Chambers, J. M. (1973). Fitting nonlinear models: numerical techniques. *Biometrika* **60,** 1–15.

Chambers, J. M. (1974). *Computational Methods for Scientific Data Analysis.* Unpublished manuscript of book.

Chen, H., Gnanadesikan, R., & Kettenring, J. R. (1974). Statistical methods for grouping corporations. *Sankhyā* B**36,** 1–28.

Chen, Hwei-Ju, Gnanadesikan, R., Kettenring, J. R., & McElroy, Marjorie B. (1970). A statistical study of groupings of corporations. *Proc. Bus. Econ. Stat. Sect. Am. Stat. Assoc.*, 447–51.

Chernoff, H. (1973). The use of faces to represent points in k-dimensional space graphically. *J. Am. Stat. Assoc.* **68,** 361–8.

Cleveland, W. S. & Kleiner, B. (1975). A graphical technique for enhancing scatterplots with moving statistics. *Technometrics* **17,** 447–54.

Cleveland, W. S., Kleiner, B., McRae, J. E., Warner, J. L., & Pasceri, R. E. (1975). The analysis of ground-level ozone data from New Jersey, New York, Connecticut, and Massachusetts: data quality assessment and temporal and geographical properties. Paper presented at the 68th annual meeting of the Air Pollution Control Association.

Coombs, C. H. (1964). *A Theory of Data.* Wiley, New York.

Cormack, R. M. (1971). A review of classification. *J. R. Stat. Soc.* A**134,** 321–67.

Cox, D. R. (1968). Notes on some aspects of regression analysis. *J. R. Stat. Soc.* A**131,** 265–79.

Cox, D. R. (1970). *The Analysis of Binary Data.* Methuen, London.

Cox, D. R. (1972). The analysis of multivariate binary data. *Appl. Stat.* **21,** 113–20.

Cox, D. R. (1973). Theories of statistical inference. Rietz Lecture, Institute of Mathematical Statistics meetings in New York.

Cox, D. R. & Hinkley, D. V. (1974). *Theoretical Statistics.* Chapman & Hall, London.

D'Agostino, R. B. (1971). An omnibus test of normality for moderate and large size samples. *Biometrika* **58,** 341–8.

D'Agostino, R. B. & Pearson, E. S. (1973). Tests for departure from normality. Empirical results for the distributions of b_2 and $\sqrt{b_1}$. *Biometrika* **60,** 613–22.

Daniel, C. (1959). The use of half-normal plots in interpreting factorial two level experiments. *Technometrics* **1,** 311–41.

David, F. N. & Johnson, N. L. (1948). The probability integral transformation when parameters are estimated from the sample. *Biometrika* **35,** 182–90.

Dempster, A. P. (1958). A high dimensional two sample significance test. *Ann. Math. Stat.* **29,** 995–1010.

Devlin, S. J., Gnanadesikan, R., & Kettenring, J. R. (1975). Robust estimation and outlier detection with correlation coefficients. *Biometrika* **62,** 531–45.

Devlin, S. J., Gnanadesikan, R., & Kettenring, J. R. (1976). Some multivariate applications of elliptical distributions. To appear in *Essays on Probability and Statistics* (S. Ikeda, ed.).

Dixon, W. J. (1953). Processing data for outliers. *Biometrics* **9,** 74–89.

Downton, F. (1966). Linear estimates with polynomial coefficients. *Biometrika* **53,** 129–41.

Draper, N. R. & Smith, H. (1966). *Applied Regression Analysis.* Wiley, New York.

Ekman, G. (1954). Dimensions of color vision. *J. Psych.* **38,** 467–74.

Everitt, B. (1974). *Cluster Analysis.* Wiley, New York.

Fisher, R. A. (1936). The use of multiple measurements in taxonomic problems. *Ann. Eugen.* **7,** 179–88.

Fisher, R. A. (1938). The statistical utilization of multiple measurements. *Ann. Eugen.* **8,** 376–86.

Fisherkeller, M. A., Friedman, J. H., & Tukey, J. W. (1974). PRIM-9. An interactive multidimensional data display and analysis system. [Also a film "PRIM-9," produced by Stanford Linear Accelerator Center (S. Steppel, ed.).] Stanford Linear Accelerator Center Pub. 1408.

Fletcher, R. & Powell, M. J. D. (1963). A rapidly convergent descent method for minimization. *Comput. J.* **2,** 163–8.

Friedman, H. P. & Rubin, J. (1967). On some invariant criteria for grouping data. *J. Am. Stat. Assoc.* **62,** 1159–78.

Gentleman, W. M. (1965). Robust estimation of multivariate location by minimizing pth power deviations. Ph.D. thesis, Princeton University.

Gnanadesikan, R. (1968). The study of multivariate residuals and methods for detecting outliers in multiresponse data. Invited paper presented at American Society for Quality Control, Chemical Division, meetings at Durham, N.C.

Gnanadesikan, R. (1972). Methods for evaluating similarity of marginal distributions. *Stat. Neerl.* **26,** No. 3, 69–78.

Gnanadesikan, R. (1973). Graphical methods for informal inference in multivariate data analysis. *Bull. Int. Stat. Inst., Proc. 39th Sess. ISI at Vienna* **45,** Book 4, 195–206.

Gnanadesikan, R. & Kettenring, J. R. (1972). Robust estimates, residuals, and outlier detection with multiresponse data. *Biometrics* **28,** 81–124.

Gnanadesikan, R. & Lee, E. T. (1970). Graphical techniques for internal comparisons amongst equal degree of freedom groupings in multiresponse experiments. *Biometrika* **57,** 229–37.

Gnanadesikan, R., Pinkham, R. S., & Hughes, L. P. (1967). Maximum likelihood estimation of the parameters of the beta distribution from smallest order statistics. *Technometrics* **9,** 607–20.

Gnanadesikan, R. & Wilk, M. B. (1966). Data analytic methods in multivariate statistical analysis. General Methodology Lecture on Multivariate Analysis, 126th Annual Meeting of the American Statistical Association, Los Angeles.

Gnanadesikan, R. & Wilk, M. B. (1969). Data analytic methods in multivariate statistical analysis. In *Multivariate Analysis* II (P. R. Krishnaiah, ed.), Academic Press, New York, pp. 593–638.

Gnanadesikan, R. & Wilk, M. B. (1970). A probability plotting procedure for general analysis of variance. *J. R. Stat. Soc.* B**32,** 88–101.

Golub, G. H. (1968). Least squares, singular values and matrix approximations. *Apl. Mat.* **13,** 44–51.

Golub, G. H. & Reinsch, C. (1970). Handbook series linear algebra: singular value decomposition and least squares solutions. *Numer. Math.* **14,** 403–20.

Gower, J. C. (1967). A comparison of some methods of cluster analysis. *Biometrics* **23,** 623–37.

Grubbs, F. E. (1950). Sample criteria for testing outlying observations. *Ann. Math. Stat.* **21,** 27–58.

Grubbs, F. E. (1969). Procedures for detecting outlying observations in samples. *Technometrics* **11,** 1–21.

Hampel, F. R. (1968). Contributions to the theory of robustness. Ph.D. thesis, University of California at Berkeley.

Hampel, F. R. (1973). Robust estimation: a condensed partial survey. *Z. Wahr. verw. Geb.* **27,** 87–104.

Hampel, F. R. (1974). The influence curve and its role in robust estimation. *J. Am. Stat. Assoc.* **69,** 383–93.

Harman, H. H. (1967). *Modern Factor Analysis,* second edition (revised). University of Chicago Press.

Hartigan, J. A. (1967). Representation of similarity matrices by trees. *J. Am. Stat. Assoc.* **62,** 1140–58.

Hartigan, J. A. (1973). Printer graphics for clustering. Unpublished manuscript. (Also see section 1.7.6. of Hartigan, 1975.)

Hartigan, J. A. (1975). *Clustering Algorithms.* Wiley, New York.

Hastings, W. K. (1970). Monte Carlo sampling methods using Markov chains and their applications. *Biometrika* **57,** 97–109

Healy, M. J. R. (1968). Multivariate normal plotting. *Appl. Stat.* **17,** 157–61.

Hoel, P. G. (1937). A significance test for component analysis. *Ann. Math. Stat.* **8,** 149–58.

Hoerl, A. E. & Kennard, R. W. (1970). Ridge regression: biased estimation for nonorthogonal problems. *Technometrics* **12,** 55–67.

Horst, P. (1965). *Factor Analysis of Data Matrices.* Holt, Rinehart & Winston, New York.

Hotelling, H. (1933). Analysis of a complex of statistical variables into principal components. *J. Educ. Psych.* **24,** 417–41, 498–520.

Hotelling, H. (1936). Relations between two sets of variates. *Biometrika* **28,** 321–77.

Hotelling, H. (1947). Multivariate quality control, illustrated by the air testing of sample bombsights. In *Selected Techniques of Statistical Analysis* (C. Eisenhart et al., eds.), McGraw-Hill, New York, pp. 111–84.

Howe, W. G. (1955). Some contributions to factor analysis. Oak Ridge National Laboratory, ORNL-1919, Oak Ridge, Tenn.

Huber, P. J. (1964). Robust estimation of a location parameter. *Ann. Math. Stat.* **35,** 73–101.

Huber, P. J. (1970). Studentizing robust estimates. In *Nonparametric Techniques in Statistical Inference* (M. L. Puri, ed.), Cambridge University Press, pp. 453–63.

Huber, P. J. (1972). Robust statistics: a review. *Ann. Math. Stat.* **43,** 1041–67.

Huber, P. J. (1973). Robust regression: asymptotics, conjectures and Monte Carlo. *Ann. Stat.* **1,** 799–821.

Imbrie, J. (1963). Factor and vector analysis programs for analysing geological data. Tech. Rept. 6, ONR Task No. 389–135.

Imbrie, J. & Kipp, N. G. (1971). A new micropaleontological method for quantitative paleoclimatology: application to a late pleistocene Caribbean core. In *Late Cenozoic Glacial Ages* (K. K. Turekian, ed.), Yale University Press, New Haven, Conn., pp. 71–181.

Imbrie, J. & Van Andel, T. H. (1964). Vector analysis of heavy-mineral data. *Bull. Geol. Soc. Am.* **75,** 1131–55.

Jackson, J. E. (1956). Quality control methods for two related variables. *Ind. Qual. Control* **12,** 2–6.

Johnson, N. L. & Leone, F. C. (1964). *Statistics and Experimental Design in Engineering and the Physical Sciences,* Vol. I. Wiley, New York.

Johnson, S. C. (1967). Hierarchical clustering schemes. *Psychometrika* **32,** 241–54.

Jöreskog, K. G. (1967). Some contributions to maximum likelihood factor analysis. *Psychometrika* **32,** 443–82.

Jöreskog, K. G. (1973). Analysis of covariance structures. In *Multivariate Analysis* III (P. R. Krishnaiah, ed.), Academic Press, New York, pp. 263–85.

Jöreskog, K. G. & Lawley, D. N. (1968). New methods in maximum likelihood factor analysis. *Br. J. Math. Stat. Psych.* **21,** 85–96.

Kempthorne, O. (1966). Multivariate responses in comparative experiments. In *Multivariate Analysis* (P. R. Krishnaiah, ed.), Academic Press, New York, pp. 521–40.

Kendall, M. G. (1968). On the future of statistics—a second look. *J. R. Stat. Soc.* A**131,** 182–92.

Kessell, D. L. & Fukunaga, K. (1972). A test for multivariate normality with unspecified parameters. Unpublished report, Purdue University School of Electrical Engineering.

Kettenring, J. R. (1969). Canonical analysis of several sets of variables. Ph.D. thesis, University of North Carolina.

Kettenring, J. R. (1971). Canonical analysis of several sets of variables. *Biometrika* **58,** 433–51.

Kruskal, J. B. (1964a). Multidimensional scaling by optimizing goodness of fit to a nonmetric hypothesis. *Psychometrika* **29,** 1–27.

Kruskal, J. B. (1964b). Nonmetric multidimensional scaling: a numerical method. *Psychometrika* **29,** 115–29.

Larsen, W. A. & McCleary, S. J. (1972). The use of partial residual plots in regression analysis. *Technometrics* **14,** 781–90.

Laue, R. V. & Morse, M. F. (1968). Simulation of traffic distribution schemes for No. 5 ACD. Bell Laboratories Memorandum.

Lawley, D. N. (1940). The estimation of factor loadings by the method of maximum likelihood. *Proc. R. Soc. Edin.* A**60,** 64–82.

Lawley, D. N. (1967). Some new results in maximum likelihood factor analysis. *Proc. R. Soc. Edin.* A**67,** 256–64.

Lawley, D. N. & Maxwell, A. E. (1963). *Factor Analysis as a Statistical Method.* Butterworth, London. (Second edition, 1971.)

Lindley, D. V. (1972). Book review (*Analysis and Design of Certain Quantitative Multiresponse Experiments* by Roy et al.) *Bull. Inst. Math. Appl.* **8,** 134.

Ling, R. F. (1971). Cluster analysis. Ph.D. thesis, Yale University.

MacQueen, J. (1965). Some methods for classification and analysis of multivariate observations. *Proc. 5th Berkeley Symp. Math. Stat. Probab.* **1,** 281–97.

Malkovich, J. F. & Afifi, A. A. (1973). On tests for multivariate normality. *J. Am. Stat. Assoc.* **68,** 176–9.

Mallows, C. L. (1973). Influence functions. Unpublished talk presented at the Working Conference on Robust Regression at National Bureau of Economics Research in Cambridge, Mass.

Mallows, C. L. & Wachter, K. W. (1970). The asymptotic configuration of Wishart eigenvalues. Abstract 126–5, *Ann. Math. Stat.* **41,** 1384.

Mardia, K. V. (1970). Measures of multivariate skewness and kurtosis with applications. *Biometrika* **57,** 519–30.

McDonald, R. P. (1962). A general approach to nonlinear factor analysis. *Psychometrika* **27,** 397–415.

McDonald, R. P. (1967). Numerical methods for polynomial models in non-linear factor analysis. *Psychometrika* **32,** 77–112.

McLaughlin, D. H. & Tukey, J. W. (1961). The variance of means of symmetrically trimmed samples from normal populations, and its estimation from such trimmed samples. (Trimming/Winsorization I.) Tech. Rept. 42, Statistical Techniques Research Group, Princeton University.

Miles, R. E. (1959). The complete amalgamation into blocks, by weighted means, of a finite set of real numbers. *Biometrika* **46,** 317–27.

Miller, G. A. & Nicely, P. E. (1955). An analysis of perceptual confusions among some English consonants. *J. Acoust. Soc. Am.* **27,** 338–52.

Miller, R. G. (1974). The jackknife—a review. *Biometrika* **61,** 1–15.

Mood, A. M. (1941). On the joint distribution of the median in samples from a multivariate population. *Ann. Math. Stat.* **12,** 268–78.

Moore, P. G. & Tukey, J. W. (1954). Answer to query 112. *Biometrics* **10,** 562–8.

Neely, P. M. (1967). Towards a theory of classification. Unpublished.

Patnaik, P. B. (1949). The non-central χ^2 and F-distributions and their approximations. *Biometrika* **36,** 202–32.

Pearson, E. S. & Hartley, H. O. (1966). *Biometrika Tables for Statisticians,* Vol. I. Cambridge University Press.

Pearson, K. (1901). On lines and planes of closest fit to systems of points in space. *Phil. Mag.* [6] **2,** 559–72.

Pillai, K. C. S. & Jayachandran, K. (1967). Power comparisons of tests of two multivariate hypotheses based on four criteria. *Biometrika* **54,** 195–210.

Puri, M. L. & Sen, P. K. (1971). *Nonparametric Methods in Multivariate Analysis.* Wiley, New York.

Rao, C. R. (1952). *Advanced Statistical Methods in Biometric Research.* Wiley, New York.

Rao, C. R. (1960). Multivariate analysis: an indispensable statistical aid in applied research. *Sankhyā* **22,** 317–38.

Rao, C. R. (1962). Use of discriminant and allied functions in multivariate analysis. *Sankhyā* A**24,** 149–54.

Rao, C. R. (1964). The use and interpretation of principal component analysis in applied research. *Sankhyā* A**26,** 329–58.

Rao, C. R. (1965). *Linear Statistical Inference and Its Applications.* Wiley, New York. (Second edition, 1973.)

Rao, C. R. (1966). Discriminant function between composite hypotheses and related problems. *Biometrika* **53,** 339–45.

Rosenblatt, M. (1952). Remarks on a multivariate transformation. *Ann. Math. Stat.* **23,** 470–2.

Rothkopf, E. Z. (1957). A measure of stimulus similarity and errors in some paired-associate learning tasks. *J. Exp. Psych.* **53,** 94–101.

Roy, S. N. (1953). On a heuristic method of test construction and its use in multivariate analysis. *Ann. Math. Stat.* **24,** 220–38.

Roy, S. N. (1957). *Some Aspects of Multivariate Analysis.* Wiley, New York.

Roy, S. N. & Bose, R. C. (1953). Simultaneous confidence interval estimation. *Ann. Math. Stat.* **24,** 513–36.

Roy, S. N. & Gnanadesikan, R. (1957). Further contributions to multivariate confidence bounds. *Biometrika* **44,** 399–410.

Roy, S. N. & Gnanadesikan, R. (1962). Two-sample comparisons of dispersion matrices for alternatives of intermediate specificity. *Ann. Math. Stat.* **33,** 432–7.

Roy, S. N., Gnanadesikan, R., & Srivastava, J. N. (1971). *Analysis and Design of Certain Quantitative Multiresponse Experiments.* Pergamon Press, Oxford.

Sarhan, A. E. & Greenberg, B. (1956). *Contributions to Order Statistics.* Wiley, New York.

Satterthwaite, F. E. (1941). Synthesis of variance. *Psychometrika* **6,** 309–16.

Scheffé, H. (1953). A method for judging all contrasts in the analysis of variance. *Biometrika* **40,** 87–104.

Scheffé, H. (1959). *The Analysis of Variance.* Wiley, New York.

Scott, A. J. & Symons, M. J. (1971). On the Edwards and Cavalli-Sforza method of cluster analysis. Note 297, *Biometrics* **27,** 217–9.

Seal, H. L. (1964). *Multivariate Statistical Analysis for Biologists.* Methuen, London.

Shapiro, S. S. & Wilk, M. B. (1965). An analysis of variance test for normality. *Biometrika* **52,** 591–611.

Shapiro, S. S., Wilk, M. B., & Chen, H. (1968). A comparative study of various tests for normality. *J. Am. Stat. Assoc.* **63,** 1343–72.

Shepard, R. N. (1962a). The analysis of proximities: multidimensional scaling with an unknown distance function—I. *Psychometrika* **27,** 125–40.

Shepard, R. N. (1962b). The analysis of proximities: multidimensional scaling with an unknown distance function—II. *Psychometrika* **27,** 219–46.

Shepard, R. N. (1963). Analysis of proximities as a study of information processing in man. *Human Factors* **5,** 33–48.

Shepard, R. N. & Carroll, J. D. (1966). Parametric representation of nonlinear data structures. In *Multivariate Analysis* (P. R. Krishnaiah, ed.), Academic Press, New York, pp. 561–92

Siotani, M. (1959). The extreme value of the generalized distances of the individual points in the multivariate sample. *Ann. Inst. Stat. Math.* **10,** 183–203.

Sneath, P. H. A. (1957). The application of computers to taxonomy. *J. Gen. Microbiol.* **17,** 201–26.

Sokal, R. R. & Sneath, P. H. A. (1963). *Principles of Numerical Taxonomy.* W. H. Freeman & Co., San Francisco.

Sørensen, T. (1948). A method of establishing groups of equal amplitude in plant sociology based on similarity of species content and its application to analyses of the vegetation on Danish commons. *Biol. Skr.* **5,** No. 4, 1–34.

Srivastava, J. N. (1966). On testing hypotheses regarding a class of covariance structures. *Psychometrika* **31,** 147–64.

Steel, R. G. D. (1951). Minimum generalized variance for a set of linear functions. *Ann. Math. Stat.* **22,** 456–60.

Stein, C. (1956). Inadmissibility of the usual estimator for the mean of a multivariate normal distribution. *Proc. 3rd Berkeley Symp. Math. Stat. Probab.* **1,** 197–206.

Stein, C. (1965). Inadmissibility of the usual estimator for the variance of a normal distribution with unknown mean. *Ann. Inst. Stat. Math.* **16,** 155–6.

Stein, C. (1969). Mimeographed lecture notes on multivariate analysis as recorded by M. Eaton. Stanford University.

Teichroew, D. (1956). Tables of expected values of order statistics and products of order statistics for samples of size twenty and less from the normal distribution. *Ann. Math. Stat.* **27,** 410–26.

Terry, M. E. (1955). On the analysis of planned experiments. *Nat. Conv. ASQC Trans.,* 553–6.

Theil, H. (1963). On the use of incomplete prior information in regression analysis. *J. Am. Stat. Assoc.* **58,** 401–14.

Thomson, G. H. (1934). Hotelling's method modified to give Spearman's g. *J. Educ. Psych.* **25,** 366–74.

Thurstone, L. L. (1931). Multiple factor analysis. *Psych. Rev.* **38,** 406–27.

Thurstone, L. L. & Thurstone, T. G. (1941). Factorial studies of intelligence. Psychometric monogr. **2.**

Tukey, J. W. (1957). On the comparative anatomy of transformations. *Ann. Math. Stat.* **28,** 602–32.

Tukey, J. W. (1958). Bias and confidence in not-quite large samples (Abstract). *Ann. Math. Stat.* **29,** 614.

Tukey, J. W. (1960). A survey of sampling from contaminated distributions. In *Contributions to Probability and Statistics* (I. Olkin et al., eds.), Stanford University Press, pp. 448–85.

Tukey, J. W. (1962). The future of data analysis. *Ann. Math. Stat.* **33,** 1–67.

Tukey, J. W. (1970). *Exploratory Data Analysis,* limited preliminary edition, Addison-Wesley, Reading, Mass.

Tukey, J. W. & McLaughlin, D. H. (1963). Less vulnerable confidence and significance procedures for location based on a single sample: trimming/Winsorization 1 *Sankhyā* A**25,** 331–52.

Tukey, J. W. & Wilk, M. B. (1966). Data analysis and statistics: an expository overview. *AFIPS Conf. Proc., Fall Joint Comput. Conf.* **29,** 695–709.

Van De Geer, J. P. (1968). Fitting a quadratic function to a two-dimensional set of points. Unpublished research note, RN 004–68, Department of Data Theory for the Social Sciences, University of Leiden, The Netherlands.

Van Eeden, C. (1957a). Maximum likelihood estimation of partially or completely ordered parameters, I. *Proc. Akad. Wet.* A**60,** 128–36.

Van Eeden, C. (1957b). Note on two methods for estimating ordered parameters of probability distributions. *Proc. Akad. Wet.* A**60,** 506–12.

Warner, J. L. (1968). An adaptation of ISODATA-POINTS, an *i*terative *s*elf-*o*rganizing *d*ata *a*nalysis *t*echnique *a*. Bell Laboratories Memorandum.

Warner, J. L. (1969). Hierarchical clustering schemes. Bell Laboratories Memorandum.

Weiss, L. (1958). A test of fit for multivariate distributions. *Ann. Math. Stat.* **29,** 595–9.

Wilk, M. B. & Gnanadesikan, R. (1961). Graphical analysis of multi-response experimental data using ordered distances. *Proc. Nat. Acad. Sci. USA* **47,** 1209–12.

Wilk, M. B. & Gnanadesikan, R. (1964). Graphical methods for internal comparisons in multiresponse experiments. *Ann. Math. Stat.* **35,** 613–31.

Wilk, M. B. & Gnanadesikan, R. (1968). Probability plotting methods for the analysis of data. *Biometrika* **55,** 1–17.

Wilk, M. B., Gnanadesikan, R., & Huyett, Miss M. J. (1962a). Probability plots for the gamma distribution. *Technometrics* **4,** 1–20.

Wilk, M. B., Gnanadesikan, R., & Huyett, M. J. (1962b). Estimation of the parameters of the gamma distribution using order statistics. *Biometrika* **49,** 525–45.

Wilk, M. B., Gnanadesikan, R., Huyett, M. J., & Lauh, Miss E. (1962). A study of alternate compounding matrices used in a graphical internal comparisons procedure. Bell Laboratories Memorandum.

Wilkinson, G. N. (1970). A general recursive procedure for analysis of variance. *Biometrika* **57,** 19–46.

Wilks, S. S. (1963). Multivariate statistical outliers. *Sankhyā* A**25,** 407–26.

Zimmer, C. E. & Larsen, R. I. (1965). Calculating air quality and its control. *J. Air Pollution Control Assoc.* **15,** 565–72.

APPENDIX

Computer Programs and Software

Almost every technique described in this book is currently available in the computing environment at Bell Telephone Laboratories. A modular set of FORTRAN subroutines, called the Statistical Computing Subroutine (SCS) Library, implemented on a Honeywell H6000 series machine, is the primary software source in this environment. The algorithms involved are of very high numerical quality. Another very appealing and central feature of this library of programs is its emphasis on a variety of graphical outputs; in fact, most of the pictorial displays in the text were obtained from the graphical facilities associated with and utilized by the SCS Library.

Unfortunately, importing this library and implementing it in its entirety on another machine and/or system is far from simple or straightforward. Indeed portability of software is still a fundamental problem that is receiving considerable attention as a research topic in computing science. Hence, in this appendix, an attempt is made to provide some references to documents wherein the interested reader might find useful information about and descriptions of algorithms or computer programs for performing several of the analyses described in this book. Although many of the references do not give actual listings of programs, they may provide sufficient information about the design of the algorithms and the software to be of value to one who is interested in developing a collection of computer programs for analyzing multivariate data.

Before listing references for specific items in the order in which the techniques were described in the text, three documents with a more extensive coverage of computing issues and programs should be mentioned:

(a) BMD-Biomedical Computer Programs (W. J. Dixon, ed.) Univ. Calif. Publ. in Autom. Comp. No. 2 (1967), No. 3 (1970) (esp. Class M and Class V programs).

(b) Cooley, W. W. & Lohnes, P. R. (1971). *Multivariate Data Analysis.* Wiley, New York. (FORTRAN listings of programs are provided throughout this book.)

(c) Chambers, J. M. (1974). *Computational Methods for Scientific Data Analysis.* Unpublished manuscript of book.

Documents concerned with software for more specific multivariate techniques are:

Principal components analysis (Sections 2.2.1 and 2.4)

(1) Businger, P. A. (1965). Algorithm 254. Eigenvalues and eigenvectors of a real symmetric matrix by the QR method. *Commun. ACM* **8,** 218–9.

(2) Businger, P. A. & Golub, G. H. (1969). Algorithm 358. Singular value decomposition of a complex matrix. *Commun. ACM* **12,** 564–5.

Factor Analysis (Section 2.2.2)

(3) Imbrie, J. H. (1963). Factor and vector analysis programs for analyzing geological data. Tech. Rept. 6, ONR Task No. 389–135.

(4) Jöreskog, K. G. (1966). UMFLA—A computer program for unrestricted maximum likelihood factor analysis. Res. Mem. 66–20, Educational Testing Service, Princeton, N.J.

(5) Jöreskog, K. G. & Gruvaeus, G. (1967). RMFLA—A computer program for restricted maximum likelihood factor analysis. Res. Mem. 67–21, Educational Testing Service, Princeton, N.J.

Multidimensional scaling and individual differences scaling (Section 2.3)

(6) Kruskal, J. B., Young, F. W., & Seery, J. B. (1973). How to use KYST, a very flexible program, to do multidimensional scaling and unfolding. Programs and user's manual available from Computing Information Service Group at Bell Laboratories.

(7) Chang, J. J. V. (1972). INDSCAL. Programs and user's manual available from Computing Information Service Group at Bell Laboratories.

Canonical correlation analysis (Section 3.2)

(8) Björck, A. & Golub, G. H. (1973). Numerical methods for computing angles between linear subspaces. *Math. Comp.* **27,** 579–94.

(9) Chen, H. J. & Kettenring, J. R. (1972). CANON: A computer program package for the multi-set canonical correlation analysis. Bell Laboratories Memorandum.

Discriminant analysis (Section 4.2)

(10) Chambers, J. M. (1974). *Op. cit.*, Section 2.6c.

Cluster analysis (Section 4.3)

(11) Hartigan, J. A. (1975). *Clustering Algorithms.* Wiley, New York. (Listings of FORTRAN programs are given at the end of each chapter.)

(12) Warner, J. L. (1968). An adaptation of ISODATA POINTS, an *i*terative *s*elf-*o*rganizing *d*ata *a*nalysis *t*echnique *a*. Bell Laboratories Memorandum.

(13) Warner, J. L. (1969). Hierarchical clustering schemes. Bell Laboratories Memorandum.

Multivariate analysis of variance (Section 5.2.1)

(14) Finn, J. (1974). MULTIVARIANCE: univariate and multivariate analysis of variance, covariance and regression. Version 5. National Educational Resources Inc., Chicago, Ill.

(15) Fowlkes, E. B. & Lee, E. T. (1971). Appendix C, Section I, in *Analysis and Design of Certain Quantitative Multiresponse Experiments* by S. N. Roy, R. Gnanadesikan, and J. N. Srivastava Pergamon Press, Oxford.

Probability plotting

(A) *Q-Q* plotting of a set of observations against the quantiles of any of 26 choices for a theoretical distribution and also a *Q-Q* plot of one sample versus another (Section 6.2)

(16) Warner, J. L. (1973). A unified probability plotting package (QQPAK). Bell Laboratories Memorandum.

(B) Single-degree-of-freedom contrast vectors (Section 6.3.1)

(17) Fowlkes, E. B. & Lee, E. T. (1971). Appendix C, Section II, in *Analysis and Design of Certain Quantitative Multiresponse Experiments* by S. N. Roy, R. Gnanadesikan and J. N. Srivastava, Pergamon Press, Oxford.

Author Index

Subject Index

Applied Probability and Statistics (Continued)

JOHNSON and KOTZ · Distributions in Statistics
Discrete Distributions
Continuous Univariate Distributions-1
Continuous Univariate Distributions-2
Continuous Multivariate Distributions

JOHNSON and LEONE · Statistics and Experimental Design: In Engineering and the Physical Sciences, Volumes I and II, *Second Edition*

KEENEY and RAIFFA · Decisions with Multiple Objectives

LANCASTER · The Chi Squared Distribution

LANCASTER · An Introduction to Medical Statistics

LEWIS · Stochastic Point Processes

MANN, SCHAFER and SINGPURWALLA · Methods for Statistical Analysis of Reliability and Life Data

MEYER · Data Analysis for Scientists and Engineers

OTNES and ENOCHSON · Digital Time Series Analysis

PRENTER · Splines and Variational Methods

RAO and MITRA · Generalized Inverse of Matrices and Its Applications

SARD and WEINTRAUB · A Book of Splines

SEAL · Stochastic Theory of a Risk Business

SEARLE · Linear Models

THOMAS · An Introduction to Applied Probability and Random Processes

WHITTLE · Optimization under Constraints

WONNACOTT and WONNACOTT · Econometrics

YOUDEN · Statistical Methods for Chemists

ZELLNER · An Introduction to Bayesian Inference in Econometrics

Tracts on Probability and Statistics

BHATTACHARYA and RAO · Normal Approximation and Asymptotic Expansions

BILLINGSLEY · Convergence of Probability Measures

CRAMER and LEADBETTER · Stationary and Related Stochastic Processes

JARDINE and SIBSON · Mathematical Taxonomy

RIORDAN · Combinatorial Identities